Lecture Notes in Earth System Sciences

More information about this series at http://www.springer.com/series/10529

Chongbin Zhao

Physical and Chemical Dissolution Front Instability in Porous Media

Theoretical Analyses and Computational Simulations

Springer

Chongbin Zhao
Computational Geosciences Research
Centre
Central South University
Changsha
China

ISSN 2193-8571 ISSN 2193-858X (electronic)
ISBN 978-3-319-08460-2 ISBN 978-3-319-08461-9 (eBook)
DOI 10.1007/978-3-319-08461-9

Library of Congress Control Number: 2014942310

Springer Cham Heidelberg New York Dordrecht London

Printed on acid-free paper

Springer is part of Springer Science+Business Media (www.springer.com)

Preface

Physical and chemical dissolution-front instability problems exist ubiquitously in many scientific and engineering fields. In geoenvironmental engineering, remediation of contaminated sites using fresh water to flush the contaminated soils involves the propagation of dissolved contaminant fronts in water-saturated porous media. In mineral mining engineering, the extraction of minerals in the deep Earth using the in-situ leaching technique may result in the propagation of dissolved mineral fronts in fluid-saturated porous media. In petroleum industry, the secondary recovery of oil by acidifying the oil field to uniformly increase porosity and hence the yield of oil is associated with the propagation of acid-dissolved material fronts in porous rocks. The common feature of physical and chemical dissolution-front propagation is that it can be mathematically treated as an instability problem involving a highly nonlinear system. When the nonlinear system is in a stable state, any small perturbation applied to the system does not cause any change in the basic characteristic of the dynamic response of the nonlinear system. However, when the nonlinear system is in an unstable state, any small perturbation applied to the system can cause a qualitative change in the basic characteristic of the dynamic response of the system. For this reason, the study of nonlinear system instability has become an important topic in many scientific and engineering fields over the past few decades.

Although dynamic mechanisms of physical and chemical dissolution-front instability phenomena in porous media have been studied and discussed for many years, there is a lack of a systematic and theoretical treatment of these phenomena in the world. Such a treatment is beneficial not only for understanding ore forming mechanisms, which are imperative to develop advanced techniques for exploring new ore deposits in the deep Earth, but also for understanding dissolved contaminant transport, which is important to develop innovative techniques for rehabilitating contaminated soils. By keeping this consideration in mind, this monograph aims to provide state-of-the-art theoretical and computational results in this particular research field, based on the author's own work during the last decade.

For this purpose, although some numerical results are provided to complement theoretical ones, the main focus of this monograph is on theoretical aspects of the topic. This means that the theoretical treatment contained in this monograph is also applicable to a wide range of scientific and engineering problems involving instability phenomena of highly nonlinear dynamic systems. To broaden the readership of this monograph, common mathematical notations are used to derive the theoretical solutions. This enables this monograph to be used either as a useful textbook for postgraduate students or as a valuable reference book for mathematicians, engineers and geoscientists. In addition, each chapter is written independently of the remainder of the monograph so that readers may read the chapter of interest separately.

In this monograph, a series of fundamental theories have been presented for dealing with physical and chemical dissolution-front instability in fluid-saturated porous media. The key of developing these theories is to mathematically establish some theoretical criteria that are used to assess whether or not physical and chemical dissolution fronts can become unstable in the dissolution systems. On the other hand, these fundamental theories can provide some benchmark solutions, which are used to verify numerical methods and algorithms that can be used to simulate complicated physical and chemical dissolution systems in the fluid-saturated porous media. Although theoretical analyses can be conducted for some physical and chemical dissolution problems with simple geometry, it is very difficult, if not impossible, to predict analytically the complicated morphological evolution of either physical or chemical dissolution front in the case of the dissolution system becoming supercritical. As an alternative, numerical methods are developed to overcome this difficulty. Specifically, the overall arrangements of this monograph are as follows: (1) a fundamental theory is presented, in Chap. 2, to deal with chemical dissolution-front instability problems in fluid-saturated porous media. The mathematical analysis is used to establish a theoretical criterion, which is employed to judge the instability of chemical dissolution fronts in a chemical dissolution system, while the computational method is used to simulate the morphological evolution of chemical dissolution fronts when the chemical dissolution system is in a supercritical state. (2) To understand the fundamental behaviors of a chemical dissolution-front instability problem, several key factors, such as the particle reactive surface area, mineral dissolution ratio, solute dispersion, medium permeability anisotropy, and medium/pore-fluid compressibility, have been deliberately considered to investigate their effects on chemical dissolution front instability in Chaps. 3–7, respectively. (3) Three-dimensional computational simulations are conducted, in Chap. 8, to investigate the three-dimensional morphological evolution of chemical dissolution fronts when the chemical dissolution system is in a supercritical state. (4) In Chap. 9, a fundamental theory is presented to deal with physical dissolution-front (known as NAPL dissolution-front) instability problems in fluid-saturated porous media. (5) Based on this theory, domain shapes and mesh discretization errors are considered, in Chap. 10, to investigate their effects on the morphological evolution of NAPL dissolution fronts in fluid-saturated porous media. In particular, the propagation theory of mesh discretization errors associated with a NAPL dissolution system is presented for both rectangular and trapezoidal domains. This further leads to the establishment of the finger-amplitude growing theory, which can be used to make the approximate error estimation of the corresponding computational simulation results. (6) A fundamental theory is presented, in Chap. 11, for dealing with acidization dissolution-front instability problems in fluid-saturated carbonate rocks. (7) Some conclusions and future research directions in this field are summarized at the end of the monograph.

March 2014 Chongbin Zhao

Acknowledgments

Dr. Chongbin Zhao expresses his sincere thanks to his wife, Ms. Peiying Xu, for her persistent support and encouragement, without which it would be impossible to write this monograph. He is very grateful to the Central South University for financial support during writing this monograph. The partial work of this monograph is also financially supported by the Natural Science Foundation of China (Grant Nos: 10872219, 10672190 and 11272359). He expresses his sincere thanks to Dr. Bruce Hobbs and Dr. Alison Ord at the University of Western Australia and the Commonwealth Science and Industrial Research Organization (CSIRO), Australia for their valuable comments and support during carrying out this research.

Contents

Nomenclature

The following symbols are commonly used with the attached definitions, unless otherwise specified in the monograph.

A	Area of a finite element
C	Species concentration
C_{eq}	Equilibrium concentration of the chemical species
C_1	Nonzero constant
C_2	Arbitrary constant
c_p	Specific heat of pore-fluid
D	Mass diffusivity
g	Acceleration due to gravity
H	Reference length
k	Medium permeability
$k_{chemical}$	Conventional rate constant of the chemical reaction
K_{eq}	Equilibrium constant of the chemical reaction
$k_{Echemical}$	Comprehensive rate constant of the chemical reaction
K_h	Reference medium permeability in the horizontal direction
L	Length of a problem domain
Le	Lewis number
p	Pressure
p_0	Hydrostatic pressure
q_C	Mass flux on the boundary of a finite element
q_T	Heat flux on the boundary of a finite element
Ra	Rayleigh number
$Ra_{critical}$	Critical Rayleigh number
S	Boundary length of a finite element
T	Temperature
t	Temporal variable
u	Darcy velocity in the x direction

U	Darcy velocity vector
v	Darcy velocity in the y direction
w	Darcy velocity in the z direction
x, y, z	Spatial coordinates
Zh	Zhao number
$Zh_{critical}$	Critical Zhao number
λ	Thermal conductivity
λ_{e0}	Reference thermal conductivity in the horizontal direction
ϕ	Porosity
ρ_0	Reference density of pore-fluid
μ	Dynamic viscosity of pore-fluid
β	Permeability ratio
σ	Stress on the boundary of a finite element
$\boldsymbol{\varphi}$	Shape function vector for the temperature, species concentration and Darcy velocity of a finite element
ζ	Thermal conductivity ratio of the underlying medium to its overlying folded layer
ε	Mineral dissolution ratio

Subscripts

f	Pertaining to pore-fluid
0	Pertaining to reference quantities

Superscripts

e	Pertaining to equivalent quantities of a porous medium
e	Pertaining to quantities in a finite element level
*	Pertaining to dimensionless quantities
s	Pertaining to solid matrix

Chapter 1
Introduction

Instability of nonlinear systems is a common phenomenon in nature. This phenomenon is the direct consequence of a nonlinear system when it reaches a qualitative change state (i.e. an unstable state) from a quantitative change state (i.e. a stable state). If a nonlinear system is in a stable state, then any small perturbation applied to the system does not cause any change in the basic characteristic of the dynamic response of the system. However, if a nonlinear system is in an unstable state, then any small perturbation applied to the system can cause a qualitative change in the basic characteristic of the dynamic response of the system. For this reason, the study of nonlinear system instability has become an important topic in many scientific and engineering fields over the past few decades.

From the driving force point of view, instability of nonlinear systems encountered in the natural science and engineering fields can be divided into the following four categories: mechanical instability, thermodynamic instability, chemical reaction instability and physical surface (including dissolution front) instability. In terms of mechanical instability, applied mechanical forces are the main driving force to cause the instability of a nonlinear system. The buckling of a column under uni-axial compression is a typical example of mechanical instability in the field of structural engineering. When the axial compressive force applied to the column is equal to or greater than the Euler critical force, which is directly proportional to the flexural rigidity of the column but inversely proportional to the square of the column length, the column losses its initial equilibrium state (i.e. an axial compression state) and reaches an unstable state (i.e. a bending state). Similarly, the folding of the upper crust of the Earth in geology (Donze et al. 1994; Strayer and Huddleston 1997; Finch et al. 2003, 2004; Zhao et al. 2007a–c; Hobbs et al. 2008), which is an analogue to the buckling of a column in the field of structural engineering, is another typical example of mechanical instability in nature.

C. Zhao, *Physical and Chemical Dissolution Front Instability in Porous Media*, Lecture Notes in Earth System Sciences,
DOI: 10.1007/978-3-319-08461-9_1

1.1 Thermodynamic Instability

In terms of thermodynamic instability, temperature gradients are the driving force to cause the instability of a nonlinear system. Convective pore-fluid flow within the upper crust of the Earth is a typical example of thermodynamic instability in geology. When the upper crust of the Earth is heated uniformly from blow, a temperature gradient is generated within the upper crust of the Earth, so that convective pore-fluid flow within the upper crust of the Earth is possible (Horton and Rogers 1945; Lapwood 1948; Nield 1968; Palm et al. 1972; Combarnous and Bories 1975; Caltagirone 1975, 1976; Buretta and Berman 1976; Kaviany 1984; Buck and Parmentier 1986; Riley and Winters 1989; Islam and Nandakumar 1990; Nield and Bejan 1992; Alavyoon 1993; Gobin and Bennacer 1994; Goyeau et al. 1996; Nithiarasu et al. 1996; Zhao et al. 1997, 1998a, b; Chevalier et al. 1999). If this temperature gradient is small, then the pore-fluid resided in the connected pores of crustal rocks is in a rest state. However, if this temperature gradient is large enough, then the pore-fluid resided in the connected pores of the crustal rocks can flow in a convective manner within the upper crust of the Earth. Since convective pore-fluid flow is an important mechanism to control ore body formation, extensive studies have been carried out to investigate, both theoretically and numerically, the thermodynamic instability of the pore-fluid within the upper crust of the Earth (Lebon and Cloot 1986; Pillatsis et al. 1987; Trevisan and Bejan 1987; Nguyen et al. 1994; Mamou et al. 1998; Zhao et al. 1999a–c, 2000a, c, 2001a–b, 2002a–b, 2003a–c, 2004, 2005a, 2006b; Lin et al. 2003; Kühn et al. 2006; Yang 2006; Yang et al. 2010; Hobbs et al. 2000). It has been demonstrated that the convective fluid flow caused by thermodynamic instability can take place in the Earth's crust not only when the pore-fluid pressure is hydrostatic (Nield and Bejan 1992; Zhao et al. 1997, 1998a–b), but also when it is lithostatic (Zhao et al. 1999a, 2008e). This resulted in a conclusion to the long-term debate, in which it was commonly regarded that the convective pore-fluid flow caused by thermodynamic instability cannot take place in the Earth's crust when the pore-fluid pressure is lithostatic (Zhao et al. 1999a). This recognition has been applied to understand the formation mechanisms of Au deposits in Yilgarn Craton, Western Australia (Sorjonen-Ward et al. 2002), Mt Isa Pb-Zn deposits in Queensland, Australia (Ord et al. 2002), Copper-Gold mineralization in the New Guinea (Gow et al. 2002). In addition, it has also been demonstrated that the convective pore-fluid flow caused by thermodynamic instability can take place in 3D fault zones (Zhao et al. 2003b, 2004, 2005a). This recognition is applied to understand the mineralization mechanisms not only in generic models (Alt-Epping and Zhao 2010), but also in realistic geological models (Zhao et al. 2008b).

1.2 Chemical Dynamic Instability

For geochemical systems, a homogeneous chemical reaction is referred to as the chemical reaction taking place within the same phase, such as between aqueous species in the fluid, while a heterogeneous chemical reaction is referred to as the chemical reaction taking place between two or more phases, such as between the fluid and the minerals. From this point of view, chemical dynamic instability can be divided into the following two types. The first type of chemical dynamic instability takes place within geochemical systems that are controlled by homogeneous chemical reactions, while the second type of chemical dynamic instability takes place within geochemical systems that are controlled by heterogeneous chemical reactions. The formation of dissipative structures of aqueous chemical species within a nonequilibrium chemical system of the Brusselator type of nonequilibrium chemical reactions (Prigogine 1980) in a fluid-saturated porous medium is a typical example of the first type of chemical dynamic instability. This instability mechanism has been used to understand the local enrichment of aqueous chemical species (such as metal ions) in ore-forming systems within the upper crust of the Earth (Fisher and Lasaga 1981; Epstein and Pojman 1998; Zhao et al. 2000b; Hobbs et al. 2011; Ord et al. 2012).

The second type of chemical dynamic instability usually involves the propagation of chemical dissolution fronts within fluid-saturated porous media, so that it is also called the chemical dissolution-front instability in the field of geochemistry. When fresh fluid (e.g. water) enters a solution-saturated porous medium, where the concentration of the solute (i.e. dissolvable solid mineral) reaches its equilibrium concentration, the concentration of the dissolvable mineral is diluted so that the dissolvable mineral is dissolved to maintain the equilibrium state of the solution. This chemical dissolution process can result in the propagation of a chemical dissolution front within the solution-saturated porous medium. Due to the dissolution of the solid mineral, the porosity of the porous medium is increased behind the chemical dissolution front. Since a change in porosity can cause a remarkable change in permeability, there is a feedback effect of the porosity change on pore-fluid flow, according to Darcy's law. Because pore-fluid flow plays an important role in the process of reactive chemical-species transport, a change in pore-fluid flow can cause a considerable change in the concentration of the dissolvable mineral within the porous medium (Steefel and Lasaga 1990, 1994; Yeh and Tripathi 1991; Raffensperger and Garven 1995; Schafer et al. 1998a–b; Xu et al. 1999, 2004; Ormond and Ortoleva 2000; Chen and Liu 2002; Zhao et al. 2005b, 2006a). This means that the problem associated with the propagation of a dissolution front is a fully coupled nonlinear problem between porosity, pore-fluid pressure and reactive chemical-species transport within the fluid-saturated porous medium. If the fresh fluid flow is slow, the feedback effect of the porosity change is weak so that the chemical dissolution front is stable. However, if the fresh fluid flow is fast enough, the feedback effect of the porosity change becomes strong so that the chemical dissolution front becomes unstable. In this case, a new morphology (i.e. dissipative structure) of the chemical dissolution front can emerge due to the self-organization of this coupled nonlinear system. This leads to the chemical dissolution front instability problem, also known as the reactive infiltration instability problem (Chadam et al. 1986, 1988; Ortoleva et al. 1987; Zhao et al. 2008a, c–e, 2009, 2010a–b, 2012a, c, 2013a), which is closely associated with mineral dissolution in a fluid-saturated porous medium. In this case, the dissolvable minerals only occupy a small portion of the porous medium, so that the porosity variation caused by the chemical dissolution does not dismantle the whole skeleton of the porous medium.

On the other hand, instability of acidization dissolution fronts (which can be viewed as a special kind of chemical dissolution front) in carbonate rocks is an important mechanism of the karst formation that is commonly observed in nature. This mechanism has been successfully used to increase the oil product in petroleum industry through injecting hydrochloric acid into carbonate rocks in the surroundings of drilling wells (Rowan 1959; Fredd and Fogler 1998; Golfier et al. 2002; Panga et al. 2005; Kalia and Balakotaiah 2007, 2009; Cohen et al. 2008; Zhao et al. 2013b–c). Due to the rapid rates of acidization dissolution, the dissolution front may become unstable under certain conditions, so that preferential flow channels such as wormholes can be generated. Generally, the acidization dissolution process of carbonate rocks involves the following three steps: (1) the acid is transported by diffusion and advection to the solid particle surfaces of the porous rock; (2) the chemical dissolution reaction takes place at the solid particle surfaces of the porous rock; and (3) the products of the chemical dissolution reaction are transported away from the solid particle surfaces of the porous rock. This means that the pore-fluid flow, which transports both the acid to and the chemical production away from the solid particle surfaces of the porous rock through advection, plays a critical role in the acidization dissolution process of carbonate rocks. In this situation, the rock itself can be completely dissolved, resulting in no resistance to pore-fluid flow. This means that both the porosity and permeability of the carbonate rock can have a dramatic change during the acidization dissolution.

1.3 Physical Surface (Including Dissolution Front) Instability

In terms of physical surface (including dissolution front) instability, the first example is to consider the propagation of an interface between two fluids of different viscosities in a fluid saturated porous medium. One of the engineering backgrounds for this example is that water is used to expel crude oil in an oil field so that the crude oil production can be raised. When a less viscous displacing fluid (such as water) is injected into a more viscous displaced fluid (such as crude oil), the original shape of the interface (e.g. a planar shape) between them can become unstable if their viscosity contrast is large enough, so that a new shape of the interface (e.g. a fingerlike shape) will emerge. This phenomenon is called the viscous fingering instability (Tan and Homsy 1986, 1987; De Wit and Homsy 1997a–b, 1999). From how the two viscous fluids interact, the viscous fingering instability can involve the following two mixing mechanisms (Tan and Homsy 1986, 1987, 1988; Zimmerman and Homsy 1991). The first is the immiscible displacement, in which the surface tension at the interface cannot be neglected, while the second is the miscible displacement, in which the surface tension at the interface can be neglected.

The second example of physical surface instability is the dissolution front instability of the nonaqueous phase liquid (NAPL), which is usually trapped at residual saturation in the pore space of a fluid-saturated porous medium and can become long-term sources of groundwater contamination as a result of its low solubility (Miller et al. 1990, 1998; Geller and Hunt 1993; Powers et al. 1994; Imhoff et al. 1994, 1996, 2002, 2003; Soerens et al. 1998; Willson et al. 1999; Seyedabbasi et al. 2008; Zhao et al. 2010c, 2011, 2012b). One of the engineering backgrounds for this example is to use water flow to flush the NAPL-type contaminants from contaminated soils in geoenvironmental engineering. When fresh water passes through a fluid-saturated porous medium that is contaminated by a NAPL, it dilutes the equilibrium concentration of the NAPL so that the solution of the NAPL is changed from an equilibrium state into a nonequilibrium state. To maintain the equilibrium state of the solution, the trapped NAPL within the pore space of the fluid-saturated porous medium is released into the aqueous phase fluid (i.e. water in this investigation) by physical elution dissolution. This indicates that as the first role played in a NAPL dissolution system, fresh water flow is the trigger to break the initial equilibrium state of the NAPL dissolution system. Since water flow can transport the dissolved NAPL from one place to another, the second role of fresh water flow is to maintain solute advection within the NAPL dissolution system. Thus, solute advection is the first important mechanism involved in a NAPL dissolution system. Another important mechanism involved in a NAPL dissolution system is the physical elution dissolution of the NAPL, which can also play two roles in the NAPL dissolution system. Firstly, the physical elution dissolution can cause a reduction in the trapped NAPL so that the NAPL can be removed from the contaminated porous medium by solute advection. Secondly, the physical elution dissolution can cause an increase in fluid flow pathways so that the permeability of the porous medium can be increased as a result of the NAPL dissolution. In addition, solute can be transported from a high concentration region to a low concentration one through solute diffusion/dispersion, which is the third important mechanism involved in a NAPL dissolution system. Due to the involvement of the above-mentioned mechanisms, a NAPL dissolution problem can be attributed to a typical reactive mass transport problem in the fluid-saturated porous medium.

Although there are some similarities between chemical dissolution-front instability problems and NAPL dissolution-front instability ones, the following considerable differences exist between them. First, in the case of chemical dissolution induced instability the dissolution process in a fluid-saturated porous medium is dominated by the chemical reaction process (i.e. the diagenetic process or water-rock chemical reaction process), while in the case of NAPL dissolution induced instability it is dominated by the physical elution process (Miller et al. 1998; Zhao et al. 2010c, 2011, 2012b). Second, the mineral dissolution ratio, which is defined as the ratio of the equilibrium concentration of the dissolvable mineral to the molar density of the solid mineral, is in a rang of 10^{-3}–10^{-9} for typical rocks (Chadam et al. 1986), while the NAPL dissolution ratio, which is similarly defined as the ratio of the equilibrium concentration of a NAPL in the aqueous phase fluid to the mass density of the NAPL itself, is of the order of 10^{-3} (Imhoff and Miller 1996). This means that the time scale of a mineral dissolution system is much larger than that of a NAPL dissolution system. For this reason, it is possible to conduct laboratory experiments to observe NAPL dissolution-front fingering phenomena (Imhoff et al. 1996), but it is very difficult, if not impossible, to observe the mineral dissolution-front instability phenomenon through laboratory experiments. Third, the pore-fluid velocity in a mineral dissolution system is often a few centimeters per year; while the aqueous phase fluid velocity in a NAPL dissolution system may reach a few meters per day. This indicates that solute dispersion (i.e. mechanical dispersion) may be neglected for a mineral dissolution system, but it must be considered in a NAPL dissolution system.

The core issue of dealing with nonlinear system instability problems is to establish a series of theoretical criteria, which are used to judge whether or not the nonlinear systems can become unstable. This goal is usually achievable through the following two mathematical ways. In the first way, the governing equations (consisting of a partial differential equation or a set of partial differential equations) of the problem are considered to derive the critical condition, at which a singular solution (or solutions) can be obtained (Horton and Rogers 1945; Lapwood 1948; Nield and Bejan 1992; Zhao et al. 1997, 1999b, 2003c, 2004, 2005a). Once this critical condition is obtained, it can be used to serve as the theoretical criteria for assessing whether or not the nonlinear system described by the considered governing equations can become unstable. In the second way, the linear stability theory is used to derive the growth rate of a small perturbation applied to the nonlinear system (Chadam et al. 1986, 1988; Ortoleva et al. 1987; Zhao et al. 2008c–e, 2009, 2010a–b, 2012a–b, 2013a). The condition at which the growth rate of the applied small perturbation is equal to zero can be used as the theoretical criteria for assessing whether or not the nonlinear system under consideration can become unstable.

1.4 The Contextual Arrangements of This Monograph

In this monograph a series of fundamental theories have been presented for dealing with physical and chemical dissolution-front instability in fluid-saturated porous media. The core of them is to mathematically establish some theoretical criteria that are used to assess whether or not physical and chemical dissolution fronts can become unstable in the dissolution systems. On the other hand, these fundamental theories can provide some benchmark solutions, which are used to verify numerical methods and algorithms that can be used to simulate complicated physical and chemical dissolution systems in fluid-saturated porous media. Although theoretical analyses can be conducted for some physical and chemical dissolution problems with simple geometry, it is very difficult, if not impossible, to predict analytically the complicated morphological evolution of either physical or chemical dissolution front in the case of the dissolution system becoming supercritical. As an alternative, numerical methods are developed to overcome this difficulty. The physical and chemical dissolution-front instability problems to be considered are closely related not only to many geological phenomena such as ore body formation and mineralization as well as the karst formation within the upper crust of the Earth, but also to many engineering problems such as chemical in-situ mining, flushing contaminated soils by water and expelling oil by water/acid.

The arrangements of the forthcoming parts of this monograph are as follows: In Chap. 2, a fundamental theory is presented to deal with chemical dissolution-front instability problems in fluid-saturated porous media. The mathematical analysis is used to establish a theoretical criterion, which is employed to judge whether or not chemical dissolution fronts become unstable, while the computational method is used to simulate the morphological evolution of chemical dissolution fronts when the chemical dissolution system is in a supercritical state. To understand the fundamental behaviors of a chemical dissolution-front instability problem, several key factors, such as the particle reactive surface area, mineral dissolution ratio, solute dispersion, medium permeability anisotropy, and medium/pore-fluid compressibility, have been deliberately considered to investigate their effects on chemical dissolution-front instability in Chaps. 3, 4, 5, 6 and 7, respectively. In Chap. 8, three-dimensional computational simulations are carried out to investigate the three-dimensional morphological evolution of chemical dissolution fronts when the chemical dissolution system is in a supercritical state. In Chap. 9, a fundamental theory is presented to deal with physical dissolution-front (known as NAPL dissolution-front) instability problems in fluid-saturated porous media. Based on this theory, domain shapes and mesh discretization error are considered, in Chap. 10, to investigate their effects on the morphological evolution of NAPL dissolution fronts in fluid-saturated porous media. In Chap. 11, a fundamental theory is presented for dealing with acidization dissolution-front instability problems in fluid-saturated carbonate rocks. Finally, summary statements are given at the end of the monograph.

References

Alavyoon F (1993) On natural convection in vertical porous enclosures due to prescribed fluxes of heat and mass at the vertical boundaries. Int J Heat Mass Transf 36:2479–2498

Alt-Epping P, Zhao C (2010) Reactive mass transport modeling of a three-dimensional vertical fault zone with a finger-like convective flow regime. J Geochem Explor 106:8–23

Buck WR, Parmentier EM (1986) Convection between young oceanic lithosphere: Implication for thermal structure and gravity. J Geophys Res 91:1961–1974

Buretta RJ, Berman AS (1976) Convective heat transfer in a liquid saturated porous layer. ASME J Appl Mech 43:249–253

Caltagirone JP (1975) Thermoconvective instabilities in a horizontal layer. J Fluid Mech 72:269–287

Caltagirone JP (1976) Thermoconvective instabilities in a porous medium bounded by two concentric horizontal cylinders. J Fluid Mech 76:337–362

Chadam J, Hoff D, Merino E, Ortoleva P, Sen A (1986) Reactive infiltration instabilities. IMA J Appl Math 36:207–221

Chadam J, Ortoleva P, Sen A (1988) A weekly nonlinear stability analysis of the reactive infiltration interface. IMA J Appl Math 48:1362–1378

Chen JS, Liu CW (2002) Numerical simulation of the evolution of aquifer porosity and species concentrations during reactive transport. Comput Geosci 28:485–499

Chevalier S, Bernard D, Joly N (1999) Natural convection in a porous layer bounded by impervious domains: from numerical approaches to experimental realization. Int J Heat Mass Transf 42:581–597

Cohen CE, Ding D, Quintard M, Bazin B (2008) From pore scale to wellbore scale: impact of geometry on wormhole growth in carbonate acidization. Chem Eng Sci 63:3088–3099

Combarnous MA, Bories SA (1975) Hydrothermal convection in saturated porous media. Adv Hydrosci 10:231–307

De Wit A, Homsy GM (1997a) Viscous fingering in periodically heterogeneous porous media. I Formulation and linear instability. J Chem Phys 107:9609–9618

De Wit A, Homsy GM (1997b) Viscous fingering in periodically heterogeneous porous media. II. Numerical simulations. J Chem Phys 107:9619–9628

De Wit A, Homsy GM (1999) Viscous fingering in reaction: diffusion systems. J Chem Phys 110:8663–8675

Donze F, Mora P, Magnier SA (1994) Numerical simulation of faults and shear zones. Geophys J Int 116:46–52

Epstein IR, Pojman JA (1998) An introduction to nonlinear chemical dynamics. Oxford, New York, p 392

Finch E, Hardy S, Gawthorpe R (2003) Discrete element modeling of contractional fault-propagation folding above rigid basement fault blocks. J Struct Geol 25:515–528

Finch E, Hardy S, Gawthorpe R (2004) Discrete element modeling of extensional fault-propagation folding above rigid basement fault blocks. Basin Res 16:489–506

Fisher GW, Lasaga AC (1981) Irreversible thermodynamics in petrology. Rev Mineral Mineral Soc Am 8:171–210

Fredd CN, Fogler HS (1998) Influence of transport and reaction on wormhole formation in porous media. AIChE J 44:1933–1949

Geller JT, Hunt JR (1993) Mass transfer from nonaqueous phase organic liquids in water-saturated porous media. Water Resour Res 29:833–845

Gobin D, Bennacer R (1994) Double diffusion convection in a vertical fluid layer: onset of the convection regime. Phys Fluids 6:59–67

Golfier F, Zarcone C, Bazin B, Lenormand R, Lasseux D, Quintard M (2002) On the ability of a Darcy-scale model to capture wormhole formation during the dissolution of a porous medium. J Fluid Mech 457:213–254

Gow P, Upton P, Zhao C, Hill K (2002) Copper-gold mineralization in the New Guinea: numerical modeling of collision, fluid flow and intrusion-related hydrothermal systems. Aust J Earth Sci 49:753–771

Goyeau B, Songbe JP, Gobin D (1996) Numerical study of double-diffusive convection in a porous cavity using Darcy-Brinkman formulation. Int J Heat Mass Transf 39:1363–1378

Hobbs BE, Zhang Y, Ord A, Zhao C (2000) Application of coupled deformation, fluid flow, thermal and chemical modelling to predictive mineral exploration. J Geochem Explor 69:505–509 (folding with thermal-mechanical feedback)

Hobbs BE, Regenauer-Lieb K, Ord A (2008) Folding with thermal-mechanical feedback. J Struct Geol 30:1572–1592

Hobbs BE, Ord A, Regenauer-Lieb K (2011) The thermodynamics of deformed metamorphic rocks: a review. J Struct Geol 33:758–818

Horton CW, Rogers FT (1945) Convection currents in a porous medium. J Appl Phys 16:367–370

Imhoff PT, Miller CT (1996) Dissolution fingering during the solubilization of nonaqueous phase liquids in saturated porous media: 1. Model predictions. Water Resour Res 32:1919–1928

Imhoff PT, Jaffe PR, Pinder GF (1994) An experimental study of complete dissolution of a nonaqueous phase liquid in saturated porous media. Water Resour Res 30:307–320

Imhoff PT, Thyrum GP, Miller CT (1996) Dissolution fingering during the solubilization of nonaqueous phase liquids in saturated porous media: 2. Experimental observations. Water Resour Res 32:1929–1942

Imhoff PT, Farthing MW, Gleyzer SN, Miller CT (2002) Evolving interface between clean and nonaqueous phase liquid (NAPL)-contaminated regions in two-dimensional porous media. Water Resour Res 38:1093–1106

Imhoff PT, Farthing MW, Miller CT (2003) Modeling NAPL dissolution fingering with upscaled mass transfer rate coefficients. Adv Water Resour 26:1097–1111

Islam MR, Nandakumar K (1990) Transient convection in saturated porous layers. Int J Heat Mass Transf 33:151–161

Kalia N, Balakotaiah V (2007) Modeling and analysis of wormhole formation in reactive dissolution in carbonate rocks. Chem Eng Sci 62:919–928

Kalia N, Balakotaiah V (2009) Effect of medium heterogeneities on reactive dissolution of carbonates. Chem Eng Sci 64:376–390

Kaviany M (1984) Thermal convective instabilities in a porous medium. ASME J Heat Transf 106:137–142

Kühn M, Dobertb F, Gessner K (2006) Numerical investigation of the effect of heterogeneous permeability distributions on free convection in the hydrothermal system at Mount Isa, Australia. Earth Planet Sci Lett 244:655–671

Lapwood ER (1948) Convection of a fluid in a porous medium. Proc Cambridge Philos Soc 44:508–521

Lebon G, Cloot A (1986) A thermodynamical modeling of fluid flows through porous media: Application to natural convection. Int J Heat Mass Transf 29:381–390

Lin G, Zhao C, Hobbs BE, Ord A, Muhlhaus HB (2003) Theoretical and numerical analyses of convective instability in porous media with temperature-dependent viscosity. Commun Numer Method Eng 19:787–799

Mamou M, Vasseur P, Bilgen E (1998) Double-diffusive convection instability in a vertical porous enclosure. J Fluid Mech 368:263–289

Miller CT, Poirier-McNeil MM, Mayer AS (1990) Dissolution of trapped nonaqueous phase liquids: mass transfer characteristics. Water Resour Res 26:2783–2796

Miller CT, Christakos TG, Imhoff PT, McBride JF, Pedit JA, Trangenstein JA (1998) Multiphase flow and transport modeling in heterogeneous porous media: challenges and approaches. Adv Water Resour 21:77–120

Nguyen HD, Pack S, Douglass RW (1994) Study of double diffusive convection in a layered anisotropic porous medium. Numer Heat Transf 26:489–505

Nield DA (1968) Onset of thermohaline convection in a porous medium. Water Resour Res 11:553–560

Nield DA, Bejan A (1992) Convection in porous media. Springer, New York
Nithiarasu P, Seetharamu KN, Sundararajan T (1996) Double-diffusive natural convection in an enclosure filled with fluid-saturated porous medium: a generalized non-Darcy approach. Numer Heat Transf 30:430–436
Ord A, Hobbs BE, Zhang Y, Broadbent GC, Brown M, Willetts G, Sorjonen-Ward P, Walshe J, Zhao C (2002) Geodynamic modelling of the century deposit, Mt Isa Province, Queensland. Aust J Earth Sci 49:1011–1039
Ord A, Hobbs BE, Lester DR (2012) The mechanics of hydrothermal systems: I. Ore systems as chemical reactors. Ore Geol Rev 49:1–44
Ormond A, Ortoleva P (2000) Numerical modeling of reaction-induced cavities in a porous rock. J Geophys Res 105:16737–16747
Ortoleva P, Chadam J, Merino E, Sen A (1987) Geochemical self-organization II: the reactive-infiltration instability. Am J Sci 287:1008–1040
Palm E, Weber JE, Kvernvold O (1972) On steady convection in a porous medium. J Fluid Mech 54:153–161
Panga MKR, Ziauddin M, Balakotaiah V (2005) Two-scale continuum model for simulation of wormholes in carbonate acidization. AIChE J 51:3231–3248
Pillatsis G, Taslim ME, Narusawa U (1987) Thermal instability of a fluid-saturated porous medium bounded by thin fluid layers. ASME J Heat Transf 109:677–682
Powers SE, Abriola LM, Weber WJ Jr (1994) An experimental investigation of nonaqueous phase liquid dissolution in saturated subsurface systems: transient mass transfer rates. Water Resour Res 30:321–332
Prigogine I (1980) From being to becoming. W. H. Freeman and Company, New York
Raffensperger JP, Garven G (1995) The formation of unconformity-type uranium ore deposits: coupled hydrochemical modelling. Am J Sci 295:639–696
Riley DS, Winters KH (1989) Modal exchange mechanisms in Lapwood convection. J Fluid Mech 204:325–358
Rowan G (1959) Theory of acid treatment of limestone formations. J Inst Petrol 45:321–334
Schafer D, Schafer W, Kinzelbach W (1998a) Simulation of reactive processes related to biodegradation in aquifers: 1. Structure of the three-dimensional reactive transport model. J Contam Hydrol 31:167–186
Schafer D, Schafer W, Kinzelbach W (1998b) Simulation of reactive processes related to biodegradation in aquifers: 2. Model application to a column study on organic carbon degradation. J Contam Hydrol 31:187–209
Seyedabbasi MA, Farthing MW, Imhoff PT, Miller CT (2008) The influence of wettability on NAPL dissolution fingering. Adv Water Resour 31:1687–1696
Soerens TS, Sabatini DA, Harwell JH (1998) Effects of flow bypassing and nonuniform NAPL distribution on the mass transfer characteristics of NAPL dissolution. Water Resour Res 34:1657–1673
Sorjonen-Ward P, Zhang Y, Zhao C (2002) Numerical modelling of orogenic processes and mineralization in the south eastern part of the Yilgarn Craton, Western Australia. Aust J Earth Sci 49:935–964
Steefel CI, Lasaga AC (1990) Evolution of dissolution patterns: permeability change due to coupled flow and reaction. In: Melchior DC, Basset RL (eds) Chemical modeling in aqueous systems II, American Chemistry Society Symposium Series, vol 416, pp 213–225
Steefel CI, Lasaga AC (1994) A coupled model for transport of multiple chemical species and kinetic precipitation/dissolution reactions with application to reactive flow in single phase hydrothermal systems. Am J Sci 294:529–592
Strayer LM, Huddleston PJ (1997) Numerical modeling of fold initiation at thrust ramps. J Struct Geol 19:551–566
Tan CT, Homsy GM (1986) Stability of miscible displacements in porous media: rectilinear flow. Phys Fluids 29(11):3549–3556
Tan CT, Homsy GM (1987) Stability of miscible displacements in porous media: radial source flow. Phys Fluids 30(5):1239–1245

Tan CT, Homsy GM (1988) Simulation of nonlinear viscous fingering in miscible displacements. Phys Fluids 31(6):1330–1338

Trevisan OV, Bejan A (1987) Mass and heat transfer by high Rayleigh number convection in a porous medium heated from below. Int J Mass Heat Transf 30:2341–2356

Willson CS, Hall JL, Miller CT, Imhoff PT (1999) Factors affecting bank formation during surfactant-enhanced mobilization of residual NAPL. Environ Sci Technol 33:2440–2446

Xu TF, Samper J, Ayora C, Manzano M, Custodio E (1999) Modelling of non-isothermal multi-component reactive transport in field scale porous media flow systems. J Hydrol 214:144–164

Xu TF, Apps JA, Pruess K (2004) Numerical simulation of CO2 disposal by mineral trapping in deep aquifers. Appl Geochem 19:917–936

Yang JW (2006) Full 3-D numerical simulation of hydrothermal fluid flow in faulted sedimentary basins: example of the McArthur Basin, Northern Australia. J Geochem Explor 89:440–444

Yang JW, Feng Z, Luo X, Chen Y (2010) Three-dimensional numerical modeling of salinity variations in driving basin-scale ore-forming fluid flow: example from Mount Isa Basin, northern Australia. J Geochem Explor 106:236–243

Yeh GT, Tripathi VS (1991) A model for simulating transport of reactive multispecies components: model development and demonstration. Water Resour Res 27:3075–3094

Zhao C, Mühlhaus HB, Hobbs BE (1997) Finite element analysis of steady-state natural convection problems in fluid-saturated porous media heated from below. Int J Numer Anal Meth Geomech 21:863–881

Zhao C, Hobbs BE, Mühlhaus HB (1998a) Finite element modelling of temperature gradient driven rock alteration and mineralization in porous rock masses. Comput Methods Appl Mech Eng 165:175–187

Zhao C, Muhlhaus HB, Hobbs BE (1998b) Effects of geological inhomogeneity on high Rayleigh number steady-state heat and mass transfer in fluid-saturated porous media heated from below. Numer Heat Transf 33:415–431

Zhao C, Hobbs BE, Baxter K, Mühlhaus HB, Ord A (1999a) A numerical study of pore-fluid, thermal and mass flow in fluid-saturated porous rock basins. Eng Comput 16:202–214

Zhao C, Hobbs BE, Mühlhaus HB (1999b) Theoretical and numerical analyses of convective instability in porous media with upward throughflow. Int J Numer Anal Meth Geomech 23:629–646

Zhao C, Hobbs BE, Mühlhaus HB (1999c) Effects of medium thermoelasticity on high Rayleigh number steady-state heat transfer and mineralization in deformable fluid-saturated porous media heated from below. Comput Methods Appl Mech Eng 173:41–54

Zhao C, Hobbs BE, Mühlhaus HB (2000a) Finite element analysis of heat transfer and mineralization in layered hydrothermal systems with upward throughflow. Comput Methods Appl Mech Eng 186:49–64

Zhao C, Hobbs BE, Mühlhaus HB, Ord A (2000b) Finite element modelling of dissipative structures for nonequilibrium chemical reactions in fluid-saturated porous media. Comput Methods Appl Mech Eng 184:1–14

Zhao C, Hobbs BE, Mühlhaus HB, Ord A, Lin G (2000c) Numerical modelling of double diffusion driven reactive flow transport in deformable fluid-saturated porous media with particular consideration of temperature-dependent chemical reaction rates. Eng Comput 17:367–385

Zhao C, Hobbs BE, Mühlhaus HB, Ord A, Lin G (2001a) Finite element modelling of three-dimensional convection problems in pore-fluid saturated porous media heated from below. Commun Numer Methods Eng 17:101–114

Zhao C, Lin G, Hobbs BE, Mühlhaus HB, Ord A, Wang Y (2001b) Finite element modelling of heat transfer through permeable cracks in hydrothermal systems with upward throughflow. Eng Comput 18:996–1011

Zhao C, Hobbs BE, Mühlhaus HB, Ord A, Lin G (2002a) Computer simulations of coupled problems in geological and geochemical systems. Comput Methods Appl Mech Eng 191:3137–3152

Zhao C, Lin G, Hobbs BE, Wang Y, Mühlhaus HB, Ord A (2002b) Finite element modelling of reactive fluids mixing and mineralization in pore-fluid saturated hydrothermal/sedimentary basins. Eng Comput 19:364–387

Zhao C, Hobbs BE, Mühlhaus HB, Ord A, Lin G (2003a) Finite element modeling of three-dimensional steady-state convection and lead/zinc mineralization in fluid-saturated rocks. J Comput Methods Sci Eng 3:73–89

Zhao C, Hobbs BE, Mühlhaus HB, Ord A, Lin G (2003b) Convective instability of three-dimensional fluid-saturated geological fault zones heated from below. Geophys J Int 155:213–220

Zhao C, Hobbs BE, Ord A, Mühlhaus HB, Lin G (2003c) Effect of material anisotropy on the onset of convective flow in three-dimensional fluid-saturated faults. Math Geol 35:141–154

Zhao C, Hobbs BE, Ord A, Peng S, Mühlhaus HB, Liu L (2004) Theoretical investigation of convective instability in inclined and fluid-saturated three-dimensional fault zones. Tectonophysics 387:47–64

Zhao C, Hobbs BE, Ord A, Peng S, Mühlhaus HB, Liu L (2005a) Double diffusion-driven convective instability of three-dimensional fluid-saturated geological fault zones heated from below. Math Geol 37:373–391

Zhao C, Hobbs BE, Ord A, Peng S, Mühlhaus HB, Liu L (2005b) Numerical modeling of chemical effects of magma solidification problems in porous rocks. Int J Numer Methods Eng 64:709–728

Zhao C, Hobbs BE, Ord A, Hornby P (2006a) Chemical reaction patterns due to fluids mixing and focusing around faults in fluid-saturated porous rocks. J Geochem Explor 89:470–473

Zhao C, Hobbs BE, Ord A, Kuhn M, Mühlhaus HB, Peng S (2006b) Numerical simulation of double-diffusion driven convective flow and rock alteration in three-dimensional fluid-saturated geological fault zones. Comput Methods Appl Mech Eng 195:2816–2840

Zhao C, Hobbs BE, Ord A, Hornby P, Peng S, Liu L (2007a) Particle simulation of spontaneous crack generation problems in large-scale quasi-static systems. Int J Numer Meth Eng 69:2302–2329

Zhao C, Hobbs BE, Ord A, Peng S, Liu L (2007b) An upscale theory of particle simulation for two-dimensional quasi-static problems. Int J Numer Meth Eng 72:397–421

Zhao C, Hobbs BE, Ord A, Robert PA, Hornby P, Peng S (2007c) Phenomenological modeling of crack generation in brittle crustal rocks using the particle simulation method. J Struct Geol 29:1034–1048

Zhao C, Hobbs BE, Hornby P, Ord A, Peng S, Liu L (2008a) Theoretical and numerical analyses of chemical-dissolution front instability in fluid-saturated porous rocks. Int J Numer Anal Meth Geomech 32:1107–1130

Zhao C, Hobbs BE, Ord A (2008b) Investigating dynamic mechanisms of geological phenomena using methodology of computational geosciences: an example of equal-distant mineralization in a fault. Sci Chin D-Ser Earth Sci 51:947–954

Zhao C, Hobbs BE, Ord A, Hornby P, Peng S (2008c) Effect of reactive surface areas associated with different particle shapes on chemical-dissolution front instability in fluid-saturated porous rocks. Transp Porous Media 73:75–94

Zhao C, Hobbs BE, Ord A, Hornby P, Peng S (2008d) Morphological evolution of three-dimensional chemical dissolution front in fluid-saturated porous media: a numerical simulation approach. Geofluids 8:113–127

Zhao C, Hobbs BE, Ord A (2008e) Convective and advective heat transfer in geological systems. Springer, Berlin

Zhao C, Hobbs BE, Ord A (2009) Fundamentals of computational geoscience: numerical methods and algorithms. Springer, Berlin

Zhao C, Hobbs BE, Ord A (2010a) Theoretical analyses of the effects of solute dispersion on chemical-dissolution front instability in fluid-saturated porous rocks. Transp Porous Media 84:629–653

Zhao C, Hobbs BE, Ord A, Peng S (2010b) Effects of mineral dissolution ratios on chemical-dissolution front instability in fluid-saturated porous media. Transp Porous Media 82:317–335

Zhao C, Hobbs BE, Ord A (2010c) Theoretical analyses of nonaqueous-phase-liquid dissolution induced instability in two-dimensional fluid-saturated porous media. Int J Numer Anal Meth Geomech 34:1767–1796

Zhao C, Hobbs BE, Regenauer-Lieb K, Ord A (2011) Computational simulation for the morphological evolution of nonaqueous-phase-liquid dissolution fronts in two-dimensional fluid-saturated porous media. Comput Geosci 15:167–183

Zhao C, Hobbs BE, Ord A (2012a) Effects of medium and pore-fluid compressibility on chemical-dissolution front instability in fluid-saturated porous media. Int J Numer Anal Meth Geomech 36:1077–1100

Zhao C, Hobbs BE, Ord A (2012b) Effects of domain shapes on the morphological evolution of nonaqueous-phase-liquid dissolution fronts in fluid-saturated porous media. J Contam Hydrol 138–139:123–140

Zhao C, Reid LB, Regenauer-Lieb K, Poulet T (2012c) A porosity-gradient replacement approach for computational simulation of chemical-dissolution front propagation in fluid-saturated porous media including pore-fluid compressibility. Comput Geosci 16:735–755

Zhao C, Hobbs BE, Ord A (2013a) Effects of medium permeability anisotropy on chemical-dissolution front instability in fluid-saturated porous rocks. Transp Porous Media 99:119–143

Zhao C, Hobbs BE, Ord A (2013b) Theoretical analyses of acidization dissolution front instability in fluid-saturated carbonate rocks. Int J Numer Anal Meth Geomech 37:2084–2105

Zhao C, Poulet T, Regenauer-Lieb K, Hobbs BE (2013c) Computational modeling of moving interfaces between fluid and porous medium domains. Comput Geosci 17:151–166

Zimmerman WB, Homsy GM (1991) Nonlinear viscous fingering in miscible displacement with anisotropic dispersion. Phys Fluids A 3(8):1859–1872

Chapter 2
Fundamental Theory for Chemical Dissolution-Front Instability Problems in Fluid-Saturated Porous Media

When fresh pore-fluid enters a solute-saturated porous medium, where the concentration of the solute (i.e. aqueous mineral) reaches its equilibrium concentration, the concentration of the aqueous mineral is diluted so that the solid part of the solute (i.e. solid mineral) is dissolved to maintain the equilibrium state of the solution. This chemical dissolution process can result in the propagation of a dissolution front within the fluid-saturated porous medium. Due to the dissolution of the solid mineral, the porosity of the porous medium is increased behind the dissolution front. Since a change in porosity can cause a remarkable change in permeability, there is a feedback effect of the porosity change on the pore-fluid flow, according to Darcy's law. Because pore-fluid flow plays an important role in the process of reactive chemical-species transport, a change in pore-fluid flow can cause a considerable change in the chemical-species concentration within the porous medium (Steefel and Lasage 1990, 1994; Yeh and Tripathi 1991; Raffensperger and Garven 1995; Schafer et al. 1998a, b; Xu et al. 1999, 2004; Ormond and Ortoleva 2000; Chen and Liu 2002; Zhao et al. 2005, 2006a). This means that the problem associated with the propagation of a dissolution front is a fully coupled nonlinear problem between porosity, pore-fluid pressure and reactive chemical-species transport within the fluid-saturated porous medium. If the fresh pore-fluid flow is slow, the feedback effect of the porosity change is weak so that the dissolution front is stable. However, if the fresh pore-fluid flow is fast enough, the feedback effect of the porosity change becomes strong so that the dissolution front becomes unstable. In this case, a new morphology (i.e. dissipative structure) of the dissolution front can emerge due to the self-organization of this coupled nonlinear system. This leads to an important scientific problem, known as the chemical dissolution front instability problem (Zhao et al. 2008a, b), which is closely associated with mineral dissolution in a fluid-saturated porous medium.

C. Zhao, *Physical and Chemical Dissolution Front Instability in Porous Media*, Lecture Notes in Earth System Sciences,
DOI: 10.1007/978-3-319-08461-9_2

This kind of chemical dissolution-front instability problem exists ubiquitously in many scientific and engineering fields. For example, in geo-environmental engineering, the rehabilitation of contaminated sites using fresh water to wash the sites involves the propagation problem of the removed contaminant material front in a water-saturated porous medium. In mineral mining engineering, the extraction of minerals in the deep Earth using the in-situ leaching technique may result in the propagation problem of the dissolved mineral front in a fluid-saturated porous medium. In the petroleum industry, the secondary recovery of oil by acidifying the oil field to uniformly increase porosity and hence the yield of oil is associated with the propagation of the acid-dissolved material front in porous rocks. More importantly, due to the ever-increasing demand for mineral resources and the likelihood of the exhaust of the existing ore deposits, it is imperative to develop advanced techniques to explore for new ore deposits. Towards this goal, there is a definite need to understand the important physical and chemical processes that control ore body formation and mineralization in the deep Earth (Raffensperger and Garven 1995; Zhao et al. 1998, 1999, 2001a, b, 2003, 2006b, 2007, 2008c; Gow et al. 2002; Schaubs and Zhao 2002). According to modern mineralization theory, ore body formation and mineralization is mainly controlled by pore-fluid flow focusing and the equilibrium concentration gradient of the concerned minerals (Phillips 1991; Zhao et al. 1998). Since the chemical dissolution front can create porosity and therefore can locally enhance the pore-fluid flow, it becomes a potentially powerful mechanism to control ore body formation and mineralization in the deep Earth.

Although analytical solutions can be obtained for some reactive transport problems with simple geometry, it is very difficult, if not impossible, to predict analytically the complicated morphological evolution of a chemical dissolution front in the case of the chemical dissolution system becoming supercritical. As an alternative, numerical methods are suitable to overcome this difficulty. Since numerical methods are approximate solution methods, they must be validated before they are used to solve any new type of scientific and engineering problem. For this reason, it is necessary to derive the analytical solution for the propagation of a planar dissolution front within a benchmark problem, the geometry of which can be accurately simulated using numerical methods such as the finite element method (Zienkiewicz 1977; Lewis and Schrefler 1998) and the finite difference method. This makes it possible to compare the numerical solution obtained from the benchmark problem with the derived analytical solution so that the proposed numerical procedure can be verified for simulating the chemical dissolution-front propagation problem in a fluid-saturated porous medium.

2.1 Mathematical Theory for Simulating Chemical Dissolution-Front Instability Problems in Fluid-Saturated Porous Media

2.1.1 A General Case of Reactive Multi-Chemical-Species Transport with Consideration of Porosity/Permeability Feedback

For a pore fluid-saturated porous medium, Darcy's law can be used to describe pore-fluid flow and Fick's law can be used to describe mass transport phenomena respectively. If both the porosity change of the porous medium is caused by chemical dissolution of soluble solid minerals within the porous medium and the feedback effect of such a change on the variation of permeability and diffusivity are taken into account, the governing equations of the coupled nonlinear problem between porosity, pore-fluid flow and reactive multi-chemical-species transport in the pore-fluid-saturated porous medium can be expressed as follows:

$$\frac{\partial}{\partial t}(\rho_f \phi) + \nabla \cdot (\rho_f \phi \vec{u}_{linear}) = 0, \tag{2.1}$$

$$\vec{u} = \phi \vec{u}_{linear} = -\frac{k(\phi)}{\mu} \nabla p, \tag{2.2}$$

$$\frac{\partial}{\partial t}(\phi C_i) + \nabla \cdot (\phi C_i \vec{u}_{linear}) = \nabla \cdot [\phi D_i(\phi) \nabla C_i] + R_i \qquad (i = 1, 2, \ldots, N), \tag{2.3}$$

where $\vec{u}_{linear}$ is the averaged linear velocity vector within the pore space of the porous medium; $\vec{u}$ is the Darcy velocity vector within the porous medium; p and C_i are pressure and the concentration (moles/pore-fluid volume) of chemical species i; μ is the dynamic viscosity of the pore-fluid; ϕ is the porosity of the porous medium; $D_i(\phi)$ is the diffusivity of chemical species i; ρ_f is the density of the pore-fluid; N is the total number of all the chemical species to be considered in the system; R_i is the source/sink term of chemical species i due to the dissolution/precipitation of solid minerals within the system; $k(\phi)$ is the permeability of the porous medium.

It is noted that in Eqs. (2.1)–(2.3), the chemical species concentration, the fluid density and averaged linear velocity of the pore-fluid are defined in the pore space, while the source/sink term and the Darcy velocity of the pore fluid are defined in the whole medium space (Phillips 1991; Nield and Bejan 1992; Zhao et al. 1994).

Since the diffusivity of each chemical species is considered as a function of porosity, a common phenomenological relation can be used for describing this function (Bear 1972; Chadam et al. 1986; Zhao et al. 2008a).

$$D_i(\phi) = D_{0i}\phi^q \quad \left(\tfrac{3}{2} \leq q \leq \tfrac{5}{2}\right), \tag{2.4}$$

where D_{0i} is the diffusivity of chemical species i in pure water.

To consider the permeability change caused by a change in porosity, an equation is needed to express the relationship between permeability and porosity. In this regard, Detournay and Cheng (1993) stated that "The intrinsic permeability k is generally a function of the pore geometry. In particular, it is strongly dependent on porosity ϕ. According to the Carman-Kozeny law (Scheidegger 1974) which is based on the conceptual model of packing of spheres, a power law relation of $k \propto \phi^3 / (1-\phi)^2$ exists. Other models based on different pore geometry give similar power laws. Actual measurements on rocks, however, often yield power law relations with exponents for ϕ significantly larger than 3." In addition, Nield and Bejan (1992) stated that "The Carman-Kozeny law is widely used since it seems to be the best simple expression available." For these reasons, the Carman-Kozeny law will be used to calculate permeability k, for a given porosity ϕ.

$$k(\phi) = \frac{k_0(1-\phi_0)^2\phi^3}{\phi_0^3(1-\phi)^2}, \tag{2.5}$$

where ϕ_0 and k_0 are the initial reference porosity and permeability of the porous medium respectively.

The source/sink term of chemical species i due to the dissolution/precipitation of solid minerals within the system can be determined in the following manner (Chadam et al. 1986; Zhao et al. 2008a). At the particle level, it is assumed that the average volume of soluble grains is $\bar{V}_p$ and that the density of the soluble grains is D_p, which is defined as the number of the soluble gains per unit medium volume. If the volume fraction of insoluble gains is denoted by $\phi_{insoluble}$, then the final (i.e. maximum) porosity of the porous medium can be denoted by $\phi_f = 1 - \phi_{insoluble}$. In this case, the average volume of soluble grains can be expressed as follows:

$$\bar{V}_p = \frac{\phi_f - \phi}{D_p}. \tag{2.6}$$

At the particle level, the rate of grain-volume change due to chemical (precipitation) reaction is denoted by R_p, so that the rate of porosity change can be expressed as:

$$\frac{\partial \phi}{\partial t} = -D_p R_p. \tag{2.7}$$

Without loss of generality, it is assumed that the solid grains are dissolved according to the following formula:

$$Solid \Rightarrow \sum_{i=1}^{N} \chi_i X_i, \tag{2.8}$$

where χ_i is the stoichiometric coefficient of the ith chemical species; X_i represents chemical species i in the pore-fluid.

It is commonly assumed that the rate of grain-volume change due to a chemical reaction can be expressed as follows (Chadam et al. 1986):

$$R_p = \frac{1}{\rho_s} k_{chemical} A_p \left(\prod_{i=1}^{N} C_i^{\chi_i} - K_{eq} \right) \tag{2.9}$$

where A_p is the averaged surface area of soluble grains; $k_{chemical}$ and K_{eq} are the conventional rate constant and equilibrium constant of the chemical reaction respectively; ρ_s is the molar density (i.e. moles per volume) of the soluble grains.

The source/sink term of chemical species i due to the dissolution/precipitation of solid minerals within the system can be expressed as follows:

$$\begin{aligned} R_i &= -\chi_i k_{chemical} D_p A_p \left(\prod_{i=1}^{N} C_i^{\chi_i} - K_{eq} \right) \\ &= -\chi_i k_{chemical} \frac{A_p}{\bar{V}_p} (\phi_f - \phi) \left(\prod_{i=1}^{N} C_i^{\chi_i} - K_{eq} \right). \end{aligned} \tag{2.10}$$

2.1.2 A Particular Case of Reactive Single-Chemical-Species Transport with Consideration of Porosity/Permeability Feedback

As a particular case, reactive transport involving single chemical-species dissolution in a fluid-saturated porous medium is first considered in this subsection. If the pore-fluid is incompressible, the governing equations of the reactive single-chemical-species transport problem in the fluid-saturated porous medium can be written as follows.

$$\frac{\partial \phi}{\partial t} - \nabla \cdot [\psi(\phi) \nabla p] = 0, \tag{2.11}$$

$$\frac{\partial}{\partial t}(\phi C) - \nabla \cdot [\phi D(\phi) \nabla C + C \psi(\phi) \nabla p] + \rho_s k_{Echemical} \frac{A_p}{\bar{V}_p} (\phi_f - \phi)(C - C_{eq}) = 0, \tag{2.12}$$

$$\frac{\partial \phi}{\partial t} + k_{Echemical} \frac{A_p}{\bar{V}_p} (\phi_f - \phi)(C - C_{eq}) = 0, \tag{2.13}$$

$$\psi(\phi) = \frac{k(\phi)}{\mu}, \tag{2.14}$$

$$k_{Echemical} = \frac{k_{chemical}}{\rho_s C_{eq}}, \tag{2.15}$$

where C and C_{eq} are the concentration and equilibrium concentration of the single chemical species; $k_{Echemical}$ is the comprehensive rate constant of the chemical reaction in the single chemical-species dissolution system. Other quantities in Eqs. (2.11)–(2.15) are of the same meanings as those defined in Eqs. (2.1)–(2.3) and (2.9).

In Eq. (2.15), $k_{chemical}$ is the conventional rate constant with the unit of mol/(m^2 · s), while ρ_s and C_{eq} have the unit of mol/m^3, $k_{Echemical}$ has the unit of m^4/(mol · s). Note that Eqs. (2.11) and (2.12) can be derived by substituting the linear average velocity into Eqs. (2.1) and (2.3) with consideration of a single-chemical species.

For this single-chemical-species system, it is very difficult, even if not impossible, to obtain a complete set of analytical solutions for the pore-fluid pressure, chemical species concentration and porosity within the fluid-saturated porous medium. However, in some special cases, it is possible to obtain analytical solutions for some variables involved in this single-chemical-species system. The first special case to be considered is a problem, in which a planar dissolution front propagates in the full space. Since the chemical dissolution front is of a planar shape, the problem, which is described by Eqs. (2.11)–(2.13) degenerates into a one-dimensional problem. For this particular case, analytical solutions can be obtained for both the propagation speed of the dissolution front and the downstream pressure gradient of the pore-fluid. The second special case to be considered is an asymptotic problem, in which the solid molar density greatly exceeds the equilibrium concentration of the chemical species, implying that the region of a considerable porosity change propagates very slowly within the fluid-saturated porous medium. In this particular case, it is possible to derive a complete set of analytical solutions for the pore-fluid pressure, chemical species concentration and porosity within the fluid-saturated porous medium. In addition, it is also possible to investigate the chemical dissolution-front instability in this particular case (Chadam et al. 1986; Zhao et al. 2008a).

2.1.2.1 The First Special Case

In this special case, the planar dissolution front is assumed to propagate in the positive x direction, so that all quantities are independent of the transverse coordinates y and z. For this reason, Eqs. (2.11)–(2.13) can be rewritten as follows:

$$\frac{\partial \phi}{\partial t} - \frac{\partial}{\partial x}\left[\psi(\phi)\frac{\partial p}{\partial x}\right] = 0, \tag{2.16}$$

$$\frac{\partial}{\partial t}(\phi C)-\frac{\partial}{\partial x}\left[\phi D(\phi)\frac{\partial C}{\partial x}+C\psi(\phi)\frac{\partial p}{\partial x}\right]+\rho_s k_{Echemical}\frac{A_p}{\bar{V}_p}(\phi_f-\phi)(C-C_{eq})=0, \tag{2.17}$$

$$\frac{\partial \phi}{\partial t}+k_{Echemical}\frac{A_p}{\bar{V}_p}(\phi_f-\phi)(C-C_{eq})=0. \tag{2.18}$$

If the chemical species is initially in an equilibrium state and fresh pore-fluid is injected at the location of x approaching negative infinity, then the boundary conditions of this special problem are expressed as

$$\lim_{x\to-\infty} C(x,t)=0, \quad \lim_{x\to-\infty}\frac{\partial p(x,t)}{\partial x}=p'_{fx} \quad \text{(upstream boundary)}, \tag{2.19}$$

$$\lim_{x\to\infty} C(x,t)=C_{eq}, \quad \lim_{x\to\infty}\frac{\partial p(x,t)}{\partial x}=p'_{0x} \quad \text{(downstream boundary)}, \tag{2.20}$$

where p'_{fx} is the pore-fluid pressure gradient as x approaching negative infinity in the upstream of the pore-fluid flow; p'_{0x} is the unknown pore-fluid pressure gradient as x approaching positive infinity in the downstream of the pore-fluid flow. Since p'_{fx} drives the pore-fluid flow continuously along the positive x direction, it has a negative algebraic value (i.e. $p'_{fx}<0$) in this analysis.

The initial condition for this theoretical problem is: $\phi(x,0)=\phi_0$ expect at the negative infinity, where $\lim_{\bar{x}\to-\infty}\phi(x,0)=\phi_f$. Note that ϕ_0 is the initial porosity of the porous medium.

If the propagation speed of the planar dissolution front is denoted by v_{front}, then it is possible to transform a moving boundary problem of the dissolution front (in an $x-t$ coordinate system) into a steady-state boundary problem of the dissolution front (in an $\xi-t$ coordinate system) using the following coordinate mapping:

$$\xi=x-v_{front}t. \tag{2.21}$$

It is necessary to relate partial derivatives with respect to ξ and t to those with respect to x and t (Turcotte and Schubert 1982).

$$\left(\frac{\partial}{\partial t}\right)_\xi=\left(\frac{\partial}{\partial t}\right)_x+\frac{\partial}{\partial x}\frac{\partial x}{\partial t}=\left(\frac{\partial}{\partial t}\right)_x+v_{front}\frac{\partial}{\partial x}, \tag{2.22}$$

$$\left(\frac{\partial}{\partial \xi}\right)_t=\left(\frac{\partial}{\partial x}\right)_t, \tag{2.23}$$

where derivatives are taken with x or t held constant as appropriate.

Since the transformed system in the $\xi - t$ coordinate system is in a steady state, the following equations can be derived from Eqs. (2.22) and (2.23).

$$\left(\frac{\partial}{\partial t}\right)_x = -v_{front}\frac{\partial}{\partial \xi}, \tag{2.24}$$

$$\left(\frac{\partial}{\partial \xi}\right)_t = \left(\frac{\partial}{\partial x}\right)_t. \tag{2.25}$$

Substituting Eqs. (2.24) and (2.25) into Eqs. (2.16)–(2.18) yields the following equations:

$$\frac{\partial}{\partial \xi}\left[\psi(\phi)\frac{\partial p}{\partial \xi} + v_{front}\phi\right] = 0, \tag{2.26}$$

$$\frac{\partial}{\partial \xi}\left[\phi D(\phi)\frac{\partial C}{\partial \xi} + C\psi(\phi)\frac{\partial p}{\partial \xi} + v_{front}(C - \rho_s)\right] = 0, \tag{2.27}$$

$$v_{front}\frac{\partial \phi}{\partial \xi} - k_{chemical}\frac{A_p}{\bar{V}_p}(\phi_f - \phi)(C - C_{eq}) = 0. \tag{2.28}$$

Integrating Eqs. (2.26) and (2.27) from negative infinity to positive infinity and using the boundary conditions [i.e. Eqs. (2.19) and (2.20)] yields the following equations:

$$C_{eq}\psi(\phi_0)p'_{ox} + v_{front}\phi_0(C_{eq} - \rho_s) + v_{front}\phi_f\rho_s = 0, \tag{2.29}$$

$$\psi(\phi_0)p'_{ox} + v_{front}\phi_0 - \psi(\phi_f)p'_{fx} - v_{front}\phi_f = 0. \tag{2.30}$$

Solving Eqs. (2.29) and (2.30) simultaneously results in the following analytical solutions:

$$v_{front} = \frac{-\psi(\phi_0)p'_{0x}C_{eq}}{\phi_0 C_{eq} + (\phi_f - \phi_0)\rho_s} = \frac{u_{0x}C_{eq}}{\phi_0 C_{eq} + (\phi_f - \phi_0)\rho_s}, \tag{2.31}$$

$$p'_{0x} = \frac{\psi(\phi_f)[\phi_0 C_{eq} + (\phi_f - \phi_0)\rho_s]}{\psi(\phi_0)[\phi_0 C_{eq} + (\phi_f - \phi_0)(\rho_s + C_{eq})]}p'_{fx}, \tag{2.32}$$

where u_{0x} is the Darcy velocity in the far downstream of the flow as x approaches positive infinity. Using Darcy's law, u_{0x} can be expressed as

$$u_{0x} = \frac{\phi_0 C_{eq} + (\phi_f - \phi_0)\rho_s}{\phi_0 C_{eq} + (\phi_f - \phi_0)(\rho_s + C_{eq})}u_{fx}, \tag{2.33}$$

where u_{fx} is the Darcy velocity in the far upstream of the flow as x approaches negative infinity.

If the finite element method is used to solve this special problem, the accuracy of the finite element simulation can be conveniently evaluated by comparing the numerical solutions with the analytical ones for both the propagation speed of the planar dissolution front (i.e. v_{front}) and the Darcy velocity in the far downstream of the flow as x approaches positive infinity (i.e. u_{0x}).

2.1.2.2 The Second Special Case (Base Solutions for a Stable State)

Since the solid molar density greatly exceeds the equilibrium concentration of the chemical species, a small parameter, which is called the mineral dissolution ratio (Zhao et al. 2010), can be defined as follows:

$$\varepsilon = \frac{C_{eq}}{\rho_s} \ll 1. \tag{2.34}$$

To facilitate the theoretical analysis in the limit case of ε approaching zero, the following dimensionless parameters and variables can be defined for a two-dimensional problem.

$$\bar{x} = \frac{x}{L^*}, \quad \bar{y} = \frac{y}{L^*}, \tag{2.35}$$

$$\bar{C} = \frac{C}{C_{eq}}, \bar{p} = \frac{p}{p^*}, \quad \bar{\vec{u}} = \frac{\vec{u}}{u^*}, \tag{2.36}$$

$$\tau = \frac{t}{t^*}\varepsilon, \tag{2.37}$$

where τ is a slow dimensionless time to describe the slowness of the chemical dissolution that takes place in the system. Other characteristic parameters used in Eqs. (2.35)–(2.37) can be expressed as follows:

$$t^* = \frac{\bar{V}_p}{k_{Echemical} A_p C_{eq}}, \quad L^* = \sqrt{\phi_f D(\phi_f) t^*}, \tag{2.38}$$

$$p^* = \frac{\phi_f D(\phi_f)}{\psi(\phi_f)}, \quad u^* = \frac{\phi_f D(\phi_f)}{L^*}, \tag{2.39}$$

$$D^*(\phi) = \frac{\phi D(\phi)}{\phi_f D(\phi_f)}, \quad \psi^*(\phi) = \frac{\psi(\phi)}{\psi(\phi_f)}. \tag{2.40}$$

Substituting Eqs. (2.35)–(2.40) into Eqs. (2.11)–(2.13) yields the following dimensionless equations:

$$\varepsilon\frac{\partial\phi}{\partial\tau} - \nabla\cdot[\psi^*(\phi)\nabla\bar{p}] = 0, \tag{2.41}$$

$$\varepsilon\frac{\partial}{\partial\tau}(\phi\bar{C}) - \nabla\cdot[D^*(\phi)\nabla\bar{C} + \bar{C}\psi^*(\phi)\nabla\bar{p}] - \frac{\partial\phi}{\partial\tau} = 0, \tag{2.42}$$

$$\varepsilon\frac{\partial\phi}{\partial\tau} + (\phi_f - \phi)(\bar{C} - 1) = 0. \tag{2.43}$$

Similarly, the boundary conditions for this special case can be expressed in a dimensionless form as follows:

$$\lim_{\bar{x}\to\infty}\bar{C}(\bar{x},\tau) = 1, \quad \lim_{\bar{x}\to\infty}\frac{\partial\bar{p}(\bar{x},\tau)}{\partial\bar{x}} = \bar{p}'_{0x} \quad \text{(downstream boundary)}, \tag{2.44}$$

$$\lim_{\bar{x}\to-\infty}\bar{C}(\bar{x},\tau) = 0, \quad \lim_{\bar{x}\to-\infty}\frac{\partial\bar{p}(\bar{x},\tau)}{\partial\bar{x}} = \bar{p}'_{fx} \quad \text{(upstream boundary)}. \tag{2.45}$$

In this case, the initial condition for this theoretical problem is: $\phi(\bar{x},0) = \phi_0$ expect at the negative infinity, where $\lim_{\bar{x}\to-\infty}\phi(\bar{x},0) = \phi_f$.

It is noted that the propagation front due to chemical dissolution divides the problem domain into two regions, an upstream region and a downstream region, relative to the propagation front. Across this propagation front, the porosity undergoes a jump from its initial value into its final value. Thus, this dissolution-front propagation problem can be considered as a Stefan moving boundary problem (Chadam et al. 1986; Zhao et al. 2008a). In the limit case of ε approaching zero, the corresponding governing equations for the dimensionless variables of the problem in both the downstream region and the upstream region can be expressed below:

$$\bar{C} = 1, \quad \nabla^2\bar{p} = 0, \quad \phi = \phi_0 \quad \text{(in the downstream region)}, \tag{2.46}$$

$$\nabla\cdot(\nabla\bar{C} + \bar{C}\nabla\bar{p}) = 0, \quad \nabla^2\bar{p} = 0, \quad \phi = \phi_f \quad \text{(in the upstream region)}. \tag{2.47}$$

If the chemical dissolution front is denoted by $S(\bar{x},\tau) = 0$, then the dimensionless pressure, chemical species concentration and mass fluxes of both the chemical species and the pore-fluid should be continuous on $S(\bar{x},\tau) = 0$. This leads to the following interface conditions for this moving-front problem:

$$\lim_{S\to0^-}\bar{C} = \lim_{S\to0^+}\bar{C}, \quad \lim_{S\to0^-}\bar{p} = \lim_{S\to0^+}\bar{p}, \tag{2.48}$$

$$\lim_{S\to0^-}\frac{\partial\bar{C}}{\partial n} = \bar{v}_{front}(\phi_f - \phi_0), \quad \lim_{S\to0^-}\frac{\partial\bar{p}}{\partial\bar{n}} = \frac{\psi(\phi_0)}{\psi(\phi_f)}\lim_{S\to0^+}\frac{\partial\bar{p}}{\partial\bar{n}}, \tag{2.49}$$

where $\bar{n} = n/L^*$; n is the normal vector of the propagating planar chemical dissolution-front; $\bar{v}_{front}$ is the dimensionless propagation speed of the planar chemical dissolution-front.

When the planar dissolution front is under stable conditions, the base solutions for this special problem can be derived from Eqs. (2.46) and (2.47) with the related boundary and interface conditions [i.e. Eqs. (2.44), (2.45), (2.48) and (2.49)]. The resulting base solutions are expressed as follows:

$$\bar{C}(\xi) = 1, \quad \bar{p}(\xi) = \bar{p}'_{0x}\xi + \bar{p}_{C1}, \quad \phi = \phi_0 \quad \text{(in the downstream region)}, \tag{2.50}$$

$$\bar{C}(\xi) = \exp(-\bar{p}'_{fx}\xi), \quad \bar{p}(\xi) = \bar{p}'_{fx}\xi + \bar{p}_{C2}, \quad \phi = \phi_f \quad \text{(in the upstream region)}, \tag{2.51}$$

where $\bar{p}_{C1}$ and $\bar{p}_{C2}$ are two constants to be determined. For example, $\bar{p}_{C1}$ can be determined by setting dimensionless pressure $\bar{p}(\xi)$ to be a constant at a prescribed location of the downstream region, while $\bar{p}_{C2}$ can be determined using the pressure continuity condition at the interface between the upstream and downstream regions. Other parameters are defined below:

$$\xi = \bar{x} - \bar{v}_{front}\tau, \quad \bar{p}'_{0x} = \frac{\psi(\phi_f)}{\psi(\phi_0)}\bar{p}'_{fx}, \quad \bar{v}_{front} = -\frac{\bar{p}'_{fx}}{\phi_f - \phi_0}. \tag{2.52}$$

Therefore, if the finite element method is used to solve the second special problem, the accuracy of the finite element simulation can be conveniently evaluated by comparing the numerical solutions with a complete set of analytical solutions including porosity, the location of the chemical dissolution front, the dimensionless chemical-species concentration and the dimensionless pore-fluid pressure.

2.1.2.3 The Second Special Case (Perturbation Solutions for an Unstable State)

When a reactive transport system represented by the above-mentioned second special problem is stable, the planar dissolution front remains planar, even though both small perturbations of the dissolution front and the feedback effect of porosity/permeability change are simultaneously considered in the analysis. However, when the reactive transport system is unstable, the planar dissolution front can change from a planar shape into a complicated one. The instability of the above-mentioned second special problem can be determined using a linear stability analysis (Chadam et al. 1986, 1988; Ortoleva et al. 1987; Zhao et al. 2008a). The main purpose of conducting such a linear stability analysis is to determine the critical condition under which the chemical dissolution front of the reactive transport system becomes unstable.

If a small time-dependent perturbation is added to the planar dissolution front, then the total solution of the system is equal to the summation of the base solution and the perturbed solution of the system.

$$S(\xi, \bar{y}, \tau) = \xi - \delta \exp(\bar{\omega}\tau) \cos(\bar{m}\bar{y}), \tag{2.53}$$

$$\bar{p}_{total}(\xi, \bar{y}, \tau) = \bar{p}(\xi, \tau) + \delta \hat{p}(\xi) \exp(\bar{\omega}\tau) \cos(\bar{m}\bar{y}), \tag{2.54}$$

$$\bar{C}_{total}(\xi, \bar{y}, \tau) = \bar{C}(\xi, \tau) + \delta \hat{C}(\xi) \exp(\bar{\omega}\tau) \cos(\bar{m}\bar{y}), \tag{2.55}$$

where $\bar{\omega}$ is the dimensionless growth rate of the perturbation; $\bar{m}$ is the dimensionless wavenumber of the perturbation; δ is the amplitude of the perturbation and $\delta \ll 1$ by the definition of a linear stability analysis.

Since $S(\xi, \bar{y}, \tau)$ is a function of coordinates $S(\xi, \bar{y}, \tau)$ and $\bar{y}$, the following derivatives exist mathematically:

$$\left(\frac{\partial}{\partial \xi}\right)_{\xi} = \frac{\partial S}{\partial \xi}\frac{\partial}{\partial S} = \left(\frac{\partial}{\partial \xi}\right)_{S}, \tag{2.56}$$

$$\left(\frac{\partial}{\partial \bar{y}}\right)_{\xi} = \frac{\partial S}{\partial \bar{y}}\frac{\partial}{\partial S} + \left(\frac{\partial}{\partial \bar{y}}\right)_{S} = \frac{\partial S}{\partial \bar{y}}\left(\frac{\partial}{\partial \xi}\right)_{S} + \left(\frac{\partial}{\partial \bar{y}}\right)_{S}, \tag{2.57}$$

$$\left(\frac{\partial^2}{\partial \xi^2}\right)_{\xi} = \left(\frac{\partial^2}{\partial \xi^2}\right)_{S}, \tag{2.58}$$

$$\left(\frac{\partial^2}{\partial \bar{y}^2}\right)_{\xi} = \frac{\partial^2 S}{\partial \bar{y}^2}\frac{\partial}{\partial \xi} + \left(\frac{\partial S}{\partial \bar{y}}\right)^2 \frac{\partial^2}{\partial \xi^2} + 2\frac{\partial S}{\partial \bar{y}}\frac{\partial^2}{\partial \xi \partial \bar{y}} + \left(\frac{\partial^2}{\partial \bar{y}^2}\right)_{S}. \tag{2.59}$$

It is noted that the total solutions expressed in Eqs. (2.54) and (2.55) must satisfy the governing equations that are expressed in Eqs. (2.46) and (2.47). With consideration of Eq. (2.59), the first-order perturbation equations of this system can be expressed as

$$\hat{C} = 0, \quad \frac{\partial^2 \hat{p}}{\partial \xi^2} - \bar{m}^2 \hat{p} + \bar{m}^2 \bar{p}'_{0x} = 0 \quad \text{(in the downstream region)}, \tag{2.60}$$

$$\frac{\partial^2 \hat{C}}{\partial \xi^2} + \bar{p}'_{fx}\frac{\partial \hat{C}}{\partial \xi} - \bar{m}^2 \hat{C} - \bar{m}^2 \bar{p}'_{fx} \exp(-\bar{p}'_{fx}\xi) - \bar{p}'_{fx} \exp(-\bar{p}'_{fx}\xi)\frac{\partial \hat{p}}{\partial \xi} = 0,$$

$$\frac{\partial^2 \hat{p}}{\partial \xi^2} - \bar{m}^2 \hat{p} + \bar{m}^2 \bar{p}'_{fx} = 0 \quad \text{(in the upstream region)}. \tag{2.61}$$

The corresponding boundary conditions of the first-order perturbation problem are:

$$\hat{C}(\xi) = 0, \quad \lim_{\xi \to \infty} \frac{\partial \hat{p}(\xi)}{\partial \xi} = 0 \quad \text{(downstream boundary)}, \tag{2.62}$$

$$\lim_{\xi \to -\infty} \hat{C}(\xi) = 0, \quad \lim_{\xi \to -\infty} \frac{\partial \hat{p}(\xi)}{\partial \xi} = 0 \quad \text{(upstream boundary)}. \tag{2.63}$$

Similarly, the interface conditions for this first-order perturbation problem can be expressed as follows:

$$\hat{C} = 0, \quad \lim_{S \to 0^-} \hat{p} = \lim_{S \to 0^+} \hat{p}, \tag{2.64}$$

$$\lim_{S \to 0^-} \frac{\partial \hat{C}}{\partial \bar{n}} = \bar{\omega}(\phi_f - \phi_0), \quad \lim_{S \to 0^-} \frac{\partial \hat{p}}{\partial \bar{n}} = \frac{\psi(\phi_0)}{\psi(\phi_f)} \lim_{S \to 0^+} \frac{\partial \hat{p}}{\partial \bar{n}}. \tag{2.65}$$

Solving Eqs. (2.60) and (2.61) with the boundary and interface conditions [i.e. Eqs. (2.62)–(2.65)] yields the following analytical results:

$$\hat{C} = 0, \quad \hat{p}(\xi) = \bar{p}'_{0x}\left[1 - \frac{1-\beta}{1+\beta}\exp(-|\bar{m}|\xi)\right] \quad \text{(in the downstream region)}, \tag{2.66}$$

$$\hat{C}(\xi) = -\bar{p}'_{fx}\left\{\exp(-\bar{p}'_{fx}\xi) - \frac{2}{1+\beta}\exp(\sigma\xi) + \frac{1-\beta}{1+\beta}\exp[(|\bar{m}| - \bar{p}'_{fx})\xi]\right\},$$

$$\hat{p}(\xi) = \bar{p}'_{fx}\left[1 + \frac{1-\beta}{1+\beta}\exp(|\bar{m}|\xi)\right] \quad \text{(in the upstream region)}, \tag{2.67}$$

where

$$\beta = \frac{\psi(\phi_0)}{\psi(\phi_f)} = \frac{k(\phi_0)}{k(\phi_f)}, \tag{2.68}$$

$$\sigma = \frac{\sqrt{(\bar{p}'_{fx})^2 + 4m^2} - \bar{p}'_{fx}}{2}. \tag{2.69}$$

Substituting Eq. (2.67) into Eq. (2.65) yields the following equation for the dimensionless growth rate of the small perturbation:

$$\bar{\omega}(\bar{m}) = \frac{-\bar{p}'_{fx}}{(1+\beta)(\phi_f - \phi_0)}\left[-\bar{p}'_{fx} - \sqrt{(\bar{p}'_{fx})^2 + 4\bar{m}^2} + (1-\beta)|\bar{m}|\right). \tag{2.70}$$

Equation (2.70) clearly indicates that the planar dissolution front of the reactive transport system, which is described by the above-mentioned second special problem, is stable to short wavelength (i.e. large dimensionless wavenumber $\bar{m}$) perturbations but it is unstable to long wavelength (i.e. small dimensionless wavenumber $\bar{m}$) perturbations.

Letting $\bar{\omega}(\bar{m}) = 0$ yields the following critical condition, under which the reactive transport system can become unstable.

$$\bar{p}'_{fx}\Big|_{critical} = -\frac{(3-\beta)(1+\beta)}{2(1-\beta)}, \tag{2.71}$$

where $\bar{p}'_{fx}\Big|_{critical}$ is the critical value of the generalized dimensionless pressure gradient in the far upstream direction as x approaching negative infinity (Zhao et al. 2008a). Since $\bar{p}'_{fx}\Big|_{critical}$ is usually of a negative value, the following critical Zhao number is defined to judge the instability of the reactive transport system:

$$Zh_{critical} = -\bar{p}'_{fx}\Big|_{critical} = \frac{(3-\beta)(1+\beta)}{2(1-\beta)}. \tag{2.72}$$

Thus, the Zhao number of the reactive transport system can be defined as follows:

$$Zh = -\bar{p}'_{fx} = -\frac{p'_{fx}L^*}{p^*} = -\frac{k(\phi_f)L^*p'_{fx}}{\phi_f \mu D(\phi_f)} = \frac{v_{flow}}{\sqrt{\phi_f D(\phi_f)}}\sqrt{\frac{\bar{V}_p}{k_{Echemical}A_p C_{eq}}}. \tag{2.73}$$

where v_{flow} is the Darcy velocity of the injected pore-fluid flow; ϕ_f is the porosity when the dissolvable minerals are completely dissolved; $D(\phi_f)$ is the molecular diffusivity of the solute in the mineral completely-dissolved region; $\bar{V}_p$ is the average volume of soluble grains; A_p is the averaged surface area of soluble grains; C_{eq} is the equilibrium concentration of the dissolvable mineral; $k_{Echemical}$ is the equivalent rate constant of the chemical reaction in the single chemical-species dissolution system [as defined in Eq. (2.15)].

Using Eqs. (2.72) and (2.73), a criterion can be established to judge the instability of a chemical dissolution front associated with the particular chemical system in this investigation. If $Zh > Zh_{critical}$, then the chemical dissolution front of the reactive transport system becomes unstable, while if $Zh < Zh_{critical}$, then the chemical dissolution front of the reactive transport system is stable. The case of $Zh = Zh_{critical}$ represents a situation where the chemical dissolution front of the reactive transport system is neutrally unstable, implying that the introduced small perturbation can be maintained but it does not grow in the corresponding reactive transport system.

Clearly, Eq. (2.73) indicates that for the reactive chemical-species transport considered in this investigation, the dissolution-enhanced permeability destabilizes the instability of the chemical dissolution front, while the dissolution-enhanced diffusivity stabilizes the instability of the chemical dissolution front. If the shape factor of soluble grains is represented by $\theta = \bar{V}_p/A_p$, then an increase in the shape factor of soluble grains can destabilize the instability of the chemical dissolution front, indicating that the instability likelihood of a porous medium comprised of irregular grains, is higher than that of a porous medium comprised of regular spherical grains. Similarly, an increase in either the equilibrium concentration of the chemical species or the chemical reaction constant of the dissolution reaction can cause the stabilization of the chemical dissolution front, for the reactive chemical-species transport considered in this investigation.

To understand the physical meanings of each term in the Zhao number, Eq. (2.73) can be rewritten in the following form:

$$Zh = F_{Advection} F_{Diffusion} F_{Chemical} F_{Shape}, \tag{2.74}$$

where $F_{Advection}$ is a term to represent the solute advection; $F_{Diffusion}$ is a term to represent the solute diffusion/dispersion; $F_{Chemical}$ is a term to represent the chemical kinetics of the dissolution reaction; F_{Shape} is a term to represent the shape factor of the soluble mineral in the fluid-rock interaction system. These terms can be expressed as follows:

$$F_{Advection} = v_{flow}, \tag{2.75}$$

$$F_{Diffusion} = \frac{1}{\sqrt{\phi_f D(\phi_f)}}, \tag{2.76}$$

$$F_{Chemical} = \sqrt{\frac{1}{k_{chemical} C_{eq}}}, \tag{2.77}$$

$$F_{Shape} = \sqrt{\frac{\bar{V}_p}{A_p}}. \tag{2.78}$$

Equations (2.74)–(2.78) clearly indicate that the Zhao number is a dimensionless number that can be used to represent the geometrical, hydrodynamic, thermodynamic and chemical kinetic characteristics of a fluid-rock system in a comprehensive manner. This dimensionless number reveals the intimate interaction between solute advection, solution diffusion/dispersion, chemical kinetics and mineral geometry in a reactive transport system.

2.2 Computational Theory for Simulating the Morphological Evolution of a Chemical Dissolution Front

Although analytical solutions can be obtained for the above-mentioned special cases, it is very difficult, if not impossible, to predict analytically the complicated morphological evolution process of a planar dissolution front in the case of the chemical dissolution system becoming supercritical. As an alternative, numerical methods are suitable to overcome this difficulty. Since numerical methods are approximate solution methods, they must be validated before they are used to solve any new type of scientific and engineering problem. For this reason, the main purpose of this section is to propose a numerical procedure for simulating how a planar dissolution front evolves into a complicated morphological front. To verify the accuracy of the numerical solution, a benchmark problem is constructed from the theoretical analysis in Sect. 2.1.2.2. As a result, the numerical solution obtained from the benchmark problem can be compared with the corresponding analytical solution. After the proposed numerical procedure is verified, it will be used to simulate the complicated morphological evolution process of a planar dissolution front in the case of the chemical dissolution system becoming supercritical.

2.2.1 Formulation of the Segregated Algorithm for Simulating the Evolution of Chemical Dissolution Fronts

In this section, Eqs. (2.41)–(2.43) are solved using the proposed numerical procedure, which is a combination of both the finite element method and the finite difference method. The finite element method is used to discretize the geometrical shape of the problem domain, while the finite difference method is used to discretize the dimensionless time. Since the system described by these equations is highly nonlinear, the segregated algorithm, in which Eqs. (2.41)–(2.43) are solved separately in a sequential manner, is used to derive the formulation of the proposed numerical procedure.

For a given dimensionless time-step, $\tau + \Delta\tau$, the porosity can be denoted by $\phi_{\tau+\Delta\tau} = \phi_\tau + \Delta\phi_{\tau+\Delta\tau}$, where ϕ_τ is the porosity at the previous time-step and $\Delta\phi_{\tau+\Delta\tau}$ is the porosity increment at the current time-step. Using the backward difference scheme, Eq. (2.43) can be written as follows:

$$\left[\frac{\varepsilon}{\Delta\tau} + (1 - \bar{C}_{\tau+\Delta\tau})\right]\Delta\phi_{\tau+\Delta\tau} = (\phi_f - \phi_\tau)(1 - \bar{C}_{\tau+\Delta\tau}), \tag{2.79}$$

where $\bar{C}_{\tau+\Delta\tau}$ is the dimensionless concentration at the current time-step; $\Delta\tau$ is the dimensionless time increment at the current time-step.

Mathematically, there exist the following relationships in the finite difference sense:

$$\varepsilon\frac{\partial(\phi\bar{C})}{\partial\tau} = \varepsilon\frac{\Delta(\phi_{\tau+\Delta\tau}\bar{C}_{\tau+\Delta\tau})}{\Delta\tau} = \varepsilon\bar{C}_{\tau+\Delta\tau}\frac{\Delta\phi_{\tau+\Delta\tau}}{\Delta\tau} + \varepsilon\phi_{\tau+\Delta\tau}\frac{\Delta(\bar{C}_{\tau+\Delta\tau})}{\Delta\tau}, \tag{2.80}$$

$$\varepsilon\frac{\partial\phi}{\partial\tau} = \varepsilon\frac{\Delta(\phi_{\tau+\Delta\tau})}{\Delta\tau} = (1-\bar{C}_{\tau+\Delta\tau})(\phi_f - \phi_{\tau+\Delta\tau}), \tag{2.81}$$

$$\nabla\cdot[D^*(\phi)\nabla\bar{C}] = \nabla\cdot[D^*(\phi_{\tau+\Delta\tau})\nabla\bar{C}_{\tau+\Delta\tau}], \tag{2.82}$$

$$\begin{aligned}\nabla\cdot[\bar{C}\psi^*(\phi)\nabla\bar{p}] &= \bar{C}\nabla\cdot[\psi^*(\phi)\nabla\bar{p}] + \nabla\bar{p}\cdot[\psi^*(\phi)\nabla\bar{C}]\\ &= \bar{C}_{\tau+\Delta\tau}\nabla\cdot[\psi^*(\phi_{\tau+\Delta\tau})\nabla\bar{p}_{\tau+\Delta\tau}] + \nabla\bar{p}_{\tau+\Delta\tau}\cdot[\psi^*(\phi_{\tau+\Delta\tau})\nabla\bar{C}_{\tau+\Delta\tau}].\end{aligned} \tag{2.83}$$

Substituting Eqs. (2.80)–(2.83) into Eq. (2.42) yields the following finite difference equation:

$$\begin{aligned}&\left[\frac{\varepsilon}{\Delta\tau}\phi_{\tau+\Delta\tau} + \frac{1}{\varepsilon}(\phi_f - \phi_{\tau+\Delta\tau})\right]\bar{C}_{\tau+\Delta\tau} - \nabla\cdot[D^*(\phi_{\tau+\Delta\tau})\nabla\bar{C}_{\tau+\Delta\tau}]\\ &\quad - \nabla\bar{p}_{\tau+\Delta\tau}\cdot[\psi^*(\phi_{\tau+\Delta\tau})\nabla\bar{C}_{\tau+\Delta\tau}] = \frac{\varepsilon}{\Delta\tau}\phi_{\tau+\Delta\tau}\bar{C}_\tau + \frac{1}{\varepsilon}(\phi_f - \phi_{\tau+\Delta\tau}).\end{aligned} \tag{2.84}$$

Similarly, Eq. (2.41) can be rewritten in the following discretized form:

$$\nabla\cdot[\psi^*(\phi)\nabla\bar{p}] = \nabla\cdot[\psi^*(\phi_{\tau+\Delta\tau})\nabla\bar{p}_{\tau+\Delta\tau}] = (1-\bar{C}_{\tau+\Delta\tau})(\phi_f - \phi_{\tau+\Delta\tau}). \tag{2.85}$$

To derive the finite element equations of the problem, the corresponding finite difference equations can be summarized as follows:

$$\left[\frac{\varepsilon}{\Delta\tau} + (1-\bar{C}_{\tau+\Delta\tau})\right]\Delta\phi_{\tau+\Delta\tau} = (\phi_f - \phi_\tau)(1-\bar{C}_{\tau+\Delta\tau}), \tag{2.86}$$

$$\begin{aligned}&\left[\frac{\varepsilon}{\Delta\tau}\phi_{\tau+\Delta\tau} + \frac{1}{\varepsilon}(\phi_f - \phi_{\tau+\Delta\tau})\right]\bar{C}_{\tau+\Delta\tau} - \nabla\cdot[D^*(\phi_{\tau+\Delta\tau})\nabla C_{\tau+\Delta\tau}]\\ &\quad - \nabla\bar{p}_{\tau+\Delta\tau}\cdot[\psi^*(\phi_{\tau+\Delta\tau})\nabla\bar{C}_{\tau+\Delta\tau}] = \frac{\varepsilon}{\Delta\tau}\phi_{\tau+\Delta\tau}\bar{C}_\tau + \frac{1}{\varepsilon}(\phi_f - \phi_{\tau+\Delta\tau}),\end{aligned} \tag{2.87}$$

$$\nabla\cdot[\psi^*(\phi)\nabla\bar{p}] = \nabla\cdot[\psi^*(\phi_{\tau+\Delta\tau})\nabla\bar{p}_{\tau+\Delta\tau}] = (1-\bar{C}_{\tau+\Delta\tau})(\phi_f - \phi_{\tau+\Delta\tau}). \tag{2.88}$$

2.2.2 Derivation of the Finite Element Equations of the Problem

Although the finite element method has been broadly used for solving many different types of scientific and engineering problems, the finite element equations related to the chemical dissolution-front propagation problem need to be given blow.

To derive the finite element equations of the problem, three unknown variables involved in the finite difference equations [i.e. Eqs. (2.86)–(2.88)] are defined as follows:

$$U_1 = \Delta\phi_{\tau+\Delta\tau}, \quad U_2 = \bar{C}_{\tau+\Delta\tau}, \quad U_3 = \bar{p}_{\tau+\Delta\tau}. \tag{2.89}$$

By using these new definitions, Eqs. (2.86)–(2.88) can be written in the following forms:

$$f_{U1}U_1 = f_{C1}, \tag{2.90}$$

$$f_{U2}U_2 - \psi^*(\phi_{\tau+\Delta\tau})(\nabla\bar{p}_{\tau+\Delta\tau} \cdot \nabla U_2) - \nabla \cdot [D^*(\phi_{\tau+\Delta\tau})\nabla U_2] = f_{C2}, \tag{2.91}$$

$$\nabla \cdot [\psi^*(\phi_{\tau+\Delta\tau})\nabla U_3] = f_{C3}, \tag{2.92}$$

where

$$f_{U1} = f_{U1}(\bar{C}_{\tau+\Delta\tau}) = \frac{\varepsilon}{\Delta\tau} + (1 - \bar{C}_{\tau+\Delta\tau}), \tag{2.93}$$

$$f_{C1} = f_{C1}(\bar{C}_{\tau+\Delta\tau}, \phi_\tau) = (\phi_f - \phi_\tau)(1 - \bar{C}_{\tau+\Delta\tau}), \tag{2.94}$$

$$f_{U2} = f_{U2}(\phi_{\tau+\Delta\tau}) = \frac{\varepsilon}{\Delta\tau}\phi_{\tau+\Delta\tau} + \frac{1}{\varepsilon}(\phi_f - \phi_{\tau+\Delta\tau}), \tag{2.95}$$

$$f_{C2} = f_{C2}(\phi_{\tau+\Delta\tau}, \bar{C}_\tau) = \frac{\varepsilon}{\Delta\tau}\phi_{\tau+\Delta\tau}\bar{C}_\tau + \frac{1}{\varepsilon}(\phi_f - \phi_{\tau+\Delta\tau}), \tag{2.96}$$

$$f_{C3} = f_{C3}(\phi_{\tau+\Delta\tau}, \bar{C}_{\tau+\Delta\tau}) = (1 - \bar{C}_{\tau+\Delta\tau})(\phi_f - \phi_{\tau+\Delta\tau}). \tag{2.97}$$

Based on the finite element method, the distribution of the above-mentioned three unknown variables in a finite element can be described as follows:

$$U_1^e = [N]\{\Delta_1\}^e, \tag{2.98}$$

$$U_2^e = [N]\{\Delta_2\}^e, \tag{2.99}$$

$$U_3^e = [N]\{\Delta_3\}^e, \tag{2.100}$$

where U_1^e, U_2^e and U_3^e are the distribution fields of the three unknown variables within the finite element; $\{\Delta_1\}^e$, $\{\Delta_2\}^e$ and $\{\Delta_3\}^e$ are the corresponding nodal vectors of the element; $[N]$ is the shape function matrix of the element.

If the finite element under consideration has Q nodes, then the following expressions exist for this element:

$$(\{\Delta_1\}^e)^T = \{\, U_{11} \quad U_{12} \quad \ldots \quad U_{1Q} \,\}, \tag{2.101}$$

$$(\{\Delta_2\}^e)^T = \{\, U_{21} \quad U_{22} \quad \ldots \quad U_{2Q} \,\}, \tag{2.102}$$

$$(\{\Delta_3\}^e)^T = \{\, U_{31} \quad U_{32} \quad \ldots \quad U_{3Q} \,\}, \tag{2.103}$$

$$[N] = [\, N_1 \quad N_2 \quad \ldots \quad N_Q \,], \tag{2.104}$$

where N_i ($i = 1, 2, \ldots, Q$) is the shape function of node i; U_{1i}, U_{2i} and U_{3i} ($i = 1, 2, \ldots, Q$) are the nodal values of the unknown variables for the finite element under consideration.

Using the Galerkin weighted-residual method, Eqs. (2.90)–(2.92) can be rewritten, at the element level, as follows:

$$\left(\iint_A [N]^T f_{U1} [N]\, dA\right)\{\Delta_1\}^e = \iint_A [N]^T f_{C1}\, dA, \tag{2.105}$$

$$\left(\iint_A [N]^T f_{U2} [N] dA\right)\{\Delta_2\}^e - \left(\iint_A \psi^*(\phi_{\tau+\Delta\tau})[N]^T (\nabla\bar{p}_{\tau+\Delta\tau} \cdot \nabla[N]) dA\right)\{\Delta_2\}^e$$
$$- \left(\iint_A [N]^T \nabla \cdot (D^*(\phi_{\tau+\Delta\tau})\nabla[N])\, dA\right)\{\Delta_2\}^e = \iint_A [N]^T f_{C2} dA, \tag{2.106}$$

$$\left(\iint_A [N]^T \nabla \cdot (\psi^*(\phi_{\tau+\Delta\tau})\nabla[N]) dA\right)\{\Delta_3\}^e = \iint_A [N]^T f_{C3} dA, \tag{2.107}$$

where A is the area of the element.

Note that the following expressions exist mathematically:

$$\left(\iint_A [N]^T \nabla \cdot (D^*(\phi_{\tau+\Delta\tau})\nabla[N]) dA\right)\{\Delta_2\}^e = \int_S [N]^T D^*(\phi_{\tau+\Delta\tau})\nabla([N]\{\Delta_2\}^e) dS$$
$$- \left(\iint_A \nabla[N]^T \cdot (D^*(\phi_{\tau+\Delta\tau})\nabla[N]) dA\right)\{\Delta_2\}^e, \tag{2.108}$$

$$\left(\iint_A [N]^T \nabla \cdot (\psi^*(\phi_{\tau+\Delta\tau}) \nabla [N])\, dA\right)\{\Delta_3\}^e = \int_S [N]^T \psi^*(\phi_{\tau+\Delta\tau}) \nabla([N]\{\Delta_3\}^e) dS - \left(\iint_A \nabla [N]^T \cdot (\psi^*(\phi_{\tau+\Delta\tau}) \nabla [N]) dA\right)\{\Delta_3\}^e, \tag{2.109}$$

where S is the length of the element boundary.

Substituting Eqs. (2.108) and (2.109) into Eqs. (2.106) and (2.107) yields the following expressions:

$$\begin{aligned} &\left(\iint_A [N]^T f_{U2} [N] dA\right)\{\Delta_2\}^e - \left(\iint_A \psi^*(\phi_{\tau+\Delta\tau}) [N]^T (\nabla \bar{p}_{\tau+\Delta\tau} \cdot \nabla [N]) dA\right)\{\Delta_2\}^e \\ &+ \left(\iint_A \nabla [N]^T \cdot (D^*(\phi_{\tau+\Delta\tau}) \nabla [N]) dA\right)\{\Delta_2\}^e \\ &= \iint_A [N]^T f_{C2} dA + \int_S [N]^T D^*(\phi_{\tau+\Delta\tau}) \nabla([N]\{\Delta_2\}^e) dS, \end{aligned} \tag{2.110}$$

$$\begin{aligned} &\left(\iint_A \nabla [N]^T \cdot (\psi^*(\phi_{\tau+\Delta\tau}) \nabla [N]) dA\right)\{\Delta_3\}^e \\ &= -\iint_A [N]^T f_{C3} dA + \int_S [N]^T \psi^*(\phi_{\tau+\Delta\tau}) \nabla([N]\{\Delta_3\}^e) dS. \end{aligned} \tag{2.111}$$

Consideration of Eqs. (2.105), (2.110) and (2.111) leads to the discretized equations of the finite element as follows:

$$[M_1]^e \{\Delta_1\}^e = \{P_1\}^e, \tag{2.112}$$

$$([M_2]^e - [H]^e + [K_1]^e)\{\Delta_2\}^e = \{P_2\}^e, \tag{2.113}$$

$$[K_2]^e \{\Delta_3\}^e = \{P_3\}^e, \tag{2.114}$$

where $[M_1]^e$, $[M_2]^e$, $[H]^e$, $[K_1]^e$ and $[K_2]^e$ are the property matrices of the finite element; $\{P_1\}^e$, $\{P_2\}^e$ and $\{P_3\}^e$ are the “load” vectors of the finite element.

These matrices and vectors can be expressed as follows:

$$[M_1]^e = \iint_A [N]^T f_{U1} [N]\, dA, \tag{2.115}$$

$$[M_2]^e = \iint_A [N]^T f_{U2} [N] dA, \tag{2.116}$$

$$[H]^e = \iint_A \psi^*(\phi_{\tau+\Delta\tau})[N]^T(\nabla\bar{p}_{\tau+\Delta\tau} \cdot \nabla[N])dA, \tag{2.117}$$

$$[K_1]^e = \iint_A \nabla[N]^T \cdot (D^*(\phi_{\tau+\Delta\tau})\nabla[N])dA, \tag{2.118}$$

$$[K_2]^e = \iint_A \nabla[N]^T \cdot (f_{U3b}\nabla[N])dA, \tag{2.119}$$

$$\{P_1\}^e = \iint_A [N]^T f_{C1} dA, \tag{2.120}$$

$$\begin{aligned} \{P_2\}^e &= \iint_A [N]^T f_{C2} dA + \int_S [N]^T D^*(\phi_{\tau+\Delta\tau})\nabla([N]\{\Delta_2\}^e)dS \\ &= \iint_A [N]^T f_{C2} dA + \int_S [N]^T D^*(\phi_{\tau+\Delta\tau})\nabla(\bar{C}_{\tau+\Delta\tau})^e dS, \end{aligned} \tag{2.121}$$

$$\begin{aligned} \{P_3\}^e &= -\iint_A [N]^T f_{C3} dA + \int_S [N]^T \psi^*(\phi_{\tau+\Delta\tau})\nabla([N]\{\Delta_3\}^e)dS \\ &= -\iint_A [N]^T f_{C3} dA + \int_S [N]^T \psi^*(\phi_{\tau+\Delta\tau})\nabla(\bar{p}_{\tau+\Delta\tau})^e dS, \end{aligned} \tag{2.122}$$

where $\nabla(\bar{C}_{\tau+\Delta\tau})^e$ and $\nabla(\bar{p}_{\tau+\Delta\tau})^e$ are the dimensionless chemical-species concentration-gradient and pore-fluid pressure-gradient at the boundary, S, of the finite element.

Assembling the property matrices and vectors of all the elements in a system yields the following discretized governing equations of the system:

$$[M_1]\{\Delta_1\} = \{P_1\}, \tag{2.123}$$

$$([M_2] - [H] + [K_1])\{\Delta_2\} = \{P_2\}, \tag{2.124}$$

$$[K_2]\{\Delta_3\} = \{P_3\}, \tag{2.125}$$

where $[M_1]$, $[M_2]$, $[M_3]$, $[H]$, $[K_1]$ and $[K_2]$ are the global property matrices of the system; $\{P_1\}$, $\{P_2\}$ and $\{P_3\}$ are the global "load" vectors of the system; $\{\Delta_1\}$, $\{\Delta_2\}$ and $\{\Delta_3\}$ are the corresponding global vectors of the system.

Clearly, Eqs. (2.123), (2.124) and (2.125) can be solved separately and sequentially for the porosity, dimensionless concentration and dimensionless pore-fluid pressure at the current time-step. Note that when Eq. (2.123) is solved using the finite element method, the dimensionless concentration at the current time-step is not known. Similarly, when Eq. (2.124) is solved using the finite element method, the dimensionless pore-fluid pressure at the current time-step remains unknown. This indicates that these three equations are fully coupled so that an iteration scheme needs to be used to solve them sequentially. At the first iteration step, the dimensionless concentration at the previous time-step is used as a reasonable guess of the dimensionless concentration at the current time-step when Eq. (2.123) is solved for the porosity. In a similar way, the dimensionless pore-fluid pressure at the previous time-step is used as a reasonable guess for the current time-step when Eq. (2.124) is solved for the dimensionless concentration. The resulting approximate porosity and dimensionless concentration can be used when Eq. (2.125) is solved for the dimensionless pore-fluid pressure. At the second iteration step, the same procedure as used in the first iteration step is followed, so that the following convergence criterion can be established after the second iteration step.

$$E = Max\left(\sqrt{\sum_{i=1}^{N_\phi}(\phi^k_{i,\tau+\Delta\tau} - \phi^{k-1}_{i,\tau+\Delta\tau})^2}, \sqrt{\sum_{i=1}^{N_{\bar{C}}}(\bar{C}^k_{i,\tau+\Delta\tau} - \bar{C}^{k-1}_{i,\tau+\Delta\tau})^2}, \sqrt{\sum_{i=1}^{N_{\bar{p}}}(\bar{p}^k_{i,\tau+\Delta\tau} - \bar{p}^{k-1}_{i,\tau+\Delta\tau})^2}\right) < \bar{E}, \tag{2.126}$$

where E and $\bar{E}$ are the maximum error at the k-th iteration step and the allowable error limit; N_ϕ, $N_{\bar{C}}$ and $N_{\bar{p}}$ are the total numbers of the degrees-of-freedom for the porosity, dimensionless concentration and dimensionless pore-fluid pressure respectively; k is the index number at the current iteration step and $k-1$ is the index number at the previous iteration step; $\phi^k_{i,\tau+\Delta\tau}$, $\bar{C}^k_{i,\tau+\Delta\tau}$ and $\bar{p}^k_{i,\tau+\Delta\tau}$ are the porosity, dimensionless concentration and dimensionless pore-fluid pressure of node i at both the current time-step and the current iteration step; $\phi^{k-1}_{i,\tau+\Delta\tau}$, $\bar{C}^{k-1}_{i,\tau+\Delta\tau}$ and $\bar{p}^{k-1}_{i,\tau+\Delta\tau}$ are the porosity, dimensionless concentration and dimensionless pore-fluid pressure of node i at the current time-step but at the previous iteration step. It is noted that $k \geq 2$ in Eq. (2.126).

The convergence criterion is checked after the second iteration step. If the convergence criterion is not met, then the iteration is repeated at the current time-step. Otherwise, the convergence solution is obtained at the current time-step and the solution procedure goes to the next time-step until the final time-step is reached.

2.3 Verification of the Proposed Numerical Algorithm for Simulating the Evolution of Chemical Dissolution Fronts

The main and ultimate purpose of a numerical simulation is to provide numerical solutions for practical problems in a real world. These practical problems are impossible and impractical to solve analytically. Since numerical methods are the basic foundation of a numerical simulation, only an approximate solution can be obtained from a computational model, which is the discretized description of a continuum mathematical model. Due to inevitable round-off errors in computation

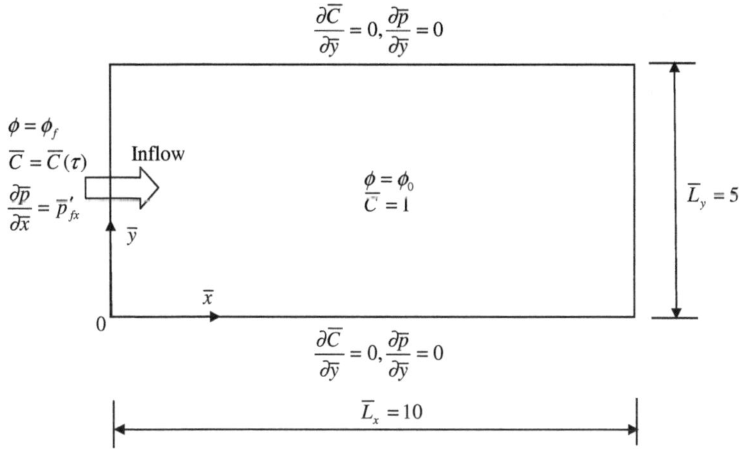

Fig. 2.1 Geometry and boundary conditions of the chemical dissolution problem

and discretized errors in temporal and spatial variables, it is necessary to verify the proposed numerical procedure so that meaningful numerical results can be obtained from a discretized computational model. For this reason, a benchmark problem, for which the analytical solutions are available, is considered in this subsection.

Figure 2.1 shows the geometry and boundary conditions of the coupled problem between porosity, pore-fluid pressure and reactive chemical-species transport within a fluid-saturated porous medium. For this benchmark problem, the dimensionless-pressure gradient (i.e. $\bar{p}'_{fx} = -1$) is applied on the left boundary, implying that there is a horizontal throughflow from the left to the right of the computational model. In this case, the Zhao number of the reactive transport system is unity. The dimensionless height and width of the computational model are 5 and 10 respectively. Except for the left boundary, the initial porosity of the porous medium is 0.1, while the initial dimensionless-concentration is one within the computational domain. The final porosity after depletion of the soluble mineral is 0.2. This final porosity is applied on the left boundary as a boundary condition of the computational domain. The permeability of the porous medium is calculated using the Carman-Kozeny formula, which has the power of 3 in the power law. The diffusivity of chemical species is calculated using the power law, which has the power of 2. Both the top and the bottom boundaries are assumed to be impermeable for the pore-fluid and chemical species. The mineral dissolution ratio of the chemical dissolution system is assumed to be 0.01, while the dimensionless time-step length is set to be 0.005 in the computation. Since the computational domain of the benchmark problem is of finite size, a time-dependent-dimensionless concentration boundary condition [i.e. $\bar{C}(\tau) = \exp(\bar{p}'_{fx}\bar{v}_{front}\tau)$] needs to be applied on the left boundary so that the numerical solutions can be compared with

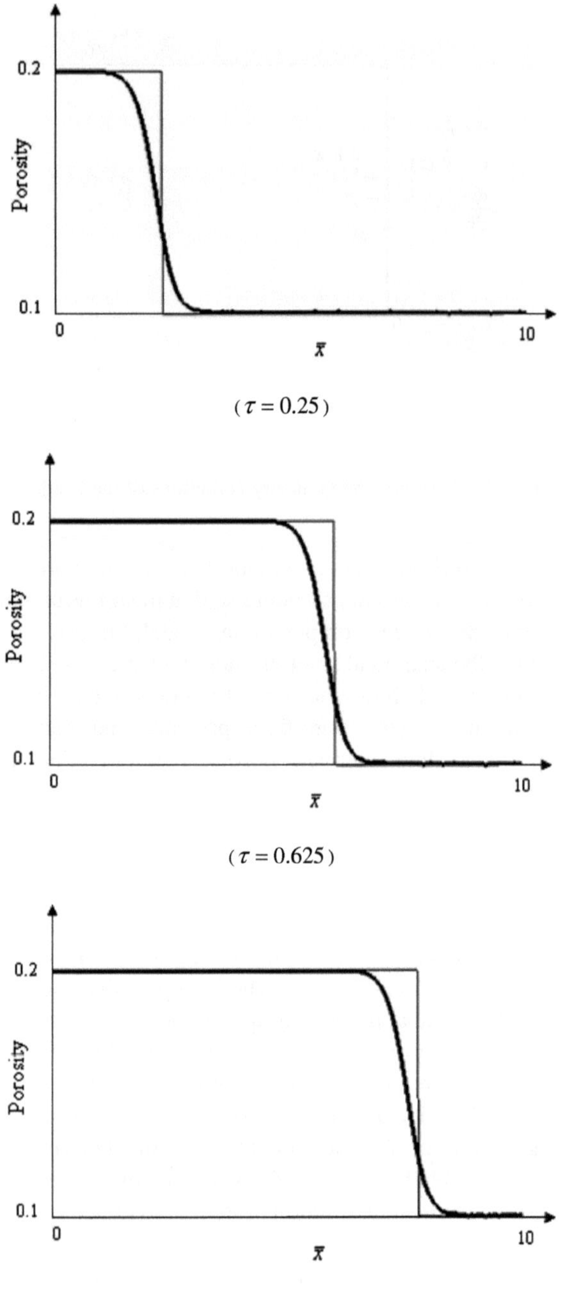

Fig. 2.2 Comparison of numerical solutions with analytical ones at different time instants (porosity): the *thick line* shows the numerical results, while the *thin line* shows the corresponding analytical solutions

Fig. 2.3 Comparison of numerical solutions with analytical ones at different time instants (dimensionless concentration): the *thick line* shows the numerical results, while the *thin line* shows the corresponding analytical solutions

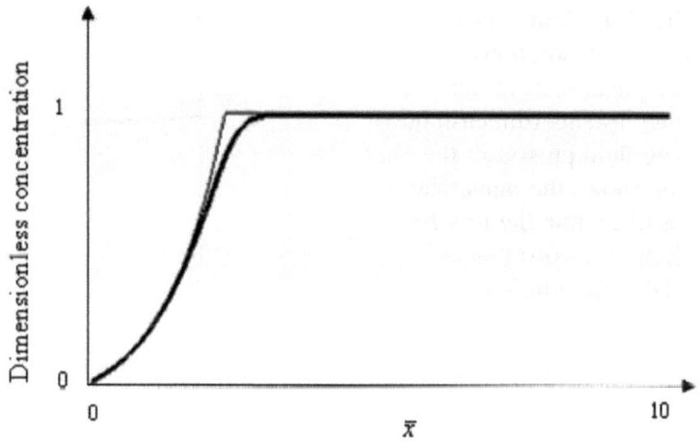

$(\tau = 0.25)$

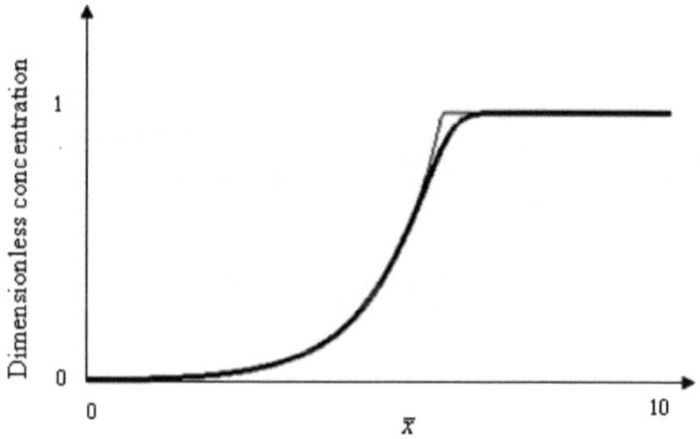

$(\tau = 0.625)$

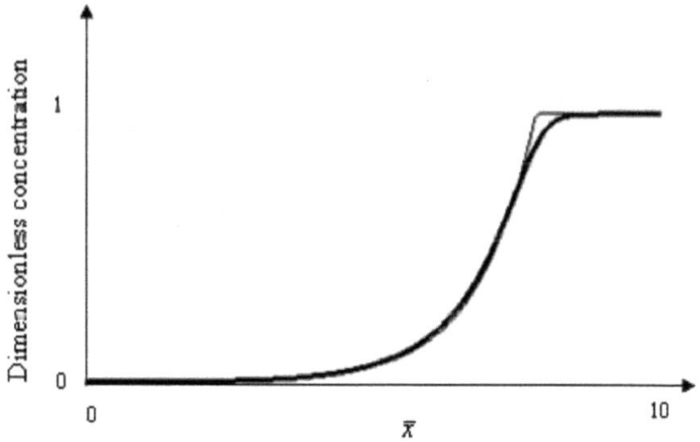

$(\tau = 0.8)$

the analytical solutions derived in the previous section. Using the above-mentioned parameters, the critical Zhao number of the system is approximately equal to 1.77. Since the Zhao number of the system is greater than its critical value, the

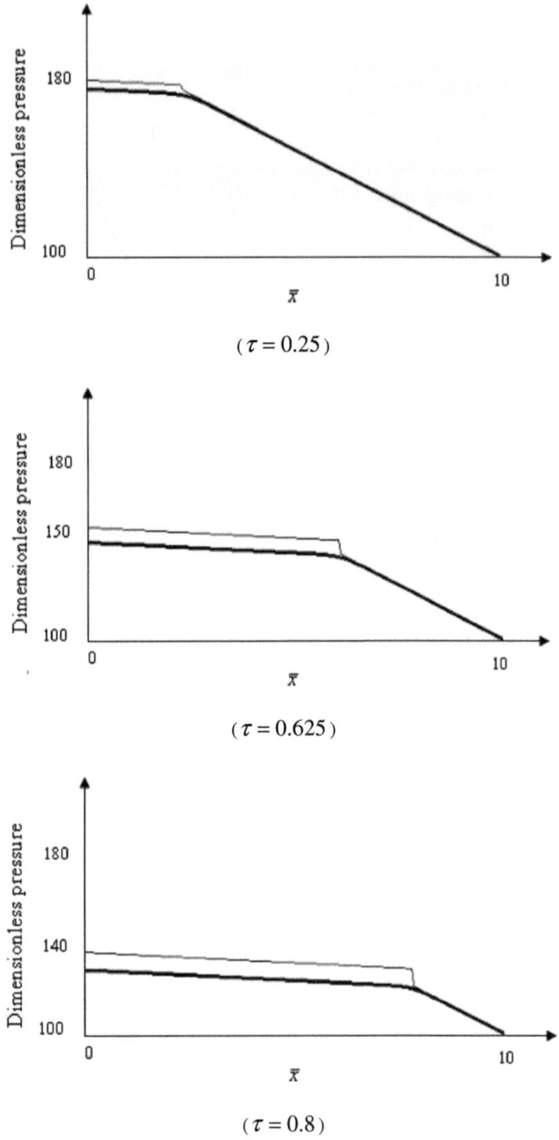

Fig. 2.4 Comparison of numerical solutions with analytical ones at different time instants (dimensionless pore-fluid pressure): the *thick line* shows the numerical results, while the *thin line* shows the corresponding analytical solutions

coupled system considered in this subsection is sub-critical so that a planar dissolution front remains planar during its propagation within the system. The dimensionless speed of the dissolution front propagation is equal to 10, which is determined using Eq. (2.52). To simulate appropriately the propagation of the dissolution front, the whole computational domain is simulated by 19701 four-node rectangular elements of 20,000 nodal points in total.

Figures 2.2, 2.3, and 2.4 show the comparison of numerical solutions with analytical ones for the porosity, dimensionless concentration and dimensionless pore-fluid pressure distributions within the computational domain at three different time instants. In these figures, the thick line shows the numerical results, while the thin line shows the corresponding analytical solutions, which can be determined from Eqs. (2.50) and (2.51) with the boundary condition of $\bar{p}(\bar{L}_x, \tau) = 100$ at the right boundary of the computational model. The resulting analytical solutions are expressed as follows:

$$\bar{C}(\bar{x}, \tau) = 1, \quad \phi(\bar{x}, \tau) = \phi_0 \quad (\bar{x} > \bar{v}_{front}\tau), \tag{2.127}$$

$$\bar{p}(\bar{x}, \tau) = -\bar{p}'_{0x}(\bar{L}_x - \bar{x}) + 100 \quad (\bar{x} > \bar{v}_{front}\tau), \tag{2.128}$$

$$\bar{C}(\bar{x}, \tau) = \exp[-\bar{p}'_{fx}(\bar{x} - \bar{v}_{front}\tau)], \quad \phi(\bar{x}, \tau) = \phi_f \quad (\bar{x} \leq \bar{v}_{front}\tau), \tag{2.129}$$

$$\bar{p}(\bar{x}, \tau) - \bar{p}'_{fx}(\bar{x} - \bar{v}_{front}\tau) - \bar{p}'_{0x}(\bar{L}_x - \bar{v}_{front}\tau) + 100 \quad (\bar{x} \leq \bar{v}_{front}\tau). \tag{2.130}$$

From these results, it can be observed that the numerical solutions agree very well with the analytical solutions, indicating that the proposed numerical procedure is capable of simulating the planar dissolution-front propagation within the fluid-saturated porous medium. As expected, the porosity propagation front is the sharpest one among the three propagation fronts, namely a porosity propagation front, a dimensionless-concentration propagation front and a dimensionless-pressure propagation front, in the computational model. Clearly, the dimensionless-pressure propagation front has the widest bandwidth, implying that it is the least sharp front in the computational model. Although there are some smoothing effects on the numerically-simulated propagation fronts as a result of numerical dispersion, the propagation speed of the numerically-simulated propagation front is in good coincidence with that of the analytically-predicted propagation front. For this benchmark problem, the overall accuracy of the numerical results is indicated by the dimensionless pore-fluid pressure. The maximum relative error of the numerically-simulated dimensionless pore-fluid pressure is 2.2, 4.6 and 5.8 % for dimensionless times of 0.25, 0.625 and 0.8 respectively. If both a small mesh size and a small time-step length are used, then the maximum relative error can be further reduced in the numerical simulation. This quantitatively demonstrates that the proposed numerical procedure can produce accurate numerical solutions for the planar dissolution-front propagation problem within a fluid-saturated porous medium.

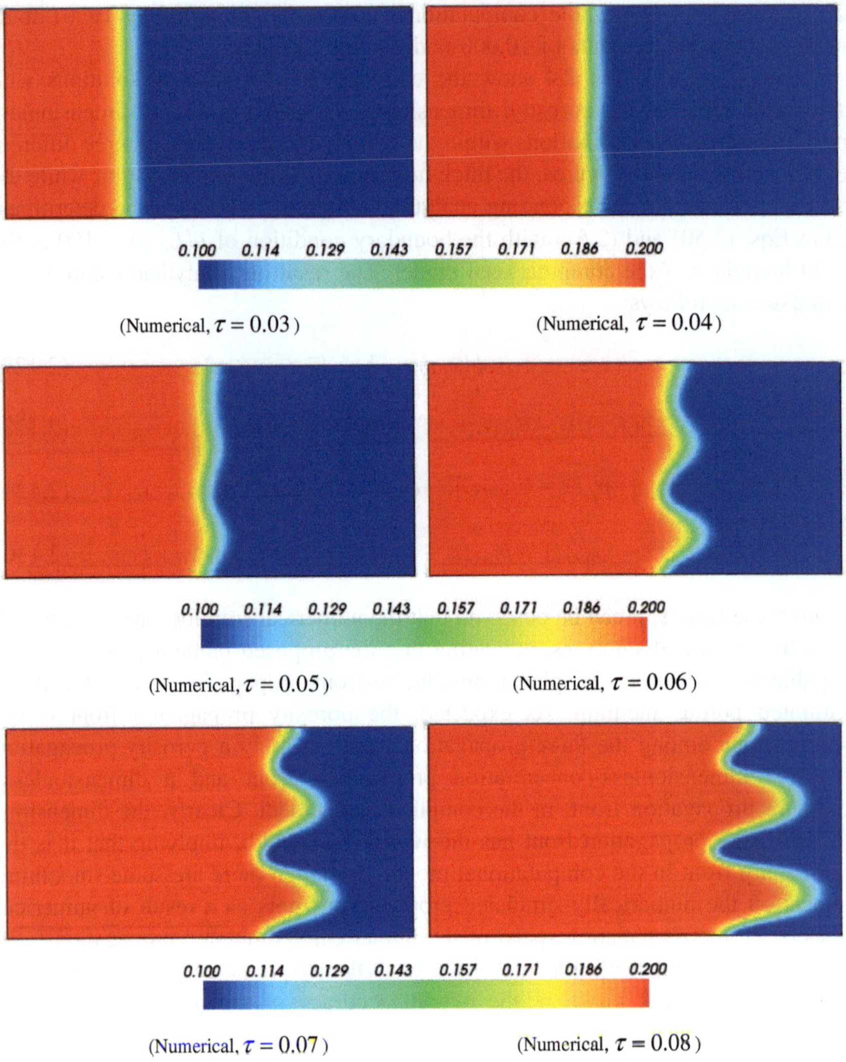

Fig. 2.5 Porosity distributions due to morphological evolution of the chemical dissolution front in the fluid-saturated porous medium

2.4 An Application Example for Simulating the Morphological Evolution of Chemical Dissolution Fronts

In this section, the proposed numerical procedure is used to simulate the morphological evolution of a chemical dissolution front in a supercritical system. For this purpose, a dimensionless-pressure gradient (i.e. $\bar{p}'_{fx} = -10$) is applied on the

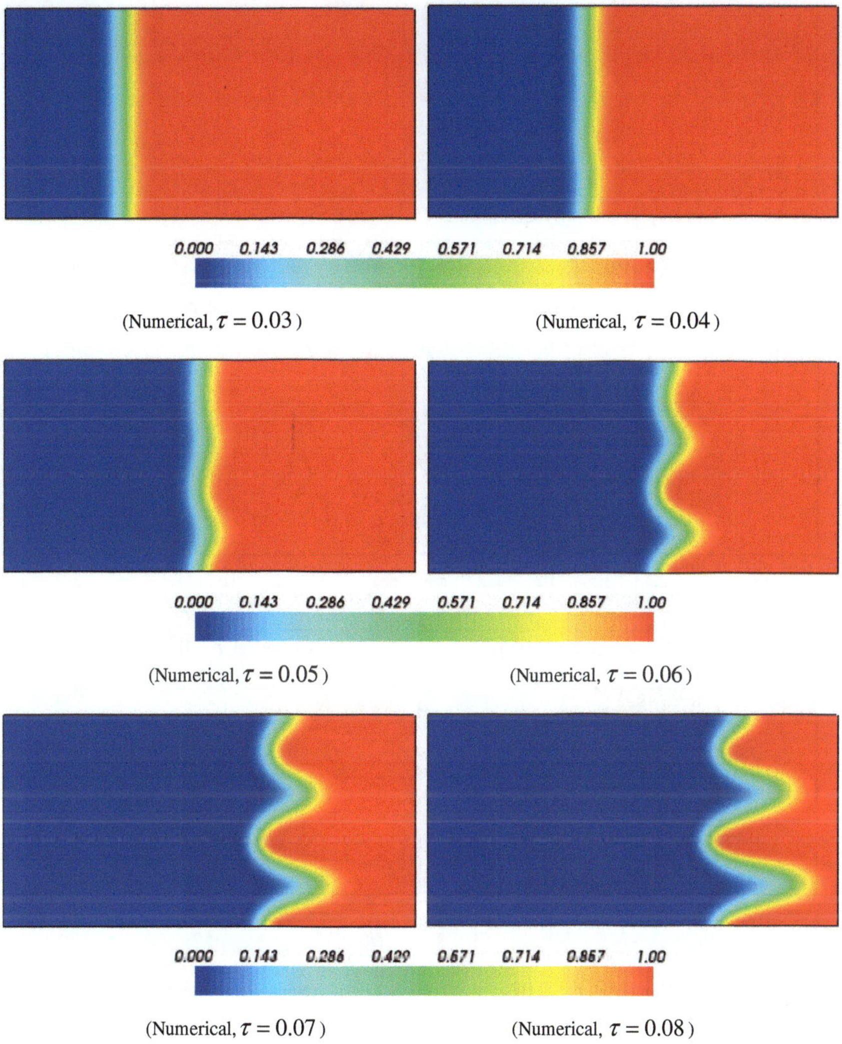

Fig. 2.6 Dimensionless concentration distributions due to morphological evolution of the chemical dissolution front in the fluid-saturated porous medium

left boundary of the computational domain so that the dimensionless speed of the dissolution front propagation is equal to 100. This means that the dissolution front propagates much faster than it does within the system considered in the previous section. Due to this change, the mineral dissolution ratio of the chemical dissolution system is assumed to be 0.001, while the dimensionless time-step length is also assumed to be 0.001 in the computation. The Zhao number of the system is increased to 10, which is greater than the critical Zhao number (i.e. approximately

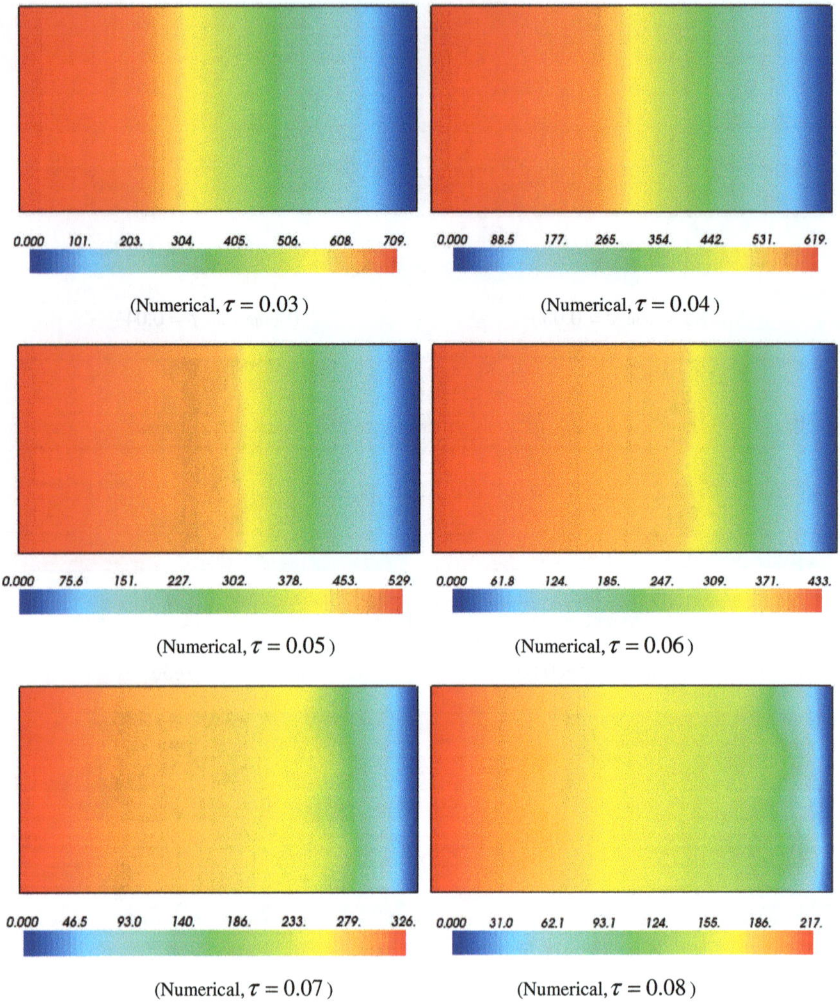

Fig. 2.7 Dimensionless pore-fluid pressure distributions due to morphological evolution of the chemical dissolution front in the fluid-saturated porous medium

1.77) of the system. The values of other parameters are exactly the same as those used in the previous section. Since the Zhao number of the system is smaller than its critical value, the coupled system considered in this section is supercritical so that a planar dissolution front evolves into a complicated morphology during its propagation within the system. In order to simulate the instability of the chemical dissolution front, a small perturbation of 1 % initial porosity is randomly added to the initial porosity field in the computational domain.

Figure 2.5 shows the porosity distributions due to the morphological evolution of the chemical dissolution front in the fluid-saturated porous medium, while

($\tau = 0.06$)

($\tau = 0.07$)

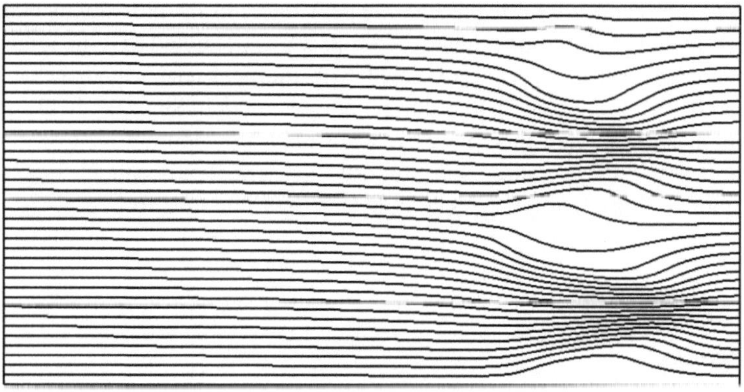

($\tau = 0.08$)

Fig. 2.8 Streamline distributions due to morphological evolution of the chemical dissolution front in the fluid-saturated porous medium

Fig. 2.6 shows the dimensionless concentration distributions due to the morphological evolution of the chemical dissolution front within the computational domain. It is observed that for the values of the dimensionless time greater than 0.03, the initial planar dissolution-front gradually changes into an irregular one. With a further increase of the dimensionless time, the amplitude of the resulting irregular dissolution-front increases significantly, indicating that the chemical dissolution front is morphologically unstable during its propagation within the computational model. Although both the porosity and the dimensionless concentration have a similar propagation front, the distribution of their maximum values along the dissolution front is clearly different. The peak value of the porosity is in good correspondence with the trough value of the dimensionless concentration due to the chemical dissolution in the system. This demonstrates that the proposed numerical procedure is capable of simulating the morphological instability of the chemical dissolution front in a fluid-saturated porous medium in the case of the coupled system being supercritical.

It is interesting to investigate how the dimensionless pore-fluid pressure and pore-fluid flow evolve with time during the propagation of the unstable dissolution front in the computational model. Figure 2.7 shows the dimensionless pore-fluid pressure distributions during the morphological evolution of the chemical dissolution front. It is noted that although the dimensionless pore-fluid pressure is continuous, there exists a clear transition for the dimensionless pressure-gradient distribution in the computational model. This phenomenon can be clearly seen at the late stages of the numerical simulation such as when the dimensionless time is equal to 0.06 and 0.07. The fluid-flow pattern evolution during the propagation of the unstable dissolution front is exhibited by the streamline evolution in the computational model. Figure 2.8 shows the streamline distributions during the morphological evolution of the chemical dissolution front within the coupled system between porosity, pore-fluid pressure and reactive chemical-species transport. Due to the growth of the amplitude of the irregular dissolution front, pore-fluid flow focusing takes place in the peak range of the porosity, which can be observed from the streamline density (in Fig. 2.8). It is noted that the width of the flow focusing zone is closely associated with the peak and trough values of the irregular dissolution front in the computational model. Since both the porosity generation and the pore-fluid flow focusing play an important role in ore body formation and mineralization, the proposed numerical procedure can provide a useful tool for simulating the related physical and chemical processes associated with the generation of giant ore deposits within the upper crust of the Earth.

References

Bear J (1972) Dynamics of fluids in porous media. American Elsevier, New York

Chadam J, Hoff D, Merino E, Ortoleva P, Sen A (1986) Reactive infiltration instabilities. IMA J Appl Math 36:207–221

Chadam J, Ortoleva P, Sen A (1988) A weekly nonlinear stability analysis of the reactive infiltration interface. IMA J Appl Math 48:1362–1378

Chen JS, Liu CW (2002) Numerical simulation of the evolution of aquifer porosity and species concentrations during reactive transport. Comput Geosci 28:485–499

Detournay E, Cheng AHD (1993) Fundamentals of poroelasticity. In: Hudson JA, Fairhurst C (eds) Comprehensive rock engineering, Vol. 2: analysis and design methods. Pergamon Press, New York

Gow P, Upton P, Zhao C, Hill K (2002) Copper-gold mineralization in the New Guinea: numerical modeling of collision, fluid flow and intrusion-related hydrothermal systems. Aust J Earth Sci 49:753–771

Lewis RW, Schrefler BA (1998) The finite element method in the static and dynamic deformation and consolidation of porous media. Wiley, New York

Nield DA, Bejan A (1992) Convection in porous media. Springer, New York

Ormond A, Ortoleva P (2000) Numerical modeling of reaction-induced cavities in a porous rock. J Geophys Res 105:16737–16747

Ortoleva P, Chadam J, Merino E, Sen A (1987) Geochemical self-organization II: the reactive-infiltration instability. Am J Sci 287:1008–1040

Phillips OM (1991) Flow and reactions in permeable rocks. Cambridge University Press, Cambridge

Raffensperger JP, Garven G (1995) The formation of unconformity-type uranium ore deposits: coupled hydrochemical modelling. Am J Sci 295:639–696

Schafer D, Schafer W, Kinzelbach W (1998a) Simulation of reactive processes related to biodegradation in aquifers: 1. Structure of the three-dimensional reactive transport model. J Contam Hydrol 31:167–186

Schafer D, Schafer W, Kinzelbach W (1998b) Simulation of reactive processes related to biodegradation in aquifers: 2. Model application to a column study on organic carbon degradation. J Contam Hydrol 31:187–209

Schaubs P, Zhao C (2002) Numerical modelling of gold-deposit formation in the Bendigo-Ballarat zone, Victoria. Aust J Earth Sci 49:1077–1096

Scheidegger AE (1974) The physics of flow through porous media. University of Toronto Press, Toronto

Steefel CI, Lasaga AC (1990) Evolution of dissolution patterns: permeability change due to coupled flow and reaction. In: Melchior DC, Basset RL (eds.) Chemical modeling in aqueous systems II, American Chemistry Society Symposium Series, vol. 416, pp. 213–225

Steefel CI, Lasaga AC (1994) A coupled model for transport of multiple chemical species and kinetic precipitation/dissolution reactions with application to reactive flow in single phase hydrothermal systems. Am J Sci 294:529–592

Turcotte DL, Schubert G (1982) Geodynamics: applications of continuum physics to geological problems. Wiley, New York

Xu TF, Samper J, Ayora C, Manzano M, Custodio E (1999) Modelling of non-isothermal multi-component reactive transport in field scale porous media flow systems. J Hydrol 214:144–164

Xu TF, Apps JA, Pruess K (2004) Numerical simulation of CO2 disposal by mineral trapping in deep aquifers. Appl Geochem 19:917–936

Yeh GT, Tripathi VS (1991) A model for simulating transport of reactive multispecies components: model development and demonstration. Water Resour Res 27:3075–3094

Zhao C, Xu TP, Valliappan S (1994) Numerical modelling of mass transport problems in porous media: a review. Comput Struct 53:849–860

Zhao C, Hobbs BE, Mühlhaus HB (1998) Finite element modelling of temperature gradient driven rock alteration and mineralization in porous rock masses. Comput Methods Appl Mech Eng 165:175–187

Zhao C, Hobbs BE, Mühlhaus HB, Ord A (1999) Finite element analysis of flow patterns near geological lenses in hydrodynamic and hydrothermal systems. Geophys J Int 138:146–158

Zhao C, Hobbs BE, Walshe JL, Mühlhaus HB, Ord A (2001a) Finite element modeling of fluid-rock interaction problems in pore-fluid saturated hydrothermal/sedimentary basins. Comput Methods Appl Mech Eng 190:2277–2293

Zhao C, Hobbs BE, Mühlhaus HB, Ord A (2001b) Finite element modelling of rock alteration and metamorphic process in hydrothermal systems. Commun Numer Methods Eng 17:833–843

Zhao C, Lin G, Hobbs BE, Ord A, Wang Y, Mühlhaus HB (2003) Effects of hot intrusions on pore-fluid flow and heat transfer in fluid-saturated rocks. Comput Methods Appl Mech Eng 192:2007–2030

Zhao C, Hobbs BE, Ord A, Peng S, Mühlhaus HB, Liu L (2005) Numerical modeling of chemical effects of magma solidification problems in porous rocks. Int J Numer Meth Eng 64:709–728

Zhao C, Hobbs BE, Ord A, Hornby P (2006a) Chemical reaction patterns due to fluids mixing and focusing around faults in fluid-saturated porous rocks. J Geochem Explor 89:470–473

Zhao C, Hobbs BE, Hornby P, Ord A, Peng S (2006b) Numerical modelling of fluids mixing, heat transfer and non-equilibrium redox chemical reactions in fluid-saturated porous rocks. Int J Numer Meth Eng 66:1061–1078

Zhao C, Hobbs BE, Ord A, Hornby P, Peng S, Liu L (2007) Mineral precipitation associated with vertical fault zones: the interaction of solute advection, diffusion and chemical kinetics. Geofluids 7:3–18

Zhao C, Hobbs BE, Hornby P, Ord A, Peng S, Liu L (2008a) Theoretical and numerical analyses of chemical-dissolution front instability in fluid-saturated porous rocks. Int J Numer Anal Meth Geomech 32:1107–1130

Zhao C, Hobbs BE, Ord A, Hornby P, Peng S (2008b) Effect of reactive surface areas associated with different particle shapes on chemical-dissolution front instability in fluid-saturated porous rocks. Transp Porous Media 73:75–94

Zhao C, Hobbs BE, Ord A, Hornby P, Mühlhaus HB, Peng S (2008c) Theoretical and numerical analyses of pore-fluid-flow focused heat transfer around geological faults and large cracks. Comput Geotech 35:357–371

Zhao C, Hobbs BE, Ord A, Peng S (2010) Effects of mineral dissolution ratios on chemical-dissolution front instability in fluid-saturated porous media. Transp Porous Media 82:317–335

Zienkiewicz OC (1977) The finite element method. McGraw-Hill, London

Chapter 3
Effects of Particle Reactive Surface Areas on Chemical Dissolution-Front Instability in Fluid-Saturated Porous Media

Particle shapes and arrangements within a porous medium can significantly affect the porosity and permeability of the porous medium (Phillips 1991; Nield and Bejan 1992; Zhao et al. 2008a). Since a change in porosity may cause a change in both the permeability and the channel of pore-fluid flow, a change in particle shapes can affect the pore-fluid flow within the porous medium. As a result, different particle shapes may affect the chemical dissolution-front instability within fluid-saturated porous media. It needs to be pointed out that for determining both the porosity and permeability in a porous medium consisting of spherical grains of equal size, the packing of grains (i.e. the orientation of an individual grain) is not relevant. However, for determining both the porosity and permeability in a porous medium consisting of cubic grains of equal size, the packing of grains (i.e. the orientation of an individual grain) must be taken into account. This means that it is the permeability that describes the shape, sorting and packing of grains, so that a system consisting of equally-sized cubic grains and having a given porosity may have a permeability that varies with the packing of the grains. Since the main purpose of this chapter is to investigate how the reactive surface area of an individual grain affects the chemical dissolution-front instability, the porosity, rather than the permeability of the porous medium, can be directly used. Although a considerable amount of research has been conducted on the instability of chemical dissolution-front propagation within fluid-saturated porous media (Chadam et al. 1986, 1988; Ortoleva et al. 1987; Chen and Liu 2002), very limited work has been conducted to analytically investigate the effect of reactive surface areas due to different particle shapes on the chemical dissolution front instability within fluid-saturated porous media (Zhao et al. 2008a). Thus, the theoretical investigation of the particle shape effect on the instability of chemical dissolution-front propagation within fluid-saturated porous media becomes the main purpose of this chapter.

From the chemical kinetics point of view, the chemical reaction rate of a heterogeneous reaction is directly proportional to the surface area of a reactive

C. Zhao, *Physical and Chemical Dissolution Front Instability in Porous Media*, Lecture Notes in Earth System Sciences,
DOI: 10.1007/978-3-319-08461-9_3

particle (Steefel and Lasage 1990, 1994). Due to the complicated feature of particle shape and size distributions in a natural porous medium, it is very difficult, if not impossible, to derive analytical solutions for directly investigating the effect of particle shape on the chemical dissolution-front instability within a fluid-saturated porous medium in nature. However, from the continuum mechanics point of view, a natural porous medium can be represented using the volume average concept. This means that if a representative elementary volume is appropriately chosen, it is possible to use the volume-averaged quantities to investigate the macroscopic behaviour of the natural porous medium (Bear 1972). Based on this understanding, the effect of reactive surface areas associated with different particle shapes on the chemical dissolution-front instability can be investigated theoretically through considering two idealized porous media consisting of two different particle shapes, namely spherical and cubic particle shapes. This makes it possible to analytically derive the critical condition, which can be used to assess whether or not the chemical dissolution front in a fully coupled system between porosity, pore-fluid flow and reactive chemical-species transport problem becomes unstable. Through examining the effect of reactive surface areas associated with different particle shapes on the derived critical condition, the effect of particle shapes on the chemical dissolution-front instability can be theoretically investigated.

In order to predict the complicated morphological evolution of a chemical dissolution front in the case of the chemical dissolution system becoming supercritical, a numerical procedure is used for solving this kind of problem. Since the proposed numerical procedure belongs to the family of approximate solution methods, it must be verified before it is used to solve any new type of scientific and engineering problem. For this reason, the analytical solution for the propagation of a planar chemical dissolution-front (Chadam et al. 1986; Zhao et al. 2008b) is used to construct a benchmark problem, the geometry of which can be accurately simulated using numerical methods such as the finite element method (Zienkiewicz 1977; Lewis and Schrefler 1998) and the finite difference method. This makes it possible to compare the numerical solution obtained from the benchmark problem with the corresponding analytical solution so that the proposed numerical procedure can be verified for simulating chemical dissolution-front propagation problems in fluid-saturated porous media. After the proposed numerical procedure is verified, it is used to simulate the morphological evolution of a chemical dissolution front in the case of the chemical dissolution system becoming supercritical. Consequently, the effect of reactive surface areas associated with different particle shapes on the morphological evolution of an unstable chemical dissolution-front can be investigated.

3.1 Theoretical Analysis of the Effect of Particle Shapes on Chemical Dissolution-Front Instability in Fluid-Saturated Porous Media

If the volume-averaged quantities over a representative elementary volume are used to describe the macroscopic behaviour of the natural porous medium (Bear 1972), the effect of reactive surface areas associated with different particle shapes on the chemical dissolution-front instability can be investigated analytically through considering two idealized porous media consisting of two different particle shapes, namely spherical and cubic particle shapes respectively. From the linear instability theory, the chemical dissolution-front instability can be determined by the critical condition, under which a fully coupled system between porosity, pore-fluid flow and reactive chemical-species transport becomes unstable. Through examining the effect of reactive surface areas associated with different particle shapes on the critical condition, the effect of particle shapes on the chemical dissolution-front instability can be investigated. Thus, the main purpose of this section is to derive the above-mentioned critical condition in a purely analytical manner.

For a fluid-saturated porous medium, Darcy's law can be used to describe pore-fluid flow and the advection-dispersion equation can be used to describe mass transport phenomena, respectively. It is assumed that the porosity change of the porous medium is caused by the chemical dissolution of a soluble solid mineral within the porous medium. Since both permeability and diffusivity are dependent on porosity, the chemically-induced porosity change has a significant feedback effect on the variation of permeability and diffusivity. If the pore-fluid is incompressible, the governing equations of the coupled nonlinear problem between porosity, pore-fluid flow and reactive single-chemical-species transport in the fluid-saturated porous medium can be expressed as follows:

$$\frac{\partial \phi}{\partial t} - \nabla \cdot [\psi(\phi) \nabla p] = 0, \tag{3.1}$$

$$\frac{\partial}{\partial t}(\phi C) - \nabla \cdot [\phi D(\phi) \nabla C + C\psi(\phi) \nabla p] + \rho_s k_{Echemical} \frac{\bar{A}_p}{\bar{V}_p}(\phi_f - \phi)(C - C_{eq}) = 0, \tag{3.2}$$

$$\frac{\partial \phi}{\partial t} + k_{Echemical} \frac{\bar{A}_p}{\bar{V}_p}(\phi_f - \phi)(C - C_{eq}) = 0, \tag{3.3}$$

$$\psi(\phi) = \frac{k(\phi)}{\mu}, \tag{3.4}$$

where p and C are the pore-fluid pressure and the concentration of the chemical species; C_{eq} is the equilibrium concentration of the chemical species; μ is the dynamic viscosity of the pore-fluid; ϕ is the porosity of the porous medium; ϕ_f is the final (i.e. maximum) porosity of the porous medium after the completion of soluble mineral dissolution; $D(\phi)$ is the diffusivity of the chemical species; $k(\phi)$ is the permeability of the porous medium. $\bar{V}_p$ is the average volume of a soluble grain; $\bar{A}_p$ is the averaged surface area of the soluble grain; $k_{Echemical}$ is the comprehensive rate constant of the chemical reaction (as defined in Eq. (2.15) previously); ρ_s is the molar density (i.e. moles per volume) of the soluble grains.

If the volumetric fraction of insoluble grains is denoted by $\phi_{insoluble}$, then the final porosity of the porous medium can be denoted by $\phi_f = 1 - \phi_{insoluble}$. In this case, the average volume of a soluble grain can be expressed as follows:

$$\bar{V}_p = \frac{\phi_f - \phi}{D_p}, \tag{3.5}$$

where D_p is the density of the soluble grains, which is defined as the total number of the soluble grains available for chemical reactions per unit medium volume.

For spherical grains (or equivalent spherical grains of equal size if their shapes are not spheres of equal size), the following relationship exists.

$$\bar{V}_p = \frac{4}{3}\pi\bar{R}^3, \tag{3.6}$$

where $\bar{R}$ is the radius of the representative grain. This radius can be considered as the characteristic length of the representative grain in the considered chemical system and expressed as

$$\bar{R} = \sqrt[3]{\frac{3(\phi_f - \phi)}{4\pi D_p}}. \tag{3.7}$$

In this case, the shape factor of the grains can be defined as follows:

$$\frac{\bar{V}_p}{\bar{A}_p} = \frac{1}{3}\bar{R} = \sqrt[3]{\frac{(\phi_f - \phi)}{36\pi D_p}} = \left(\sqrt[3]{\frac{1}{36\pi}}\right)\sqrt[3]{\frac{(\phi_f - \phi)}{D_p}} = \alpha_{sphere}\sqrt[3]{\frac{(\phi_f - \phi)}{D_p}}, \tag{3.8}$$

where $\alpha_{sphere} = \sqrt[3]{1/(36\pi)}$ is the shape coefficient of the spherical grains.

Substituting Eq. (3.8) into Eqs. (3.2) and (3.3) yields the following equations.

$$\frac{\partial}{\partial t}(\phi C) - \nabla \cdot [\phi D(\phi)\nabla C + C\psi(\phi)\nabla p] + \rho_s k_{Echemical}\frac{\sqrt[3]{D_p}}{\alpha_{sphere}}(\phi_f - \phi)^{\frac{2}{3}}(C - C_{eq}) = 0, \tag{3.9}$$

$$\frac{\partial \phi}{\partial t} + k_{Echemical}\frac{\sqrt[3]{D_p}}{\alpha_{sphere}}(\phi_f - \phi)^{\frac{2}{3}}(C - C_{eq}) = 0. \tag{3.10}$$

Suppose the chemical species is initially in an equilibrium state within the whole system and that the fresh pore-fluid is injected at the location of x approaching negative infinity, the following boundary conditions of the problem can be used in this theoretical investigation.

$$\lim_{x \to -\infty} C(x, t) = 0, \quad \lim_{x \to -\infty} \frac{\partial p(x, t)}{\partial x} = p'_{fx} \quad \text{(upstream boundary)}, \tag{3.11}$$

$$\lim_{x \to \infty} C(x, t) = C_{eq}, \quad \lim_{x \to \infty} \frac{\partial p(x, t)}{\partial x} = p'_{0x} \quad \text{(downstream boundary)}, \tag{3.12}$$

where p'_{fx} is the pore-fluid pressure gradient as x approaching negative infinity in the upstream of the pore-fluid flow; p'_{0x} is the unknown pore-fluid pressure gradient as x approaching positive infinity in the downstream of the pore-fluid flow. Since p'_{fx} drives the pore-fluid flow continuously along the positive x direction, it has a negative algebraic value (i.e. $p'_{fx} < 0$) in this analysis.

The initial condition for this theoretical problem is: $\phi(x, 0) = \phi_0$ expect at the negative infinity, where $\lim_{\bar{x} \to -\infty} \phi(x, 0) = \phi_f$. Note that ϕ_0 is the initial porosity of the porous medium.

In most geochemical systems consisting reactive rocks, the solid molar density of a soluble mineral is usually much greater than the corresponding equilibrium concentration of the soluble mineral. In such a case, a small parameter, which is known as the mineral dissolution ratio (Zhao et al. 2010), can be defined below:

$$\varepsilon = \frac{C_{eq}}{\rho_s} << 1. \tag{3.13}$$

To analytically derive the critical condition, which can be used to examine whether or not a chemical dissolution front in the coupled system between porosity, pore-fluid flow and reactive chemical-species transport within the fluid-saturated porous medium becomes unstable, the following dimensionless parameters and variables need to be defined for a two-dimensional problem.

$$\bar{x} = \frac{x}{L^*}, \quad \bar{y} = \frac{y}{L^*}, \quad \bar{C} = \frac{C}{C_{eq}}, \quad \bar{p} = \frac{p}{p^*}, \quad \tau = \frac{t}{t^*}\varepsilon \tag{3.14}$$

where τ is a slow dimensionless time to describe the slowness of the chemical dissolution that takes place in the system. Other characteristic parameters used in Eq. (3.14) can be defined as follows:

$$t^* = \frac{\alpha_{sphere}}{(k_{Echemical} C_{eq}) \sqrt[3]{D_p}}, \quad L^* = \sqrt{\phi_f D(\phi_f) t^*}, \quad p^* = \frac{\phi_f D(\phi_f)}{\psi(\phi_f)},$$
$$D^*(\phi) = \frac{\phi D(\phi)}{\phi_f D(\phi_f)}, \quad \psi^*(\phi) = \frac{\psi(\phi)}{\psi(\phi_f)}. \tag{3.15}$$

If Eqs. (3.14) and (3.15) are inserted into Eqs. (3.1), (3.9) and (3.10), the following dimensionless equations can be obtained.

$$\varepsilon\frac{\partial\phi}{\partial\tau} - \nabla\cdot[\psi^*(\phi)\nabla\bar{p}] = 0, \tag{3.16}$$

$$\varepsilon\frac{\partial}{\partial\tau}(\phi\bar{C}) - \nabla\cdot[D^*(\phi)\nabla\bar{C} + \bar{C}\psi^*(\phi)\nabla\bar{p}] - \frac{\partial\phi}{\partial\tau} = 0, \tag{3.17}$$

$$\varepsilon\frac{\partial\phi}{\partial\tau} + (\phi_f - \phi)^{\frac{2}{3}}(\bar{C} - 1) = 0. \tag{3.18}$$

Obviously, the dimensionless variables, such as the dimensionless pore-fluid pressure and dimensionless chemical species concentration, need to be used to describe the following boundary conditions for this dimensionless problem.

$$\lim_{\bar{x}\to\infty}\bar{C}(\bar{x},\tau) = 1, \quad \lim_{\bar{x}\to\infty}\frac{\partial\bar{p}(\bar{x},\tau)}{\partial\bar{x}} = \bar{p}'_{0x} \quad \text{(downstream boundary)}, \tag{3.19}$$

$$\lim_{\bar{x}\to-\infty}\bar{C}(\bar{x},\tau) = 0, \quad \lim_{\bar{x}\to-\infty}\frac{\partial\bar{p}(\bar{x},\tau)}{\partial\bar{x}} = \bar{p}'_{fx} \quad \text{(upstream boundary)}. \tag{3.20}$$

In this case, the initial condition for this problem is: $\phi(\bar{x}, 0) = \phi_0$ expect at the negative infinity, where $\lim_{\bar{x}\to-\infty}\phi(\bar{x}, 0) = \phi_f$.

In the limit case of ε approaching zero, the corresponding governing equations for the dimensionless variables of the problem in both the downstream region and the upstream region can be derived from Eqs. (3.16)–(3.18) and expressed as follows (Chadam et al. 1986; Zhao et al. 2008b):

$$\bar{C} = 1, \quad \nabla^2\bar{p} = 0, \quad \phi = \phi_0 \quad \text{(in the downstream region)}, \tag{3.21}$$

$$\nabla\cdot(\nabla\bar{C} + \bar{C}\nabla\bar{p}) = 0, \quad \nabla^2\bar{p} = 0, \quad \phi = \phi_f \quad \text{(in the upstream region)}. \tag{3.22}$$

If the chemical dissolution front is denoted by $S(\bar{x}, \tau) = 0$, then the dimensionless pore-fluid pressure, dimensionless chemical species concentration and mass fluxes of both the chemical species and the pore-fluid should be continuous on $S(\bar{x}, \tau) = 0$. With this factor taking into account, the following interface conditions for this chemical dissolution-front propagation problem exist mathematically.

$$\lim_{S\to0^-}\bar{C} = \lim_{S\to0^+}\bar{C}, \quad \lim_{S\to0^-}\bar{p} = \lim_{S\to0^+}\bar{p}, \tag{3.23}$$

$$\lim_{S\to0^-}\frac{\partial\bar{C}}{\partial\bar{n}} = \bar{v}_{front}(\phi_f - \phi_0), \quad \lim_{S\to0^-}\frac{\partial\bar{p}}{\partial\bar{n}} = \frac{\psi(\phi_0)}{\psi(\phi_f)}\lim_{S\to0^+}\frac{\partial\bar{p}}{\partial\bar{n}}. \tag{3.24}$$

where $\bar{n} = n/L^*$; n is the normal vector of the propagating planar chemical dissolution-front; $\bar{v}_{front}$ is the dimensionless propagation speed of the chemical dissolution front.

Equations (3.21)–(3.24) clearly indicate that in the limit case of ε approaching zero, the dimensionless governing equations and the boundary conditions are independent of the reactive surface area associated with the soluble grain shapes. For this reason, the same procedure as used in the previous investigation (Chadam et al. 1986; Zhao et al. 2008b) can be followed to derive the critical condition, which is used to assess whether or not a chemical dissolution front in the coupled system between porosity, pore-fluid flow and reactive chemical-species transport becomes unstable. Following the same procedures as those in Chap. 2, the resulting critical condition is given below:

$$Zh_{critical} = \frac{(3-\beta)(1+\beta)}{2(1-\beta)}, \quad \beta = \frac{\psi(\phi_0)}{\psi(\phi_f)} = \frac{k(\phi_0)}{k(\phi_f)}, \tag{3.25}$$

where $Zh_{critical}$ is the critical Zhao number of the chemical dissolution system. The corresponding Zhao number of the chemical dissolution system in a fluid-saturated porous medium consisting of spherical grains can be defined as follows:

$$Zh = -\frac{p'_{fx}L^*}{p^*} = -\frac{k(\phi_f)L^*p'_{fx}}{\phi_f\mu D(\phi_f)} = -\frac{k(\phi_f)p'_{fx}}{\mu\sqrt{\phi_f D(\phi_f)}}\sqrt{\frac{\alpha_{sphere}}{(k_{Echemical}C_{eq})\sqrt[3]{D_p}}} \tag{3.26}$$

Using Eqs. (3.25) and (3.26), a theoretical criterion can be established to judge the instability of a chemical dissolution front in the fluid-saturated porous medium consisting of spherical grains.

It is noted that the *critical* Zhao number and the Zhao number of a chemical dissolution system are two totally different concepts. The *critical* Zhao number of a chemical dissolution system with the effect of porosity feedback represents the critical condition, which can be used to judge whether or not the chemical dissolution front in the chemical dissolution system becomes unstable. In this regard, the *critical* Zhao number of a chemical dissolution system represents the common characteristic of a kind of chemical dissolution system. On the other hand, the Zhao number of a chemical dissolution system represents the particular characteristic of the individual chemical dissolution system, which is one particular case of the above-mentioned kind of chemical dissolution system. As indicated by Eq. (3.25), the *critical* Zhao number of a chemical dissolution system with the effect of porosity feedback is independent of the soluble grain shapes. However, Eq. (3.26) clearly indicated that the shape coefficient of soluble grains can affect the Zhao number of the chemical dissolution system. As a result, an increase in the shape coefficient of soluble grains can destabilize the instability of the chemical dissolution front within the fluid-saturated porous medium.

If the soluble grains are of cubic shapes, then the characteristic length of the representative grain in the chemical dissolution system can be expressed as follows:

$$\bar{L} = \sqrt[3]{\frac{(\phi_f - \phi)}{D_p}}. \tag{3.27}$$

In this case, the shape factor of the grains can be defined as

$$\frac{\bar{V}_p}{\bar{A}_p} = \frac{1}{6}\bar{L} = \sqrt[3]{\frac{(\phi_f - \phi)}{216 D_p}} = (\sqrt[3]{\frac{1}{216}})\sqrt[3]{\frac{(\phi_f - \phi)}{D_p}} = \alpha_{cubic}\sqrt[3]{\frac{(\phi_f - \phi)}{D_p}}. \tag{3.28}$$

where $\alpha_{cubic} = \sqrt[3]{1/216}$ is the shape coefficient of cubic grains.

Following the same procedures as used to define the Zhao number of a chemical dissolution system in the fluid-saturated porous medium consisting of spherical grains, the corresponding Zhao number of a chemical dissolution system in the fluid-saturated porous medium consisting of cubic grains can be defined as follows:

$$Zh = -\frac{p'_{fx} L^*}{p^*} = -\frac{k(\phi_f) L^* p'_{fx}}{\phi_f \mu D(\phi_f)} = -\frac{k(\phi_f) p'_{fx}}{\mu\sqrt{\phi_f D(\phi_f)}}\sqrt{\frac{\alpha_{cubic}}{(k_{Echemical} C_{eq})\sqrt[3]{D_p}}}. \tag{3.29}$$

Since the shape coefficient of spherical grains is greater than that of cubic grains, the chemical dissolution system in a fluid-saturated porous medium consisting of spherical grains is more unstable than that consisting of cubic grains, implying that the instability likelihood of a natural porous medium, which is comprised of irregular grains, is smaller than that of an idealized porous medium, which is comprised of regular spherical grains. The reason for this is due to the physical fact that if reaction rates are fast enough (relative to flow rates), then the Zhao number of the chemical dissolution system is small (relative to the corresponding critical Zhao number), so that the chemical dissolution front in the chemical dissolution system is stable. It is clearly demonstrated [from Eqs. (3.26) and (3.29)] that a chemical dissolution system containing grains with a larger reactive surface area (i.e. a smaller shape coefficient), such as cubic grains over spherical grains, is more reactive and has the tendency to produce a more stable reaction front.

3.2 Numerical Modeling of the Chemical Dissolution-Front Evolution Within Fluid-Saturated Porous Media Consisting of Different Particle Shapes

Even though the effects of reactive surface areas associated with different particle shapes on the critical condition, which can be used to assess chemical dissolution-front instability in a chemical dissolution system, can be investigated analytically, it is necessary to computationally simulate how a chemical dissolution-front evolves when the chemical dissolution system, which can be treated as a fully-coupled system between porosity, pore-fluid flow and reactive chemical-species transport, reaches a supercritical state. Therefore, the main purpose of this section is to propose a numerical procedure for simulating chemical dissolution-front evolution in a fluid-saturated porous medium, which is comprised of either spherical grains or cubic grains, depending on the specific parameters used in the related equations. To verify the accuracy of the numerical solution, a benchmark problem is established on the basis of the previous theoretical analysis (Chadam et al. 1986; Zhao et al. 2008b). This makes it possible to compare the numerical solution obtained from the benchmark problem with the corresponding analytical one. After the proposed numerical procedure is verified, it is used to simulate chemical dissolution-front instability problems in fluid-saturated porous media, so that the effects of reactive surface areas associated with different particle shapes on the morphological evolution of an unstable chemical dissolution-front can be investigated.

To obtain numerical solutions, Eqs. (3.16)–(3.18) are solved using the proposed numerical procedure, which is based on a combination of the finite element and finite difference methods. The finite element method is used to discretize the geometrical shape of the problem domain, while the finite difference method is used to discretize the dimensionless time. Because of high nonlinearity of these three equations, the segregated algorithm is used to solve them both separately and iteratively in a sequential manner.

In terms of porosity, Eq. (3.18) is a nonlinear equation, which can be linearized using the Taylor expansion as follows:

$$[\phi_f - (\phi + \Delta\phi)]^{\frac{2}{3}} = [(\phi_f - \phi) - \Delta\phi)]^{\frac{2}{3}} = (\phi_f - \phi)^{\frac{2}{3}} - \frac{2}{3}(\phi_f - \phi)^{-\frac{1}{3}}\Delta\phi. \quad (3.30)$$

For a given dimensionless time-step, $\tau + \Delta\tau$, the porosity at the current time-step can be denoted by $\phi_{\tau+\Delta\tau} = \phi_\tau + \Delta\phi_{\tau+\Delta\tau}$, where ϕ_τ is the porosity at the previous time-step and $\Delta\phi_{\tau+\Delta\tau}$ is the porosity increment at the current time-step. Using the backward difference scheme, Eq. (3.18) can be written as follows:

$$\left[\frac{\varepsilon}{\Delta\tau}(\phi_f - \phi_\tau)^{\frac{1}{3}} + \frac{2}{3}(1 - \bar{C}_{\tau+\Delta\tau})\right]\Delta\phi_{\tau+\Delta\tau} = (\phi_f - \phi_\tau)(1 - \bar{C}_{\tau+\Delta\tau}), \quad (3.31)$$

where $\bar{C}_{\tau+\Delta\tau}$ is the dimensionless chemical species concentration at the current time-step; $\Delta\tau$ is the dimensionless time increment at the current time-step.

In the finite difference sense, Eq. (3.17) can be mathematically expressed in the following form:

$$\begin{aligned}&[\tfrac{\varepsilon}{\Delta\tau}\phi_{\tau+\Delta\tau}+\tfrac{1}{\varepsilon}(\phi_f-\phi_{\tau+\Delta\tau})^{\frac{2}{3}}]\bar{C}_{\tau+\Delta\tau}-\nabla\cdot[D^*(\phi_{\tau+\Delta\tau})\nabla\bar{C}_{\tau+\Delta\tau}]\\&-\nabla\bar{p}_{\tau+\Delta\tau}\cdot[\psi^*(\phi_{\tau+\Delta\tau})\nabla\bar{C}_{\tau+\Delta\tau}]\\&=\tfrac{\varepsilon}{\Delta\tau}\phi_{\tau+\Delta\tau}\bar{C}_{\tau}+\tfrac{1}{\varepsilon}(\phi_f-\phi_{\tau+\Delta\tau})^{\frac{2}{3}}\end{aligned} \quad . \tag{3.32}$$

Similarly, Eq. (3.16) can be rewritten as follows:

$$\nabla\cdot[\psi^*(\phi)\nabla\bar{p}]=\nabla\cdot[\psi^*(\phi_{\tau+\Delta\tau})\nabla\bar{p}_{\tau+\Delta\tau}]=(1-\bar{C}_{\tau+\Delta\tau})(\phi_f-\phi_{\tau+\Delta\tau})^{\frac{2}{3}}. \tag{3.33}$$

To avoid the nonlinear term of the porosity at the current time-step, the porosity increment, instead of the porosity itself, is used for the finite difference discretization of the porosity evolution equation [i.e. Eq. (3.18)]. On the other hand, the dimensionless chemical species concentration and dimensionless pore-fluid pressure are used for the finite difference discretization of both the mass transport equation [i.e. Eq. (3.17)] and the continuity equation [i.e. Eq. (3.16)]. This means that a combination of the incremental variable method and the full variable method is used, in this investigation, to solve the fully coupled problem between porosity, pore-fluid flow and reactive chemical-species transport within the fluid-saturated porous medium. Since the finite element method (Zienkiewicz 1977; Lewis and Schrefler 1998) can be straightforwardly used to discretize Eqs. (3.31)–(3.33) in space, as demonstrated in Chap. 2, the details of the related finite element formulation are not given here.

Using the proposed segregated scheme and finite element method, Eqs. (3.31)–(3.33) are solved separately and sequentially for the porosity, dimensionless chemical-species concentration and dimensionless pore-fluid pressure at the current time-step. Note that when Eq. (3.31) is solved using the finite element method, the dimensionless chemical-species concentration at the current time-step is not known. Similarly, when Eq. (3.32) is solved using the finite element method, the dimensionless pore-fluid pressure at the current time-step remains unknown. This indicates that these three equations are fully coupled so that an iteration scheme needs to be used to solve them sequentially. At the first iteration step, the dimensionless chemical-species concentration at the previous time-step is used as a reasonable guess for the dimensionless chemical-species concentration at the current time-step when Eq. (3.31) is solved for the porosity. In the similar way, the dimensionless pore-fluid pressure at the previous time-step is used as a reasonable guess at the current time-step when Eq. (3.32) is solved for the dimensionless chemical-species concentration. The resulting approximate porosity and dimensionless chemical-species concentration can be used when Eq. (3.33) is solved for the dimensionless pore-fluid pressure. At the second iteration step, the same procedure as used in the first iteration step is followed, so that the following convergence criterion can be established after the second iteration step.

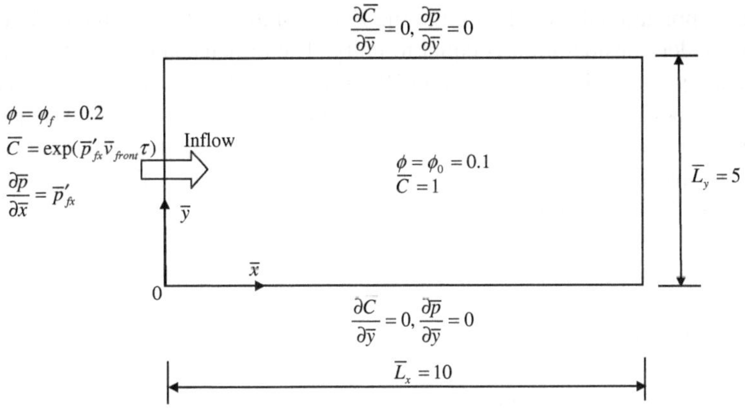

Fig. 3.1 Geometry and boundary conditions of the benchmark problem

$$E = Max\left(\sqrt{\sum_{i=1}^{N_\phi} (\phi_{i,\tau+\Delta\tau}^{k} - \phi_{i,\tau+\Delta\tau}^{k-1})^2}, \sqrt{\sum_{i=1}^{N_{\bar{C}}} (\bar{C}_{i,\tau+\Delta\tau}^{k} - \bar{C}_{i,\tau+\Delta\tau}^{k-1})^2}, \sqrt{\sum_{i=1}^{N_{\bar{p}}} (\bar{p}_{i,\tau+\Delta\tau}^{k} - \bar{p}_{i,\tau+\Delta\tau}^{k-1})^2}\right) < \bar{E}, \tag{3.34}$$

where E and $\bar{E}$ are the maximum error at the k-th iteration step and the allowable error limit; N_ϕ, $N_{\bar{C}}$ and $N_{\bar{p}}$ are the total numbers of the degree-of-freedom for the porosity, dimensionless chemical-species concentration and dimensionless pore-fluid pressure respectively; k is the index number at the current iteration step and $k-1$ is the index number at the previous iteration step; $\phi_{i,\tau+\Delta\tau}^{k}$, $\bar{C}_{i,\tau+\Delta\tau}^{k}$ and $\bar{p}_{i,\tau+\Delta\tau}^{k}$ are the porosity, dimensionless chemical-species concentration and dimensionless pore-fluid pressure of node i at both the current time-step and the current iteration step; $\phi_{i,\tau+\Delta\tau}^{k-1}$, $\bar{C}_{i,\tau+\Delta\tau}^{k-1}$ and $\bar{p}_{i,\tau+\Delta\tau}^{k-1}$ are the porosity, dimensionless chemical-species concentration and dimensionless pore-fluid pressure of node i at the current time-step but at the previous iteration step. It is noted that $k \geq 2$ in Eq. (3.34).

The convergence criterion is checked after the second iteration step. If the convergence criterion is not met, then the iteration is repeated at the current time-step. Otherwise, the convergent solution is obtained at the current time-step and the solution procedure goes to the next time-step until the final time-step is reached.

The proposed numerical procedure can be verified using a benchmark problem consisting of spherical grains, for which the analytical solutions are available (Chadam et al. 1986; Zhao et al. 2008b). Figure 3.1 shows the geometry and boundary conditions of the benchmark problem, which represents a fully coupled problem between porosity, pore-fluid flow and reactive chemical-species transport within a fluid-saturated porous medium. For this benchmark problem, the dimensionless pore-fluid pressure-gradient (i.e. $\bar{p}'_{fx} = -1$) is applied on the left boundary, implying that there is a horizontal throughflow from the left to the right of the computational model and that the Zhao number of the chemical dissolution system under consideration is equal to unity. The dimensionless height and width of the computational model are 5 and 10 respectively. Except for the left boundary, the initial porosity of the porous medium is 0.1, while the initial dimensionless chemical-species concentration is unity within the computational domain. The final porosity after the depletion of the soluble mineral is 0.2. This final porosity is applied on the left boundary as a boundary condition of the computational domain. The permeability of the porous medium is calculated using the Carman-Kozeny formula (Scheidegger 1974; Nield and Bejan 1992), which has the power of 3 in the power law. The diffusivity of chemical species is calculated using the power law, which has the power of 2. Both the top and the bottom boundaries are assumed to be impermeable for the pore-fluid and chemical species. The mineral dissolution ratio is assumed to be 0.01, while the dimensionless time-step length is set to be 0.005 in the computation. Since the computational domain of the benchmark problem is of finite size, a time-dependent boundary condition of the dimensionless chemical-species concentration [i.e. $\bar{C}(\tau) = \exp(\bar{p}'_{fx}\bar{v}_{front}\tau)$] needs to be applied on the left boundary so that the numerical solutions can be compared with the analytical solutions (Chadam et al. 1986; Zhao et al. 2008b). Using the above-mentioned parameters, the critical Zhao number of the chemical dissolution system is equal to 1.77. Since the Zhao number of the chemical dissolution system is smaller than the corresponding critical Zhao number, the coupled system under consideration is in a subcritical state, so that a planar chemical dissolution-front remains planar during its propagation within the chemical dissolution system. The dimensionless speed of the chemical dissolution-front propagation is equal to 10. To appropriately simulate the propagation of the chemical dissolution front, the whole computational domain is simulated by 19,701 four-node rectangular elements of 20,000 nodal points in total.

Figures 3.2, 3.3 and 3.4 show the comparison of numerical solutions with analytical ones for the porosity, dimensionless chemical-species concentration and dimensionless pore-fluid pressure distributions within the computational domain at three different time instants. In these figures, the left column shows the numerical results, while the right column shows the corresponding analytical solutions. From these results, it can be observed that the numerical solutions agree very well with the analytical ones, indicating that the proposed numerical procedure is capable of simulating the planar dissolution-front propagation within the fluid-saturated porous medium. Although there are some smooth effects on the numerically-simulated propagation fronts due to numerical dispersion, the propagation speed of the numerically-simulated propagation front is in good coincidence with that of analytically-predicted propagation front. For this benchmark problem, the overall accuracy of the numerical results is indicated by the dimensionless pore-fluid pressure. The maximum relative error of the numerically-simulated dimensionless pore-fluid pressure is 2.2, 3.7 and 6.3 % for the dimensionless time being 0.25, 0.5 and 0.75 respectively. It should be pointed out that if necessary, this relative error can be further reduced by: (1) using smaller elements and time-step length;

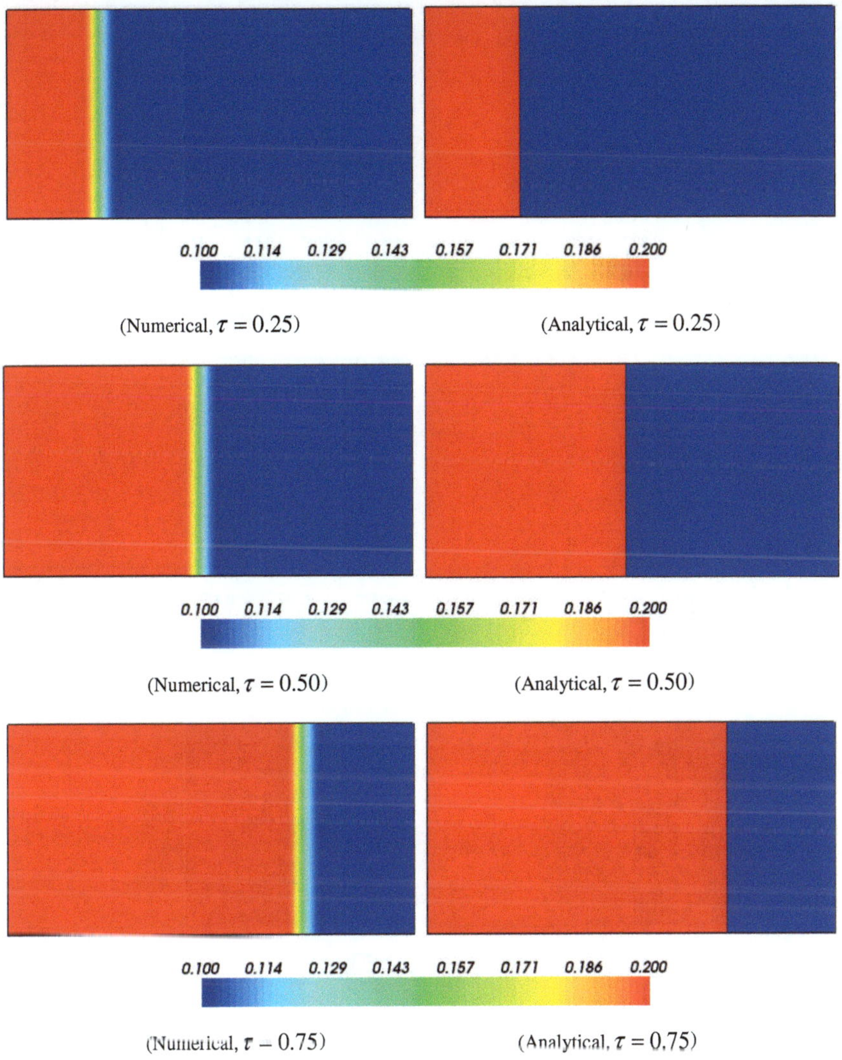

Fig. 3.2 Comparison of numerical solutions with analytical ones at different time instants (Porosity)

(2) using higher-order transport schemes such as the total variation diminishing (TVD) schemes (Harten 1983; Sweby 1985); and (3) developing a stepwise element to simulate the discontinuity of the chemical dissolution front. This quantitatively demonstrates that the proposed numerical procedure can produce accurate numerical solutions for the planar dissolution-front propagation problem within a fluid-saturated porous medium.

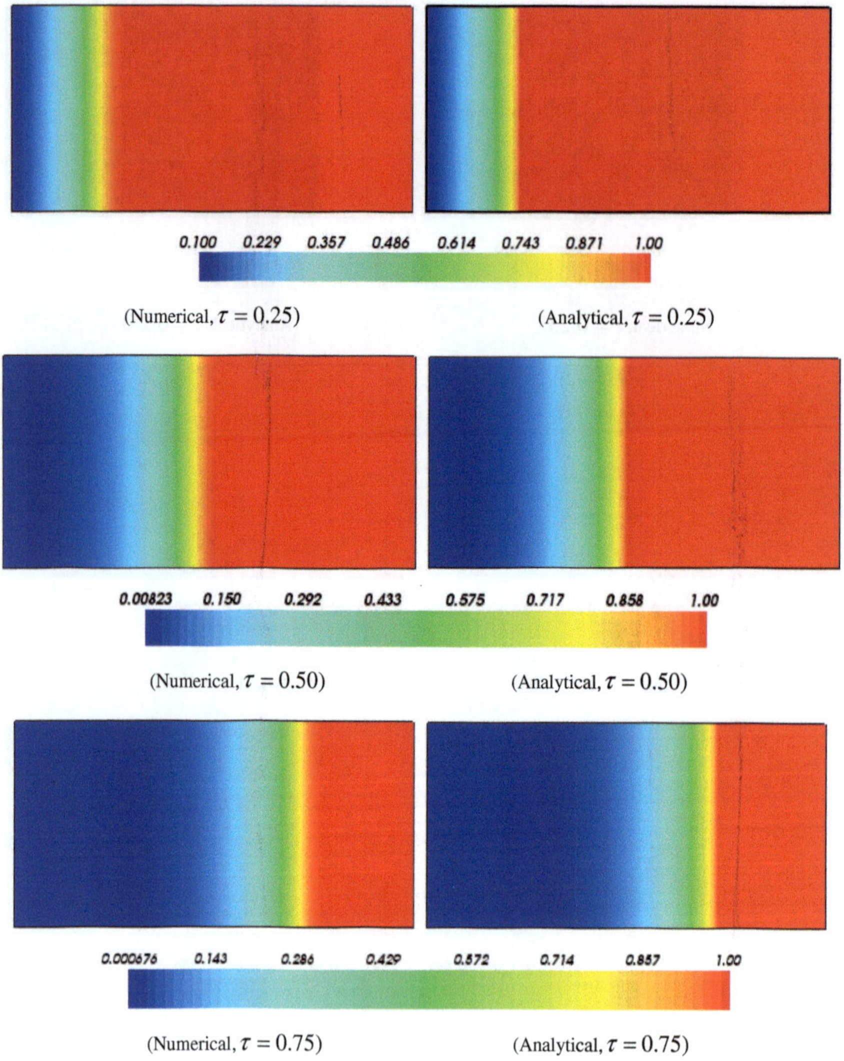

Fig. 3.3 Comparison of numerical solutions with analytical ones at different time instants (Dimensionless chemical-species concentration)

After the proposed numerical procedure is verified, it is used to simulate the morphological evolution of a chemical dissolution front within a supercritical system consisting of spherical grains, which has a dimensionless pore-fluid pressure-gradient (i.e. $\bar{p}'_{fx} = -10$) applied on the left boundary of the computational domain. This means that the Zhao number of the chemical dissolution system under consideration now is equal to 10, so that the chemical dissolution system is in a supercritical state. In this case, the dimensionless speed of the chemical dissolution-front

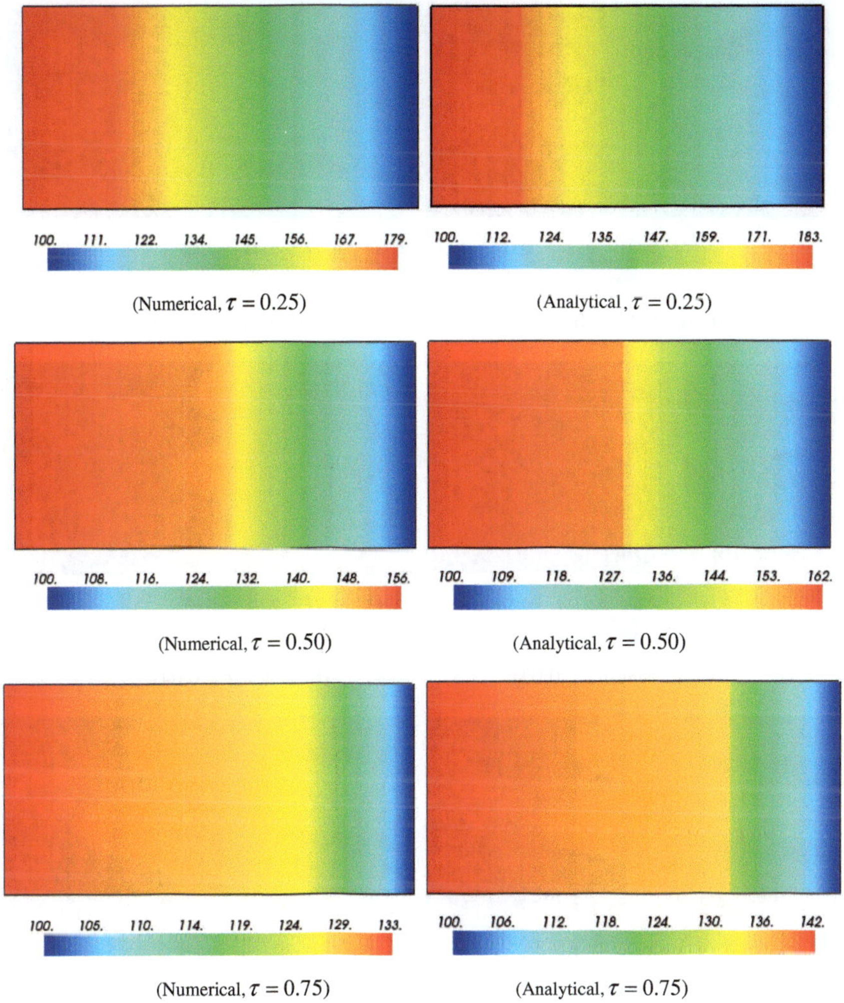

Fig. 3.4 Comparison of numerical solutions with analytical ones at different time instants (Dimensionless pore-fluid pressure)

propagation is equal to 100. The mineral dissolution ratio is 0.001, while the dimensionless time-step length is also 0.001 in the computation. The values of other parameters are exactly the same as those used when the Zhao number of the chemical dissolution system is equal to unity. In order to simulate the instability of the chemical dissolution front, a small perturbation of 1 % initial porosity is randomly added to the initial porosity field in the computational domain.

Figures 3.5 and 3.6 show the porosity and dimensionless chemical-species-concentration distributions associated with the morphological evolution of the chemical dissolution front in the fluid-saturated porous medium consisting of

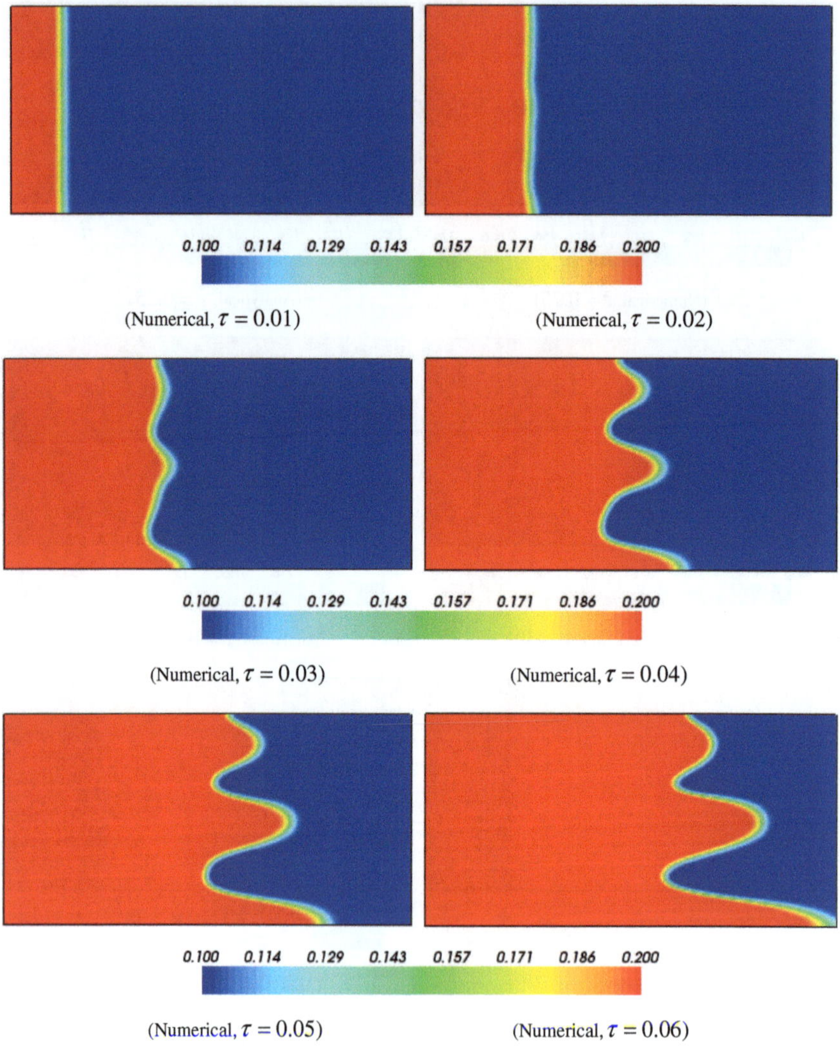

Fig. 3.5 Porosity distributions due to morphological evolution of the chemical dissolution front in the fluid-saturated porous medium

spherical grains. It is observed that after the dimensionless time is greater than 0.02, the initially-planar dissolution front is gradually changed into an irregular one. With the increase of the dimensionless time, the amplitude of the resulting irregular dissolution front increases significantly. For example, the finger-like dissolution front can be clearly formed after the dimensionless time is greater than 0.04. Although both the porosity and the dimensionless chemical-species concentration have the similar propagation front, the distribution of their maximum values along the dissolution front is in a complementary manner. This can be

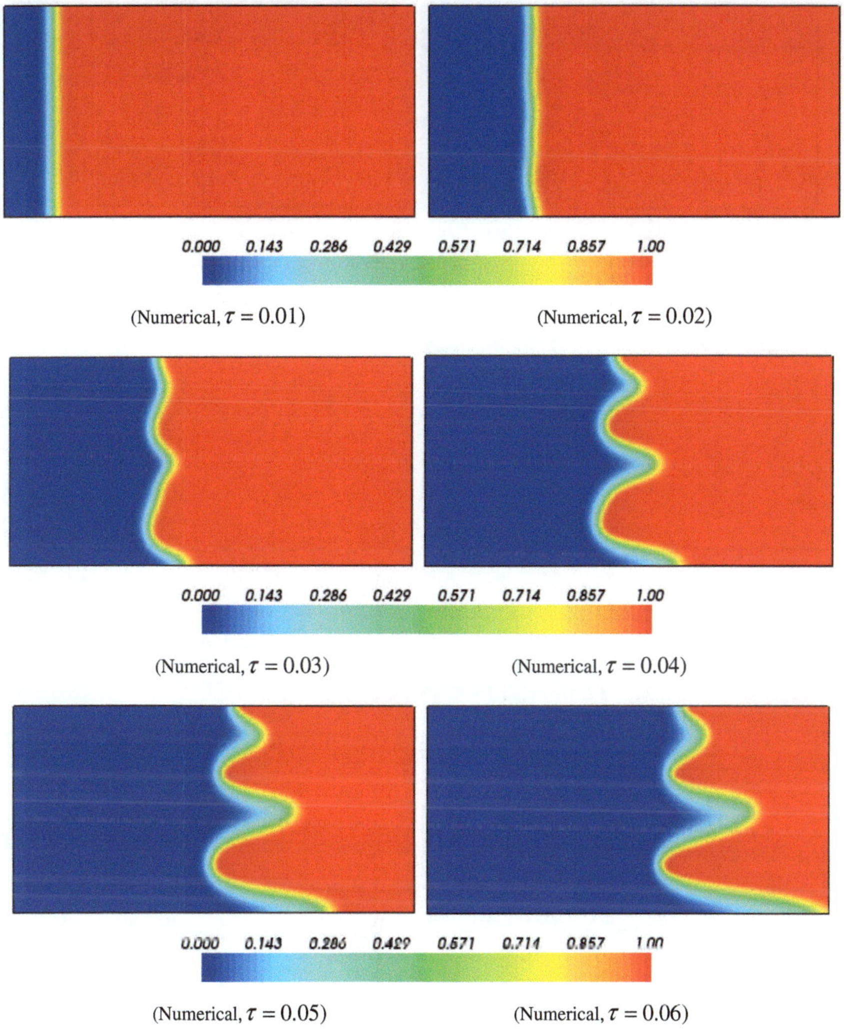

Fig. 3.6 Dimensionless chemical-species concentration distributions due to morphological evolution of the chemical dissolution front in the fluid-saturated porous medium

observed from the fact that the peak value of the porosity is in good correspondence with the trough value of the dimensionless chemical-species concentration due to the chemical dissolution reaction in the system. In this regard, it can be concluded that the proposed numerical procedure is capable of simulating the morphological evolution of the chemical dissolution front in a fluid-saturated porous medium when the coupled system becomes supercritical.

Except for simulating the evolution patterns of both the porosity and the dimensionless chemical-species concentration, the dimensionless pore-fluid-pressure

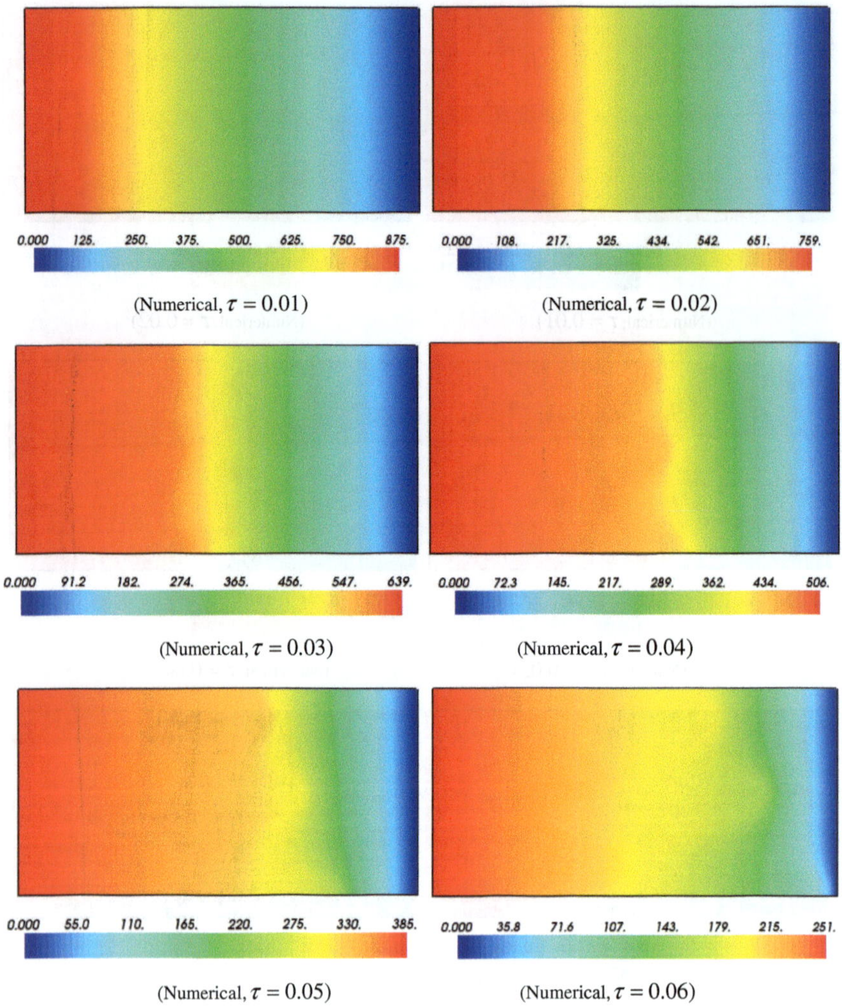

Fig. 3.7 Dimensionless pore-fluid pressure distributions due to morphological evolution of the chemical dissolution front in the fluid-saturated porous medium

and pore-fluid-flow evolution patterns during the propagation of the unstable chemical dissolution-front can be also simulated in the computational model. Figure 3.7 shows the dimensionless pore-fluid-pressure evolution pattern during the propagation of the chemical dissolution front in the computational model consisting of spherical grains. It is observed that even though the dimensionless pore-fluid pressure is continuous, there exists a clear transition for the dimensionless pore-fluid-pressure-gradient distribution in the computational model. The pore-fluid-flow pattern evolution during the propagation of the unstable chemical dissolution-front is displayed by the streamline evolution pattern in the computational model.

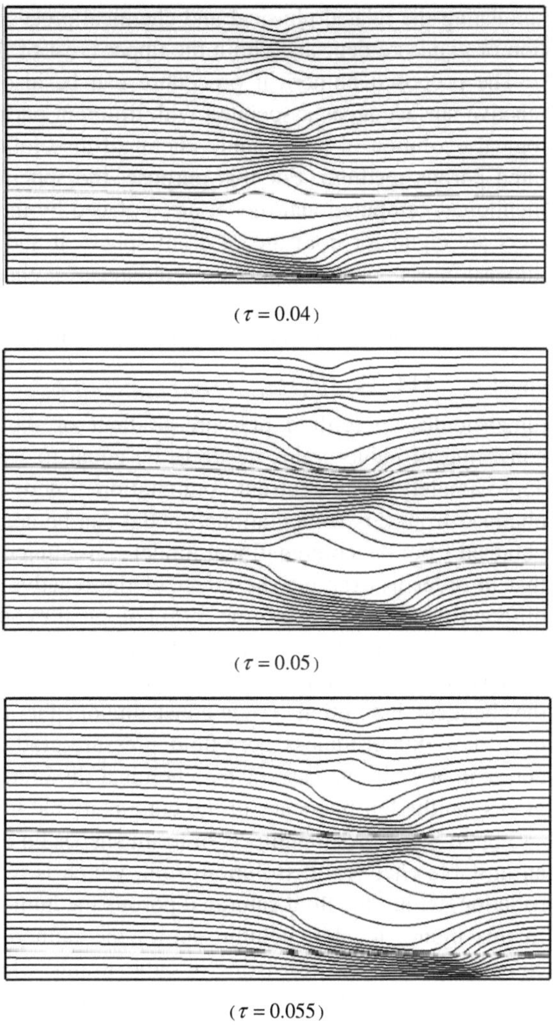

Fig. 3.8 Streamline distributions due to morphological evolution of the chemical dissolution front in the fluid-saturated porous medium

Figure 3.8 shows the streamline evolution pattern during the propagation of the unstable chemical dissolution-front within the fully coupled system between porosity, pore-fluid flow and reactive chemical-species transport within the fluid-saturated porous medium. With the amplitude increase of the irregular chemical dissolution-front, the pore-fluid flow focusing takes place in the peak region of the porosity, which can be clearly observed from the streamline density (in Fig. 3.8). It is noted that the width of the flow focusing zone is closely associated with the peak value zone of the porosity in the computational model. Because the porosity creation and pore-fluid flow focusing play an important role in ore body formation and mineralization, the proposed numerical procedure can provide a useful tool for simulating the related physical and chemical processes associated with the generation of giant ore deposits within the upper crust of the Earth.

The proposed numerical procedure can be also used to investigate how the reactive surface areas associated with different particle shapes affect the morphological evolution of an unstable chemical dissolution-front within the fluid-saturated porous medium. For this purpose, the spherical grains used in the above-mentioned computational model are replaced by the cubic grains. If other parameters are equal for both the cases of spherical and cubic grains, the ratio of the dimensionless pore-fluid pressure-gradient of a chemical dissolution system consisting of spherical grains to that consisting of cubic grains can be determined by considering Eqs. (3.26) and (3.29) simultaneously. Since this ratio is approximately equal to 1.114 (i.e. $(\bar{p}'_{fx})_{sphere}\big/(\bar{p}'_{fx})_{cubic} = \sqrt{\alpha_{sphere}\big/\alpha_{cubic}} = \sqrt[6]{6/\pi} \approx 1.114$), the dimensionless pore-fluid pressure-gradient of the chemical dissolution system consisting of cubic grains needs to be equal to -8.98 (i.e. $-10/1.114 = -8.98$ approximately) in the current computational model, so that the numerical results obtained from this computational model can be compared with the previous ones obtained from the computational model consisting of spherical grains. This means that in order to examine the effect of the reactive surface areas associated with different particle shapes on the morphological evolution of an unstable chemical dissolution-front within the fluid-saturated porous medium, the previously-applied dimensionless pore-fluid pressure-gradient, which has a value of -10 in the case of the chemical dissolution system consisting of spherical grains, needs to be replaced by a new dimensionless pore-fluid pressure-gradient, which is equal to -8.98 approximately in the case of the chemical dissolution system consisting of cubic grains, on the left boundary of the computational model.

Figure 3.9 shows the effects of grain shapes on the morphological evolution of an unstable chemical dissolution-front, which is represented by the porosity distribution, within the fluid-saturated porous medium at three different time instants. In this figure, the left column shows the numerical results obtained from the computational model consisting of cubic grains, while the right column shows the numerical results obtained from the computational model consisting of spherical grains. It is noted that since the Zhao number of the chemical dissolution system consisting of spherical grains is smaller than that consisting of cubic grains, the chemical dissolution front in the case of the chemical dissolution system consisting of spherical grains propagates faster than that in the case of the chemical dissolution system consisting of cubic grains. Because the Zhao numbers of two systems are different, the morphological evolution patterns of the chemical dissolution front are clearly different within these two computational models, indicating that the reactive surface area associated with different particle shapes can have a significant influence on the morphological evolution of an unstable chemical dissolution-front within the fluid-saturated porous medium.

In summary, the related theoretical and numerical results have demonstrated that: (1) since the shape coefficient of spherical grains is greater than that of cubic grains, the chemical dissolution system consisting of spherical grains is more

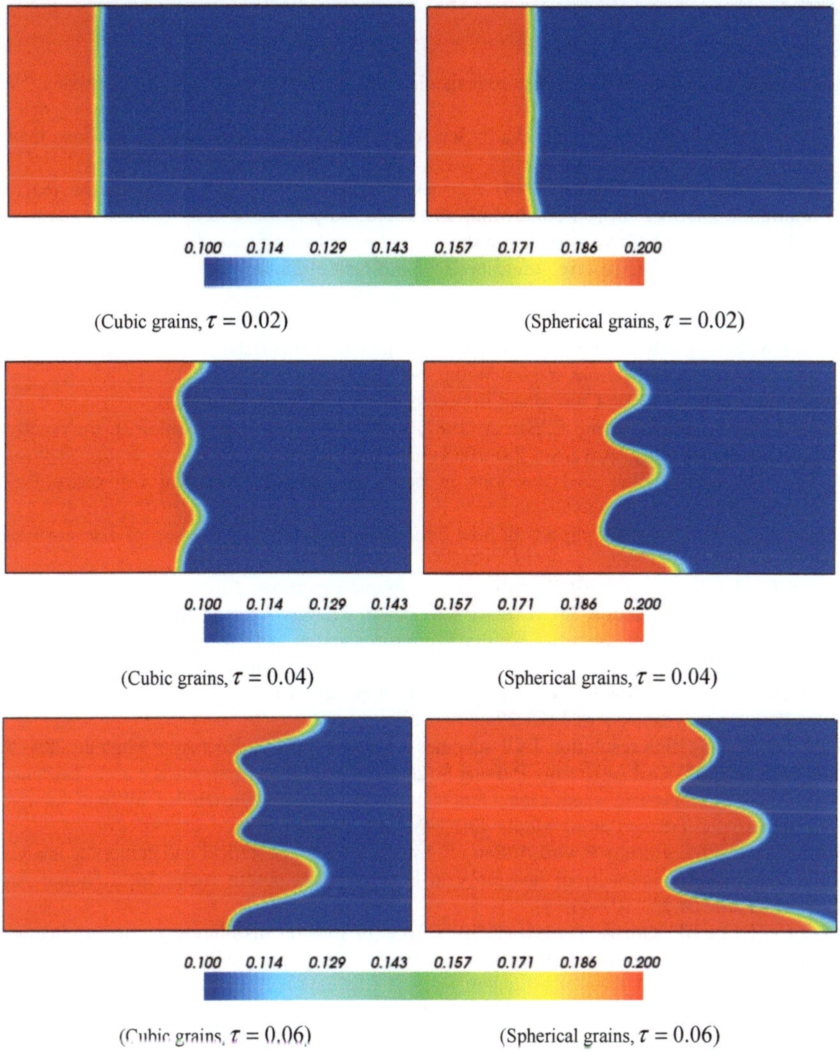

Fig. 3.9 Effect of grain shapes on the morphological evolution of the chemical dissolution front in the fluid-saturated porous medium

unstable than that consisting of cubic grains; (2) the instability likelihood of a natural porous medium, which is comprised of irregular grains, is smaller than that of an idealized porous medium, which is comprised of regular spherical grains; (3) reactive surface areas associated with different particle shapes can have a significant influence on the morphological evolution of an unstable chemical dissolution-front within the fluid-saturated porous medium.

References

Bear J (1972) Dynamics of fluids in porous media. American Elsevier Publishing Company, New York

Chadam J, Hoff D, Merino E, Ortoleva P, Sen A (1986) Reactive infiltration instabilities. IMA J Appl Math 36:207–221

Chadam J, Ortoleva P, Sen A (1988) A weekly nonlinear stability analysis of the reactive infiltration interface. IMA J Appl Math 48:1362–1378

Chen JS, Liu CW (2002) Numerical simulation of the evolution of aquifer porosity and species concentrations during reactive transport. Comput Geosci 28:485–499

Harten A (1983) High resolution schemes for hyperbolic conservation laws. J Comput Phys 49:357–393

Lewis RW, Schrefler BA (1998) The finite element method in the static and dynamic deformation and consolidation of porous media. Wiley, New York

Nield DA, Bejan A (1992) Convection in porous media. Springer, New York

Ortoleva P, Chadam J, Merino E, Sen A (1987) Geochemical self-organization II: the reactive-infiltration instability. Am J Sci 287:1008–1040

Phillips OM (1991) Flow and reactions in permeable rocks. Cambridge University Press, Cambridge

Scheidegger AE (1974) The physics of flow through porous media. University of Toronto Press, Toronto

Steefel CI, Lasaga AC (1990) Evolution of dissolution patterns: permeability change due to coupled flow and reaction. In: Melchior DC, Basset RL (eds) Chemical modeling in Aqueous systems II. American chemistry society symposium series, vol 416. pp 213–225

Steefel CI, Lasaga AC (1994) A coupled model for transport of multiple chemical species and kinetic precipitation/dissolution reactions with application to reactive flow in single phase hydrothermal systems. Am J Sci 294:529–592

Sweby PK (1985) High resolution TVD schemes using flux limiters. Lect Appl Math 22:289–309

Zhao C, Hobbs BE, Ord A, Hornby P, Peng S (2008a) Effect of reactive surface areas associated with different particle shapes on chemical-dissolution front instability in fluid-saturated porous rocks. Transp Porous Media 73:75–94

Zhao C, Hobbs BE, Hornby P, Ord A, Peng S, Liu L (2008b) Theoretical and numerical analyses of chemical-dissolution front instability in fluid-saturated porous rocks. Int J Numer Anal Meth Geomech 32:1107–1130

Zhao C, Hobbs BE, Ord A, Peng S (2010) Effects of mineral dissolution ratios on chemical-dissolution front instability in fluid-saturated porous media. Transp Porous Media 82:317–335

Zienkiewicz OC (1977) The finite element method. McGraw-Hill, London

Chapter 4
Effects of Mineral Dissolution Ratios on Chemical Dissolution-Front Instability in Fluid-Saturated Porous Media

Considcration of mineral dissolution due to chemical reactions plays an important role in the computational simulation of reactive transport problems involving fluid-saturated porous media (Steefel and Lasage 1990, 1994; Yeh and Tripathi 1991; Raffensperger and Garven 1995; Schafcr ct al. 1998a, b; Xu et al. 1999; Ormond and Ortoleva 2000; Alt-Epping and Smith 2001; Chen and Liu 2002; Zhao et al. 1998, 2001, 2003, 2005, 2006). Under certain circumstances, the resulting chemical dissolution-front can become unstable during its propagation in the fluid-saturated medium so that the pore-fluid flow is strongly enhanced in some regions of the problem domain. This kind of chemical dissolution problem, known as the chemical dissolution-front instability problem in the scientific term, was initially studied in the later eighties of the last century (Chadam et al. 1986, 1988; Ortoleva et al. 1987). Although some preliminary theoretical work was provided, the overall progress on the theoretical respect of the chemical dissolution-front instability problem has been limited for many years as a result of the complicated nature of the problem itself. Generally, when a chemical dissolution-front instability problem in a fluid-saturated porous medium is considered, it is necessary to deal with a fully coupled nonlinear problem between porosity evolution, pore-fluid flow, mass transport and chemical reactions in the fluid-saturated porous medium. From the mathematical point of view, it is necessary to solve a set of simultaneous nonlinear partial differential equations so as to get a better theoretical understanding of the chemical dissolution-front instability problem. This is the main reason why the only limited progress has been achieved on the theoretical respect of this problem until the recent years.

Under the stimulus of understanding the physical and chemical processes associated with mineral enrichment phenomena within the upper crust of the Earth, the chemical dissolution-front instability problem has received further attention in the recent years. In particular, Zhao et al. have revisited the theoretical respect of the problem (Zhao et al. 2008a, b) and systematically reported the current status of the related theoretical work in this particular research field (Zhao et al. 2009).

C. Zhao, *Physical and Chemical Dissolution Front Instability in Porous Media*, Lecture Notes in Earth System Sciences,
DOI: 10.1007/978-3-319-08461-9_4

To assess whether or not a planar chemical dissolution-front is stable during its propagation in the fluid-saturated porous medium, both the Zhao number and the critical Zhao number concepts have been proposed. The Zhao number is a dimensionless number, which can be used to represent the fundamental characteristics including the solute diffusion/dispersion, solute advection, chemical kinetics and mineral shapes of the system, while the critical Zhao number is used to represent the critical condition, which can be used to examine whether or not a planar chemical dissolution-front can become unstable during its propagation in the fluid-saturated porous medium. Based on the Zhao number and critical Zhao number, a theoretical criterion has been established. If the Zhao number of a chemical dissolution system is smaller than the corresponding critical Zhao number of the system, then the chemical dissolution front is stable during its propagation in the fluid-saturated porous medium. If the Zhao number is equal to the corresponding critical Zhao number, then the chemical dissolution front is neutrally unstable, implying that any small perturbation to the chemical dissolution front can be maintained without growing during its propagation. However, if the Zhao number is greater than the corresponding critical Zhao number, then the chemical dissolution front is unstable during its propagation. In this supercritical case, a small perturbation can enable a planar chemical dissolution-front to evolve into different morphologies.

Despite the above-mentioned progress on the theoretical respect of the chemical dissolution-front instability problem, the critical conditions associated with chemical dissolution-front instability problems are derived under the assumption that the mineral dissolution ratio, which is defined as the ratio of the dissolved-mineral equilibrium concentration (in the pore-fluid) to the molar concentration of the dissolvable mineral in the solid matrix of the fluid-saturated porous medium, is an infinitesimal quantity. If the mineral dissolution ratio is a finite number, then the following two fundamental questions need to be answered: (1) how does this ratio affect the propagation speed of a planar chemical dissolution-front in the fluid-saturated porous medium? (2) how does this ratio affect the evolution pattern of the planar chemical dissolution-front when the chemical dissolution system becomes supercritical. The main purpose of this chapter is to answer them through both theoretical analyses and computational simulations.

4.1 Theoretical Considerations of the Problem

For a fluid-saturated porous medium, Darcy's law is often used to describe pore-fluid flow, while Fick's law and the related chemical reaction kinetics (Steefel and Lasaga 1994) are commonly used to describe reactive mass transport phenomena. If the pore-fluid (i.e. water in this investigation) is incompressible and the dissolvable mineral is comprised of spheres, the governing equations of the coupled nonlinear problem between porosity evolution, pore-fluid flow and reactive dissolved-mineral transport in the fluid-saturated porous medium can be expressed as follows:

$$\frac{\partial \phi}{\partial t} - \nabla \cdot [\psi(\phi)\nabla p] = 0, \tag{4.1}$$

$$\frac{\partial}{\partial t}(\phi C) - \nabla \cdot [\phi \mathbf{D}(\phi) \cdot \nabla C + C\psi(\phi)\nabla p] + \rho_s k_{Echemical} \frac{\sqrt[3]{D_p}}{\alpha_{sphere}} (\phi_f - \phi)^{\frac{2}{3}} (C - C_{eq}) = 0, \tag{4.2}$$

$$\frac{\partial \phi}{\partial t} + k_{Echemical} \frac{\sqrt[3]{D_p}}{\alpha_{sphere}} (\phi_f - \phi)^{\frac{2}{3}} (C - C_{eq}) = 0, \tag{4.3}$$

$$\psi(\phi) = \frac{k(\phi)}{\mu}, \tag{4.4}$$

where p and C are the pore-fluid pressure and the concentration of the dissolved mineral; C_{eq} is the equilibrium concentration of the dissolved mineral; μ is the dynamic viscosity of the pore-fluid; ϕ is the porosity of the porous medium; ϕ_f is the final (i.e. maximum) porosity of the porous medium after the completion of dissolvable mineral dissolution; $\mathbf{D}(\phi)$ is the general dispersion tensor of the dissolved mineral; $k(\phi)$ is the permeability of the porous medium; $k_{Echemical}$ is the comprehensive rate constant of the chemical reaction (as defined in Eq. (2.15) previously); ρ_s is the molar density (i.e. moles per volume) of the dissolvable grains; $\alpha_{sphere} = \sqrt[3]{1/(36\pi)}$ is the shape coefficient of the spherical grains; D_p is the density of the dissolvable grains, which is defined as the number of the dissolvable gains per unit medium volume.

For a two-dimensional reactive transport problem in a fluid-saturated porous medium, the general form of the dispersion tensor (Scheidegger 1961; Bear 1972, Holzbecher 1998) can be expressed as follows:

$$\mathbf{D}(\phi) = \begin{bmatrix} D(\phi) + \left(\alpha_T \bar{u}_A + (\alpha_L - \alpha_T)\frac{u_x^2}{\bar{u}_A}\right) & (\alpha_L - \alpha_T)\frac{u_x u_y}{\bar{u}_A} \\ (\alpha_L - \alpha_T)\frac{u_x u_y}{\bar{u}_A} & D(\phi) + \left(\alpha_T \bar{u}_A + (\alpha_L - \alpha_T)\frac{u_y^2}{\bar{u}_A}\right) \end{bmatrix}, \tag{4.5}$$

where $D(\phi)$ is the diffusivity of the dissolved mineral; α_T and α_L are the transversal and longitudinal dispersivities of the dissolved mineral; $\bar{u}_A$ is the absolute magnitude of the mean averaged linear velocity; u_x and u_y are the averaged linear velocity of the pore-fluid in the x and y directions, respectively.

Based on Darcy's law, the averaged linear velocity vector of the pore-fluid can be expressed as follows:

$$\vec{u}_{linear} = \begin{Bmatrix} u_1 \\ u_2 \end{Bmatrix} = \frac{\vec{u}}{\phi} = -\frac{k(\phi)}{\phi\mu}\nabla p, \tag{4.6}$$

where $\vec{u}$ is the Darcy velocity vector of the pore-fluid; μ is the dynamic viscosity of the pore-fluid; ϕ is the porosity of the porous medium; $k(\phi)$ is the permeability of the porous medium.

As the dissolved-mineral diffusivity is usually considered as a function of porosity in the chemical dissolution-front instability problems (Bear 1972; Chadam et al. 1986), it can be expressed using the following equation:

$$D(\phi) = D_0 \phi^q \quad \left(\frac{3}{2} \leq q \leq \frac{5}{2}\right), \tag{4.7}$$

where D_0 is the chemical species diffusivity in pure water.

The permeability change caused by a change in porosity can be described by the Carman-Kozeny law (Scheidegger 1974; Nield and Bejan 1992) as follows:

$$k(\phi) = \frac{k_0 (1 - \phi_0)^2 \phi^3}{\phi_0^3 (1 - \phi)^2}, \tag{4.8}$$

where ϕ_0 and k_0 are the initial reference porosity and permeability of the porous medium respectively.

If the dissolved mineral is initially in an equilibrium state within the whole system and the fresh pore-fluid is injected at the location of x approaching negative infinity in an infinite space, then the following boundary conditions can be used for the chemical dissolution-front instability problem under the theoretical consideration.

$$\lim_{x \to -\infty} C = 0, \quad \lim_{x \to -\infty} \frac{\partial p}{\partial x} = p'_{fx} \quad \text{(upstream boundary)}, \tag{4.9}$$

$$\lim_{x \to \infty} C = C_{eq}, \quad \lim_{x \to \infty} \frac{\partial p}{\partial x} = p'_{0x} \quad \text{(downstream boundary)}, \tag{4.10}$$

where p'_{fx} is the pore-fluid pressure gradient as x approaching negative infinity in the upstream of the pore-fluid flow; p'_{0x} is the unknown pore-fluid pressure gradient as x approaching positive infinity in the downstream of the pore-fluid flow. Since p'_{fx} drives the pore-fluid flow continuously along the positive x direction, it has a negative algebraic value (i.e. $p'_{fx} < 0$) in this investigation.

The initial condition for this theoretical problem is: $\phi(x, 0) = \phi_0$ expect at the negative infinity, where $\lim_{\bar{x} \to -\infty} \phi(x, 0) = \phi_f$. Note that ϕ_0 is the initial porosity of the porous medium.

To get a theoretical understanding of how a planar chemical dissolution-front propagates in the infinite space, Eqs. (4.1)–(4.3) can be rewritten into the following one-dimensional form:

$$\frac{\partial \phi}{\partial t} - \frac{\partial}{\partial x}\left[\psi(\phi)\frac{\partial p}{\partial x}\right] = 0, \tag{4.11}$$

$$\frac{\partial}{\partial t}(\phi C) - \frac{\partial}{\partial x}\left\{\left[\phi D(\phi) + \alpha_L \psi(\phi)\left|\frac{\partial p}{\partial x}\right|\right]\frac{\partial C}{\partial x} + C\psi(\phi)\frac{\partial p}{\partial x}\right\}, \\ + \rho_s k_{Echemical}\frac{\sqrt[3]{D_p}}{\alpha_{sphere}}(\phi_f - \phi)^{\frac{2}{3}}(C - C_{eq}) = 0 \tag{4.12}$$

$$\frac{\partial \phi}{\partial t} + k_{Echemical}\frac{\sqrt[3]{D_p}}{\alpha_{sphere}}(\phi_f - \phi)^{\frac{2}{3}}(C - C_{eq}) = 0. \tag{4.13}$$

Suppose the propagation speed of the planar chemical dissolution-front is denoted by v_{front}, the moving boundary problem of the chemical dissolution front (in the $x-t$ coordinate system) can be transformed into a steady-state boundary-value problem (in the $\xi - t$ coordinate system) using the following coordinate mapping:

$$\xi = x - v_{front}t. \tag{4.14}$$

The resulting mapped governing equations in the $\xi - t$ coordinate system can be expressed as follows:

$$\frac{\partial}{\partial \xi}\left[\psi(\phi)\frac{\partial p}{\partial \xi} + v_{front}\phi\right] = 0, \tag{4.15}$$

$$\frac{\partial}{\partial \xi}\left\{\left[\phi D(\phi) + \alpha_L \psi(\phi)\left|\frac{\partial p}{\partial x}\right|\right]\frac{\partial C}{\partial \xi} + C\psi(\phi)\frac{\partial p}{\partial \xi} + v_{front}(C - \rho_s)\phi\right\} = 0, \tag{4.16}$$

$$v_{front}\frac{\partial \phi}{\partial \xi} - k_{Echemical}\frac{\sqrt[3]{D_p}}{\alpha_{sphere}}(\phi_f - \phi)^{\frac{2}{3}}(C - C_{eq}) = 0. \tag{4.17}$$

Integrating Eqs. (4.15) and (4.16) from negative infinity to positive infinity and using the boundary conditions (i.e. Eqs. (4.9) and (4.10)) yield the following equations:

$$C_{eq}\psi(\phi_0)p'_{ox} + v_{front}\phi_0(C_{eq} - \rho_s) + v_{front}\phi_f\rho_s = 0, \tag{4.18}$$

$$\psi(\phi_0)p'_{ox} + v_{front}\phi_0 - \psi(\phi_f)p'_{fx} - v_{front}\phi_f = 0. \tag{4.19}$$

Solving Eqs. (4.18) and (4.19) simultaneously results in the following analytical solutions.

$$v_{front} = \frac{-\psi(\phi_f)p'_{fx}C_{eq}}{\phi_f C_{eq} + (\phi_f - \phi_0)\rho_s}, \tag{4.20}$$

$$p'_{0x} = \frac{\psi(\phi_f)[\phi_0 C_{eq} + (\phi_f - \phi_0)\rho_s]}{\psi(\phi_0)[\phi_0 C_{eq} + (\phi_f - \phi_0)(\rho_s + C_{eq})]} p'_{fx}. \tag{4.21}$$

Since the mineral dissolution ratio is defined as the ratio of the dissolved-mineral equilibrium concentration (in the pore-fluid) to the molar density of the dissolvable mineral in the solid matrix of the fluid-saturated porous medium, it can be mathematically expressed as follows:

$$\varepsilon = \frac{C_{eq}}{\rho_s}, \tag{4.22}$$

where C_{eq} is the equilibrium concentration of the dissolved mineral (in the pore-fluid); ρ_s is the molar density (i.e. moles per volume) of the dissolvable mineral in the solid matrix of the fluid-saturated porous medium.

Inserting Eq. (4.22) into Eqs. (4.20) and (4.21) yields the following equations:

$$v_{front} = \frac{-\psi(\phi_f)p'_{fx}}{\phi_f + (\phi_f - \phi_0)/\varepsilon} = \frac{u_{fx}}{\phi_f + (\phi_f - \phi_0)/\varepsilon}, \tag{4.23}$$

$$p'_{0x} = \frac{\psi(\phi_f)\left[\phi_0\varepsilon + (\phi_f - \phi_0)\right]}{\psi(\phi_0)\left[\phi_0\varepsilon + (\phi_f - \phi_0)(1 + \varepsilon)\right]} p'_{fx}, \tag{4.24}$$

where u_{fx} is the Darcy velocity in the far upstream of the pore-fluid flow as x approaching negative infinity.

If ε is an infinitesimal quantity, then Eq. (4.23) can be approximately written as follows:

$$v_{front} \approx \frac{-\psi(\phi_f)p'_{fx}\varepsilon}{\phi_f - \phi_0}. \tag{4.25}$$

It needs to be pointed out that Eq. (4.23) clearly states that the propagation speed of the planar chemical dissolution front is only dependent on the mineral dissolution ratio (ε), initial porosity (ϕ_0), final porosity (ϕ_f) and upstream Darcy velocity (u_{fx}) of the chemical dissolution system, but independent of both the diffusion and the dispersion mechanisms of the dissolved mineral in the chemical dissolution system.

4.2 Numerical Simulation of the Problem

Since the mineral dissolution ratio is usually a small quantity in most chemical dissolution problems involving fluid-saturated porous media, it is desirable to solve a chemical dissolution problem using dimensionless quantities (Chadam et al. 1986; Zhao et al. 2008a, b), so that highly-accurate numerical results can be obtained from the computational models. Toward this end, the following dimensionless quantities are used in the rest of this investigation.

$$\bar{x} = \frac{x}{L^*}, \quad \bar{y} = \frac{y}{L^*}, \quad \bar{C} = \frac{C}{C_{eq}}, \quad \bar{p} = \frac{p}{p^*}, \quad \bar{v}_{front} = \frac{v_{front}}{v^*_{front}}, \tag{4.26}$$

$$\bar{\alpha}_L = \frac{\alpha_L}{L^*}, \quad \bar{\alpha}_T = \frac{\alpha_T}{L^*}, \quad \tau = \frac{t}{t^*}\varepsilon, \quad \varepsilon = \frac{C_{eq}}{\rho_s} \ll 1, \tag{4.27}$$

where τ is a slow dimensionless time to describe the slowness of the chemical dissolution that takes place in the chemical dissolution system. Other characteristic parameters used in Eqs. (4.26) and (4.27) can be expressed as follows:

$$\begin{gathered} t^* = \frac{\alpha_{aphere}}{k_{Echemical} C_{eq} \sqrt[3]{D_p}}, \quad L^* = \sqrt{\phi_f D(\phi_f) t^*}, \quad p^* = \frac{\phi_f D(\phi_f)}{\psi(\phi_f)}, \quad u^* = \frac{\phi_f D(\phi_f)\varepsilon}{L^*}, \\ D^*(\phi) = \frac{\phi D(\phi)}{\phi_f D(\phi_f)}, \quad \psi^*(\phi) = \frac{\psi(\phi)}{\psi(\phi_f)}, \quad v^*_{front} = \frac{\phi_f D(\phi_f)\varepsilon}{L^*}. \end{gathered} \tag{4.28}$$

By substituting Eqs. (4.26)–(4.28) into Eqs. (4.1)–(4.3), the following dimensionless partial differential equations can be obtained.

$$\varepsilon \frac{\partial \phi}{\partial \tau} + (\phi_f - \phi)^{\frac{2}{3}} (\bar{C} - 1) = 0, \tag{4.29}$$

$$\varepsilon \frac{\partial}{\partial \tau}(\phi \bar{C}) - \nabla \cdot \left\{ \left[D^*(\phi) \mathbf{I} + \mathbf{D}^*_{dispersion}(\phi) \right] \cdot \nabla \bar{C} + \bar{C} \psi^*(\phi) \nabla \bar{p} \right\} - \frac{\partial \phi}{\partial \tau} = 0, \tag{4.30}$$

$$\varepsilon \frac{\partial \phi}{\partial \tau} - \nabla \cdot [\psi^*(\phi) \nabla \bar{p}] = 0, \tag{4.31}$$

where $\mathbf{I}$ is a unit second-order tensor; $\mathbf{D}^*_{dispersion}(\phi)$ is the dimensionless second-order tensor due to pure dispersion. As the pore-fluid flow is dominant in the x direction for the chemical dissolution problem under consideration, the dimensionless second-order tensor due to pure dispersion can be expressed as follows:

$$\mathbf{D}^*_{dispersion}(\phi) = \begin{bmatrix} \bar{\alpha}_L \psi^*(\phi) \left| \frac{\partial \bar{p}}{\partial \bar{x}} \right| & 0 \\ 0 & \bar{\alpha}_T \psi^*(\phi) \left| \frac{\partial \bar{p}}{\partial \bar{x}} \right| \end{bmatrix}. \tag{4.32}$$

Using the above-defined dimensionless quantities, the boundary conditions for the chemical dissolution problem can be rewritten in the dimensionless form:

$$\lim_{\bar{x}\to\infty} \bar{C} = 1, \qquad \lim_{\bar{x}\to\infty} \frac{\partial \bar{p}}{\partial \bar{x}} = \bar{p}'_{0x} \qquad \text{(downstream boundary)}, \tag{4.33}$$

$$\lim_{\bar{x}\to-\infty} \bar{C} = 0, \qquad \lim_{\bar{x}\to-\infty} \frac{\partial \bar{p}}{\partial \bar{x}} = \bar{p}'_{fx} \qquad \text{(upstream boundary)}. \tag{4.34}$$

Similarly, the initial condition for this problem is: $\phi(\bar{x}, 0) = \phi_0$ expect at the negative infinity, where $\lim_{\bar{x}\to-\infty} \phi(\bar{x}, 0) = \phi_f$.

Considering Eqs. (4.23), (4.26) and (4.28) yields the following expression for the dimensionless propagation speed of the planar chemical dissolution-front in the fluid-saturated porous medium.

$$\bar{v}_{front} = \frac{Zh}{\phi_f \varepsilon + (\phi_f - \phi_0)}, \tag{4.35}$$

where Zh is the Zhao number of the chemical dissolution system and $Zh = -\bar{p}'_{fx}$ by definition (Zhao et al. 2009).

If ε approaches zero, Eq. (4.35) degenerates into the following form:

$$\lim_{\varepsilon\to 0} \bar{v}_{front} = \frac{Zh}{(\phi_f - \phi_0)}. \tag{4.36}$$

It is worth pointing out that in this limit case, Eq. (4.36) is identical to that derived in the previous investigations (Chadam et al. 1986, Zhao et al. 2008a, b). By comparing Eqs. (4.35) with (4.36), it can be concluded that the consideration of a non-zero mineral-dissolution ratio (ε) results in a smaller dimensionless propagation speed of the planar chemical dissolution-front in the fluid-saturated porous medium. To demonstrate this point, the theoretical curves of the dimensionless propagation speed (i.e. $\bar{v}_{front}$) of the planar chemical dissolution-front versus the mineral dissolution ratio (i.e. ε) have been obtained for two different sets of initial and final porosities in the fluid-saturated porous medium. Figure 4.1 shows the effects of the mineral dissolution ratio on the dimensionless propagation speed of a planar chemical dissolution-front in the fluid-saturated porous medium due to different Zhao numbers of the chemical dissolution system. It is noted that for all the Zhao numbers under consideration, an increase in the value of the mineral dissolution ratio results in a remarkable decrease in the value of the dimensionless propagation speed of the planar chemical dissolution-front in the fluid-saturated porous medium. As expected, for both a given Zhao number and a given mineral dissolution ratio, the value of the dimensionless propagation speed of the planar chemical dissolution-front decreases with an increase in the final porosity of the fluid-saturated porous medium. This theoretical understanding clearly reveals that the mineral dissolution ratio plays an important role in determining the

Fig. 4.1 Effect of the mineral dissolution ratio on the dimensionless propagation speed of a planar chemical dissolution-front in the fluid-saturated porous medium

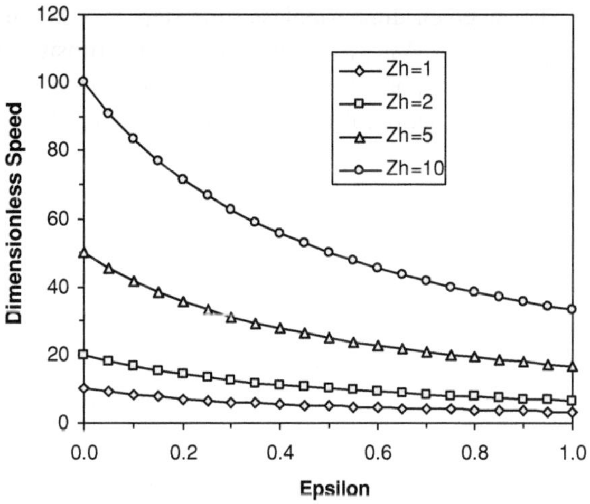

(a) $(\phi_0 = 0.1, \phi_f = 0.2)$

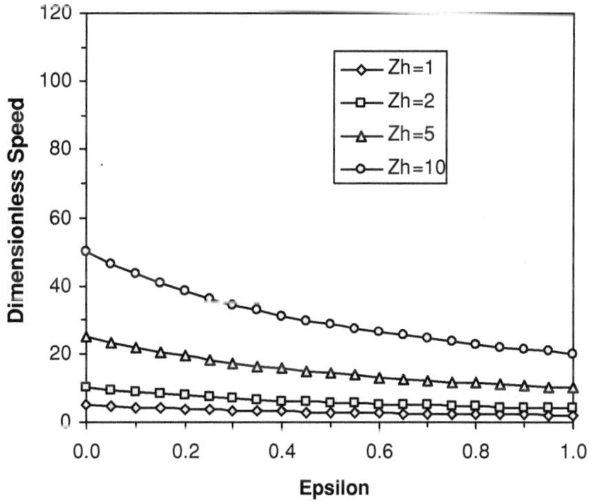

(b) $(\phi_0 = 0.1, \phi_f = 0.3)$

dimensionless propagation speed of the planar chemical dissolution-front in the fluid-saturated porous medium.

Equations (4.29)–(4.31) can be solved using a combination of both the finite element method and the finite difference method (Zhao et al. 2008a, b). By means of this numerical procedure, the finite element method is used to discretize the geometrical shape of the problem domain, while the finite difference method is employed to discretize the dimensionless time. Since the system described by these equations is highly nonlinear, the segregated algorithm, in which Eqs. (4.29)–(4.31) are solved separately in a sequential manner, is used to derive the formulation of the proposed numerical procedure.

For a given dimensionless time-step, $\tau + \Delta\tau$, the porosity can be denoted by $\phi_{\tau+\Delta\tau} = \phi_\tau + \Delta\phi_{\tau+\Delta\tau}$, where ϕ_τ is the porosity at the previous time-step and $\Delta\phi_{\tau+\Delta\tau}$is the porosity increment at the current time-step. Since Eq. (4.29) is a nonlinear equation, it can be linearized using the Taylor expansion as follows:

$$[\phi_f - (\phi + \Delta\phi)]^{\frac{2}{3}} = [(\phi_f - \phi) - \Delta\phi]^{\frac{2}{3}} = (\phi_f - \phi)^{\frac{2}{3}} - \frac{2}{3}(\phi_f - \phi)^{-\frac{1}{3}}\Delta\phi \quad (4.37)$$

If the backward difference scheme is used, Eq. (4.29) can be written in the following finite difference form.

$$\left[\frac{\varepsilon}{\Delta\tau}(\phi_f - \phi_\tau)^{\frac{1}{3}} + \frac{2}{3}(1 - \bar{C}_{\tau+\Delta\tau})\right]\Delta\phi_{\tau+\Delta\tau} = (\phi_f - \phi_\tau)(1 - \bar{C}_{\tau+\Delta\tau}) \quad (4.38)$$

where $\bar{C}_{\tau+\Delta\tau}$ is the dimensionless concentration of the dissolved mineral at the current time-step; $\Delta\tau$ is the dimensionless time increment at the current time-step.

In the finite difference sense, Eq. (4.30) can be mathematically expressed as follows:

$$\begin{aligned}&\left[\frac{\varepsilon}{\Delta\tau}\phi_{\tau+\Delta\tau} + \frac{1}{\varepsilon}(\phi_f - \phi_{\tau+\Delta\tau})^{\frac{2}{3}}\right]\bar{C}_{\tau+\Delta\tau} - \nabla\cdot\left\{\left[D^*(\phi_{\tau+\Delta\tau})\mathbf{I} + \mathbf{D}^*_{dispersion}(\phi_{\tau+\Delta\tau})\right]\cdot\nabla\bar{C}_{\tau+\Delta\tau}\right\} \\ &\quad - \nabla\bar{p}_{\tau+\Delta\tau}\cdot\left[\psi^*(\phi_{\tau+\Delta\tau})\nabla\bar{C}_{\tau+\Delta\tau}\right] = \frac{\varepsilon}{\Delta\tau}\phi_{\tau+\Delta\tau}\bar{C}_\tau + \frac{1}{\varepsilon}(\phi_f - \phi_{\tau+\Delta\tau})^{\frac{2}{3}}.\end{aligned} \quad (4.39)$$

Similarly, Eq. (4.31) can be rewritten in the following discretized form:

$$\nabla\cdot[\psi^*(\phi)\nabla\bar{p}] = \nabla\cdot\left[\psi^*(\phi_{\tau+\Delta\tau})\nabla\bar{p}_{\tau+\Delta\tau}\right] = (1 - \bar{C}_{\tau+\Delta\tau})(\phi_f - \phi_{\tau+\Delta\tau})^{\frac{2}{3}}. \quad (4.40)$$

Using the segregated scheme and finite element method (Zhao et al. 2008a, b), Eqs. (4.38)–(4.40) are solved separately and sequentially for the porosity, dimensionless concentration of the dissolved mineral and dimensionless pore-fluid pressure at the current time-step.

4.3 Effects of Mineral Dissolution Ratios on the Evolution Patterns of Chemical Dissolution Fronts During Propagation in Fluid-Saturated Porous Media

Since the first fundamental question raised in the introduction has been theoretical answered in Sect. 4.2, the focus of this section is on the numerical investigation of how the mineral dissolution ratio affects the evolution patterns of a planar chemical dissolution-front during its propagation in the fluid-saturated porous medium. It is expected that the computational simulation results from this

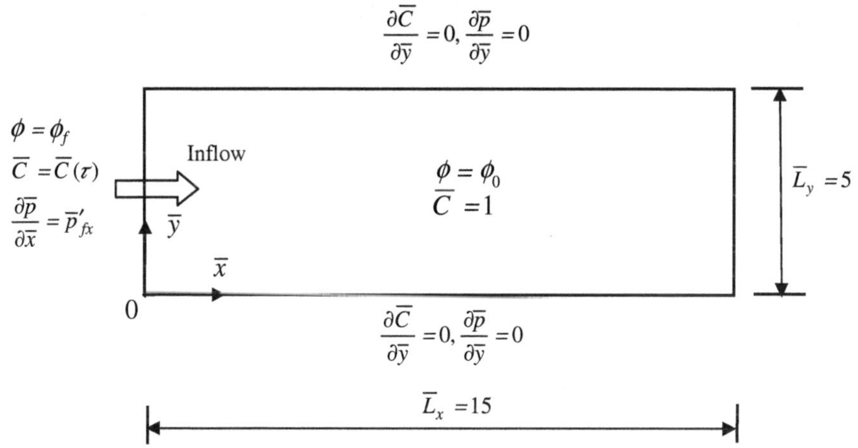

Fig. 4.2 Geometry and boundary conditions for the computational model

investigation can be used to answer the second fundamental question raised previously. For this purpose, the proposed numerical procedure, which is comprised of a combination of the finite element and finite difference methods, is used to simulate the morphological evolution of a planar chemical dissolution-front in the fluid-saturated porous medium under the condition of supercritical Zhao numbers. Since the proposed numerical procedure can only provide approximate solutions, it is necessary to verify it through some benchmark problems, the analytical solutions of which are ready for comparison with the computational simulation results obtained from the proposed numerical procedure. After the proposed numerical procedure is verified, it is used to simulate the morphological evolution process of a planar chemical dissolution-front in the case of the chemical dissolution system becoming supercritical.

Figure 4.2 shows the geometry and boundary conditions of the proposed benchmark problem, for which analytical solutions have been derived in the previous publications (Zhao et al. 2008a, b). As indicated in this figure, the dimensionless height and width of the computational model are 5 and 15 respectively. Except for the left boundary, the initial porosity of the porous medium prior to perturbation is 0.1, while the initial dimensionless concentration of the dissolved mineral is unity within the computational domain. The final porosity with a value of 0.2 is applied to the left boundary of the computational domain. Since the analytical solutions are only suitable for subcritical situations, the Zhao number of a value of unity is used for the chemical dissolution system under consideration. This means that a dimensionless pore-fluid pressure gradient (i.e. $\bar{p}'_{fx} = -1$) is applied to the left boundary, so that there is a horizontal throughflow from the left to the right of the computational model. To compare the resulting computational simulation results with the analytical solutions, a time-dependent-dimensionless-concentration boundary condition (i.e. $\bar{C}(\tau) = \exp(\bar{p}'_{fx}\bar{v}_{front}\tau)$) is applied to the left boundary of

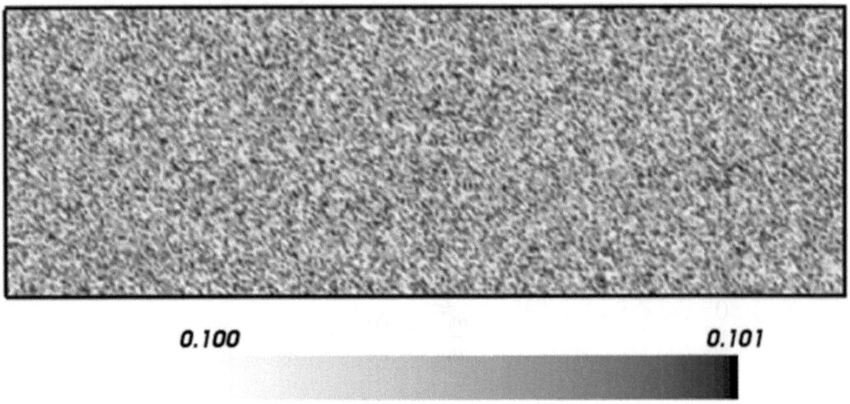

Fig. 4.3 The initial porosity distribution after the small perturbation in the fluid-saturated porous medium

the computational model. Since the value of the dimensionless pore-fluid pressure can be chosen arbitrarily for the benchmark problem (Zhao et al. 2008a), it is assumed to be 100 (i.e. $\bar{p}(\bar{L}_x, \tau) = 100$) at the right boundary of the computational domain.

To consider whether or not a small perturbation affects the morphological evolution of a planar chemical dissolution-front during its propagation in the fluid-saturated porous medium, the initial porosity field is randomly perturbed by a small amount of 1 % of the originally-input initial porosity (i.e. $\phi_0 = 0.1$) before running the computational model. Figure 4.3 shows the porosity distribution after the small perturbation in the computational model. This kind of porosity distribution is more suitable for simulating the porous rock in nature. The permeability of the porous medium is calculated using the Carman-Kozeny formula, which has the power of 3 in the power law, while the diffusivity of the solute is calculated using the power law, which has the power of 2. The mineral dissolution ratio is assumed to be 0.01, while the dimensionless time-step length is set to be 0.001 in the computation. To consider the effect of dissolved-mineral dispersion, the dimensionless longitudinal dispersivity is assumed to be 0.2, while the dimensionless transversal dispersivity is assumed to be 0.02 in the computational model. Using the above parameters, the critical Zhao number of the chemical dissolution system is approximately equal to 2.001. Since the Zhao number of the system is smaller than its critical value, the chemical dissolution system simulated here is in a subcritical state. The theoretical value for the dimensionless propagation speed of the chemical dissolution front is approximately equal to 10, which is determined from Eq. (4.35). To appropriately simulate the propagation of the chemical dissolution front, the whole computational domain is simulated by 29,601 four-node square elements of 30,000 nodal points in total, resulting in a uniform mesh size of $5/99$ for both $\Delta\bar{x}$ and $\Delta\bar{y}$ in the computational model.

Fig. 4.4 Comparison of the computational simulation results with analytical solutions at three different time instants (Porosity)

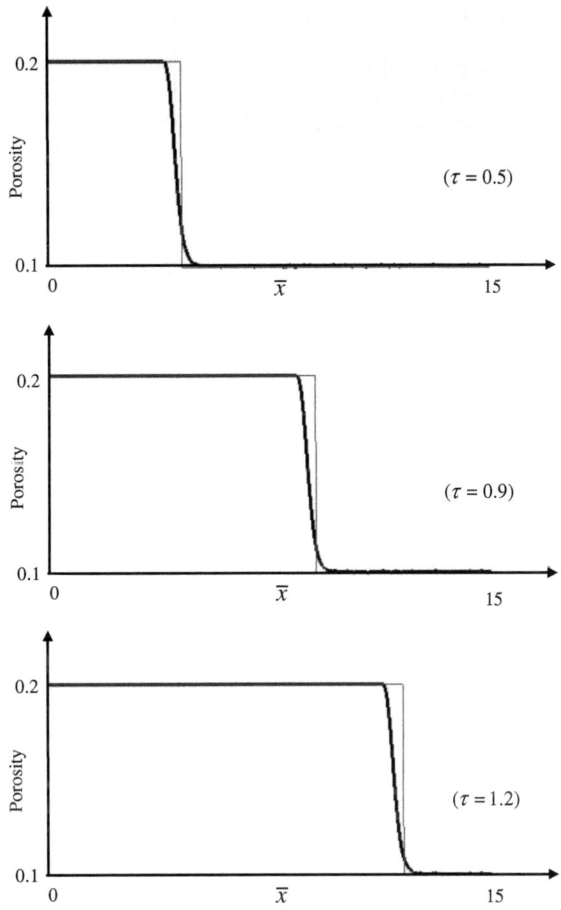

Figures 4.4 and 4.5 show the comparison of the computational simulation results with analytical solutions for the porosity and dimensionless concentration of the dissolved mineral in the computational domain at three different time instants. In these figures, the thick lines are used to represent the computational simulation results, while the thin lines are used to represent the analytical solutions. Although the initial porosity field after a small perturbation (shown in Fig. 4.3) is used as an initial porosity condition for the computational model, the planar chemical dissolution-front remains planar throughout the whole computational simulation process as a result of the subcritical Zhao number of the chemical dissolution system. Note that the computational simulation results for both the porosity and the dimensionless concentration of the dissolved mineral agree very well with the analytical solutions. This fact indicates that the proposed numerical procedure is capable of simulating the planar dissolution-front propagation within the fluid-saturated porous medium. Compared with the theoretical dimensionless

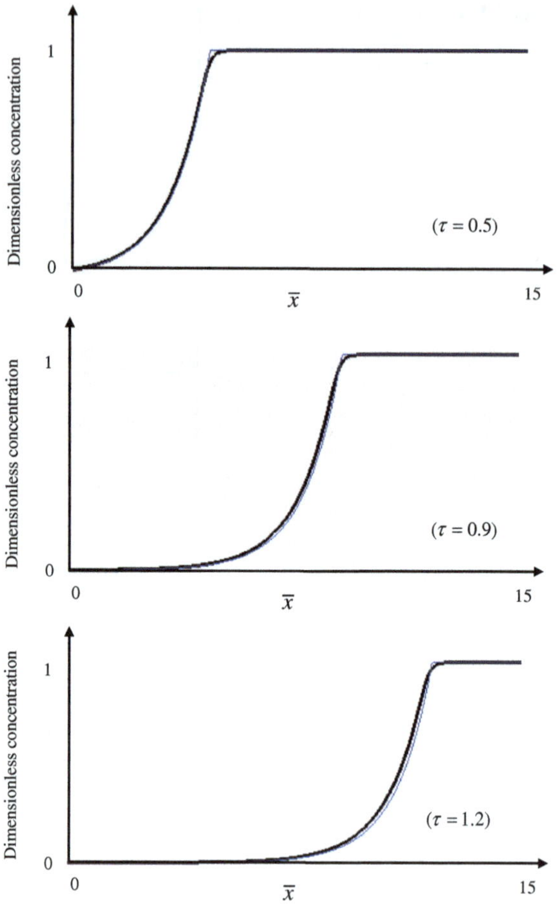

Fig. 4.5 Comparison of the computational simulation results with analytical solutions at three different time instants (Dimensionless concentration)

propagation speed of the planar chemical dissolution-front in the case of the mineral dissolution ratio approaching zero (i.e. $\varepsilon \rightarrow 0$), the simulated dimensionless propagation speed of the planar chemical dissolution-front in the case of the mineral dissolution ratio having a finite value (i.e. $\varepsilon = 0.01$) is slightly slower in the computational model, as expected from the previous theoretical analysis. Since the analytical solutions are derived under the assumption that the mineral dissolution ratio is an infinitesimal quantity, there is a discrepancy between the computational model that is used to produce the computational simulation results and the theoretical model that is used to derive the analytical solutions. Nevertheless, such a discrepancy has only a limited influence on the sharpness of the chemical dissolution front when it is represented by the porosity. This influence becomes negligible if the chemical dissolution front is represented by the dimensionless concentration of the dissolved mineral in the computational model.

To examine whether the proposed numerical procedure is suitable for simulating the complicated morphological evolution process of a planar chemical dissolution-front under supercritical Zhao number conditions, the Zhao number used in the benchmark problem is increased to a value of 10, implying that the dimensionless pore-fluid pressure gradient (i.e. $\bar{p}'_{fx} = -10$), which is applied to the left boundary of the computational domain, is changed from -1 into -10 in the computational model. In this situation, the dimensionless propagation speed of the corresponding planar chemical dissolution-front is approximately equal to 100. This indicates that in the case of a chemical dissolution system with a supercritical Zhao number, the chemical dissolution front propagates faster than it does in the case of the system with a subcritical Zhao number. Due to the change in the Zhao number that is used in the benchmark problem, the mineral dissolution ratio needs to be changed from 0.01 into 0.005, while the dimensionless time-step length is assumed to be 0.001 in the current computational model. Other parameters are of exactly the same values as those used in the above benchmark problem. Since the Zhao number of the chemical dissolution system is greater than its corresponding critical value (i.e. $Zh_{critical} = 2.001$ in this situation), the chemical dissolution system under consideration is supercritical and therefore, a planar chemical dissolution-front can evolve into a complicated morphology during its propagation in the computational model. The same initial porosity field after the small perturbation as that used in the benchmark problem is employed as the initial porosity condition for the computational model.

Figure 4.6 shows porosity distributions in the computational model as a result of the morphological evolution of the chemical dissolution front in the fluid-saturated porous medium, while Fig. 4.7 shows dimensionless concentration distributions within the computational domain. Since the small perturbation grows with time for a supercritical system, the initial planar chemical dissolution-front is gradually changed into an irregular one after the dimensionless time is greater than 0.07. With an increase in the dimensionless time, there is a significant increase in the amplitude of the irregular chemical dissolution-front. This indicates that the chemical dissolution front is morphologically unstable during its propagation within the computational domain. Note that the chemical dissolution front represented by the dimensionless concentration of the dissolved mineral is more diffusive than that represented by the porosity. Nevertheless, both the porosity and the dimensionless concentration of the dissolved mineral have the similar propagation front morphology, even though the location of the maximum value of the porosity is clearly different from that of the dimensionless concentration along the chemical dissolution front. The peak value of the porosity is in good correspondence with the trough value of the dimensionless concentration. This demonstrates that the proposed numerical procedure is also capable of simulating the morphological evolutions of chemical dissolution fronts in the supercritical chemical dissolution systems.

At this stage, the proposed numerical procedure can be convincingly used to investigate how the mineral dissolution ratio affects the morphological evolution

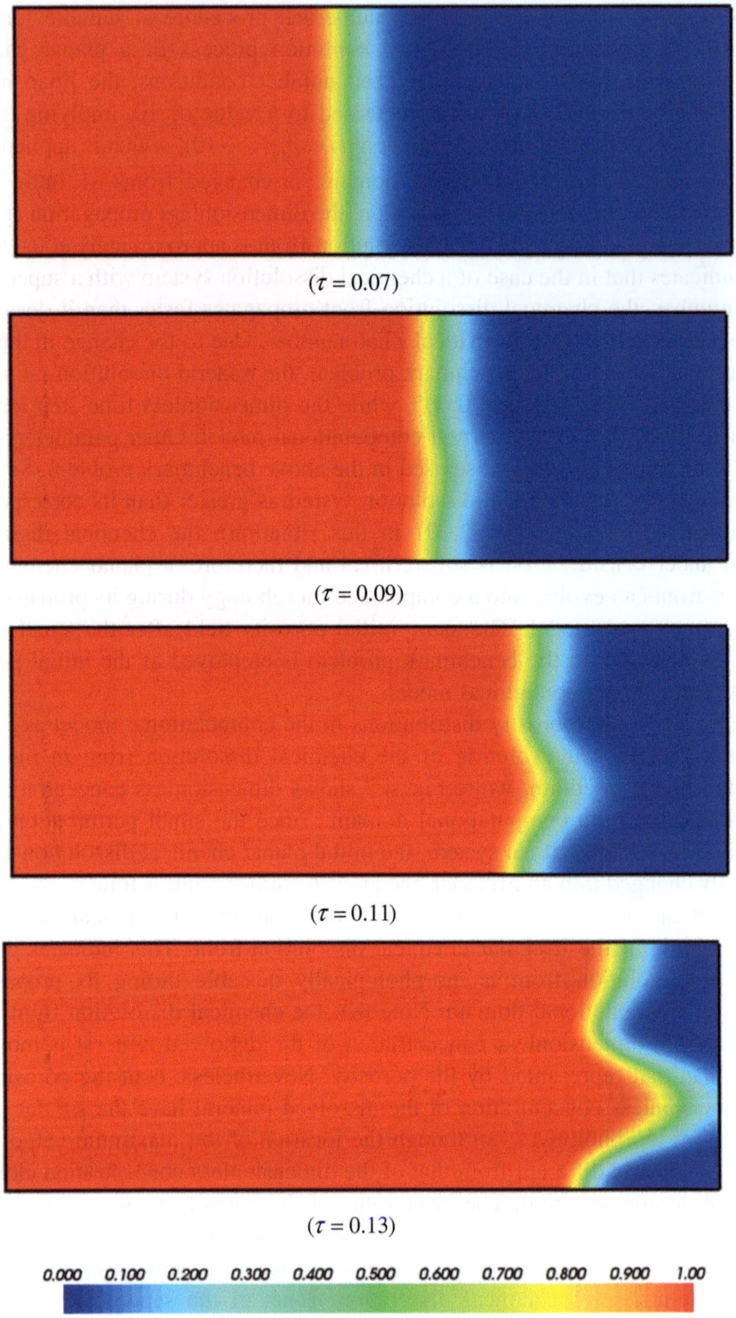

Fig. 4.6 Porosity distributions due to the morphological evolution of the chemical dissolution front in the computational model ($\varepsilon = 0.01$)

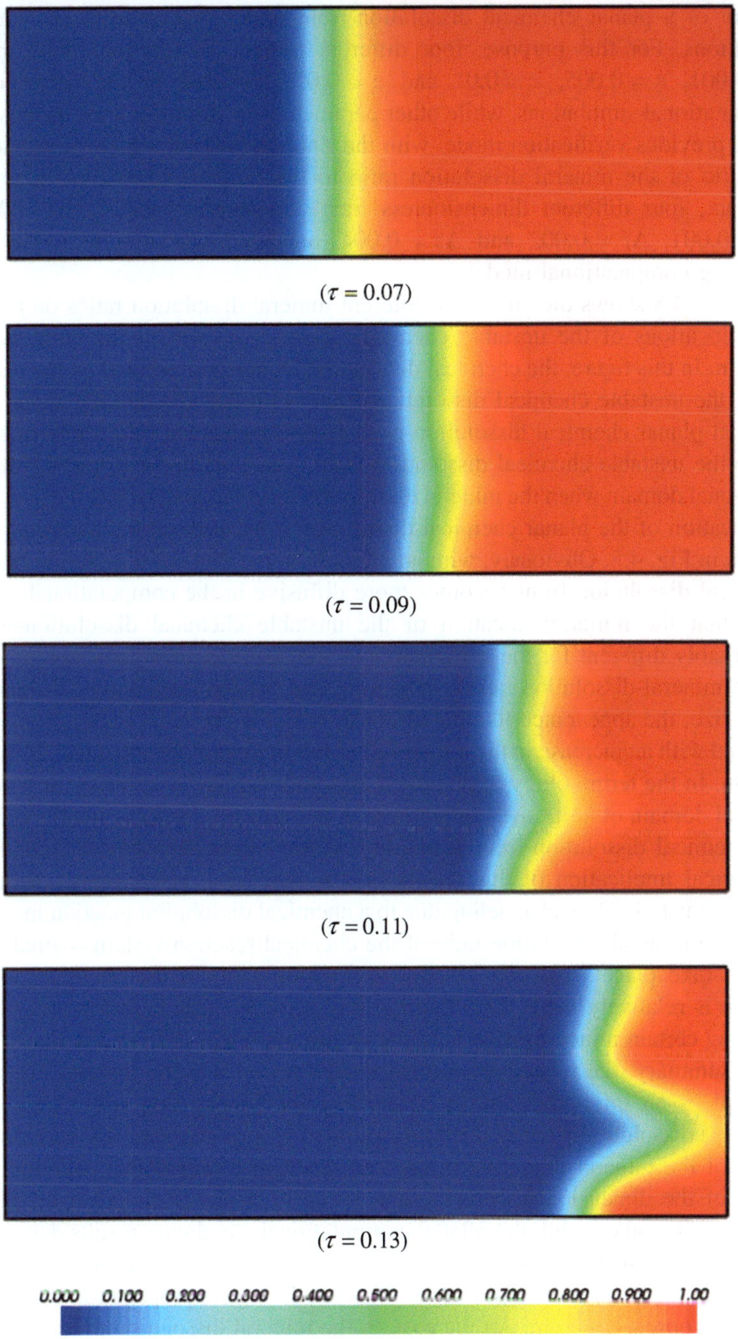

Fig. 4.7 Dimensionless concentration distributions due to the morphological evolution of the chemical dissolution front in the computational model ($\varepsilon = 0.01$)

pattern of a planar chemical dissolution-front under supercritical Zhao number conditions. For this purpose, four different mineral dissolution ratios, namely $\varepsilon = 0.001$, $\varepsilon = 0.005$, $\varepsilon = 0.01$ and $\varepsilon = 0.02$, are used in the corresponding computational simulations, while other parameters are kept the same as those used in the previous verification model with the Zhao number of 10. However, to keep the ratio of the mineral dissolution ratio to the dimensionless time-step length constant, four different dimensionless time-step lengths, namely $\Delta\bar{\tau} = 0.0002$, $\Delta\bar{\tau} = 0.001$, $\Delta\bar{\tau} = 0.002$ and $\Delta\bar{\tau} = 0.004$, are used accordingly in the corresponding computational models.

Figure 4.8 shows the effects of different mineral dissolution ratios on the initiation locations of the unstable chemical dissolution-front in the computational domain. In this figure, the chemical dissolution front is represented by the porosity, while the unstable chemical dissolution-front is produced when the shape of the original planar chemical dissolution-front is firstly changed into different shapes. Since the unstable chemical dissolution-front does not appear in the whole computational domain when the mineral dissolution ratio is equal to 0.02 (i.e. $\varepsilon = 0.02$), the location of the planar chemical dissolution-front at the last simulation step is shown in Fig. 4.8. Obviously, with an increase in the mineral dissolution ratio, the chemical dissolution front becomes more diffusive in the computational domain. Note that the initiation location of the unstable chemical dissolution-front is remarkably different for the computational simulation results associated with different mineral-dissolution ratios, indicating that for the computational domain of finite size, the appearance likelihood of the unstable chemical dissolution-front is reduced with an increase in the mineral dissolution ratio of the chemical dissolution system. In the terminology of chemical dissolution-front instability, for a computational domain of finite size, an increase in the mineral dissolution ratio stabilizes the chemical dissolution-front instability within the computational domain. The geological implication of this recognition is that for a given region within the Earth's crust, the flow channeling due to a chemical dissolution reaction may occur when the mineral dissolution ratio of the chemical reaction system is small, but it cannot take place when the mineral dissolution ratio of the chemical reaction system is relatively large. This finding may provide some help for better understanding certain mineralization patterns within the upper crust of the Earth.

In summary, the related theoretical results reveal that the mineral dissolution ratio plays an important role in controlling the propagation speed of a planar chemical dissolution-front in the fluid-saturated porous medium. An increase in the value of the mineral dissolution ratio can result in a remarkable decrease in the value of the dimensionless propagation speed of a planar chemical dissolution-front. On the other hand, the related computational simulation results demonstrate that the mineral dissolution ratio has a considerable influence on the evolution pattern of a planar chemical dissolution-front during its propagation in the fluid-saturated porous medium. An increase in the mineral dissolution ratio can reduce the likelihood for a planar chemical dissolution-front to evolve from the initial planar shape into different morphologies within the fluid-saturated porous medium of finite size.

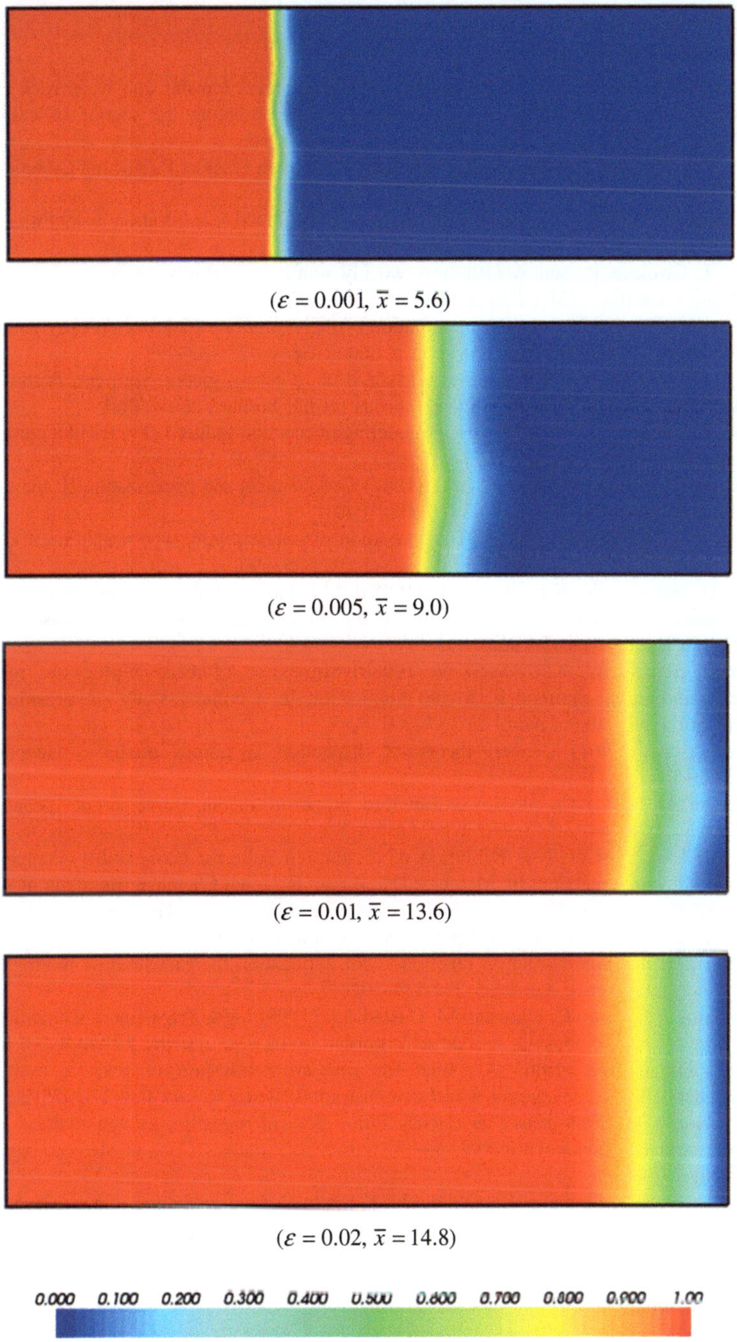

Fig. 4.8 Effects of mineral dissolution ratios on the initiation locations of unstable chemical dissolution-fronts in the computational domain (Porosity)

References

Alt-Epping P, Smith L (2001) Computing geochemical mass transfer and water/rock ratios in submarine hydrothermal systems: implications for estimating the vigour of convection. Geofluids 1:163–181

Bear J (1972) Dynamics of fluids in porous media. American Elsevier Publishing Company, New York

Chadam J, Hoff D, Merino E, Ortoleva P, Sen A (1986) Reactive infiltration instabilities. IMA J Appl Math 36:207–221

Chadam J, Ortoleva P, Sen A (1988) A weekly nonlinear stability analysis of the reactive infiltration interface. IMA J Appl Math 48:1362–1378

Chen JS, Liu CW (2002) Numerical simulation of the evolution of aquifer porosity and species concentrations during reactive transport. Comput Geosci 28:485–499

Holzbecher EO (1998) Modeling density-driven flow in porous media. Springer, Berlin

Nield DA, Bejan A (1992) Convection in porous media. Springer, New York

Ormond A, Ortoleva P (2000) Numerical modeling of reaction-induced cavities in a porous rock. J Geophys Res 105:16737–16747

Ortoleva P, Chadam J, Merino E, Sen A (1987) Geochemical self-organization II: the reactive-infiltration instability. Am J Sci 287:1008–1040

Raffensperger JP, Garven G (1995) The formation of unconformity-type uranium ore deposits: coupled hydrochemical modelling. Am J Sci 295:639–696

Schafer D, Schafer W, Kinzelbach W (1998a) Simulation of reactive processes related to biodegradation in aquifers: 1. Structure of the three-dimensional reactive transport model. J Contam Hydrol 31:167–186

Schafer D, Schafer W, Kinzelbach W (1998b) Simulation of reactive processes related to biodegradation in aquifers: 2. Model application to a column study on organic carbon degradation. J Contam Hydrol 31:187–209

Scheidegger AE (1961) General theory of dispersion in porous media. J Geophys Res 66:3273–3278

Scheidegger AE (1974) The physics of flow through porous media. University of Toronto Press, Toronto

Steefel CI, Lasaga AC (1990) Evolution of dissolution patterns: permeability change due to coupled flow and reaction. In: Melchior DC, Basset RL (eds) Chemical Modeling in Aqueous Systems II, American Chemistry Society Symposium Series, vol 416, pp 213–225

Steefel CI, Lasaga AC (1994) A coupled model for transport of multiple chemical species and kinetic precipitation/dissolution reactions with application to reactive flow in single phase hydrothermal systems. Am J Sci 294:529–592

Xu TF, Samper J, Ayora C, Manzano M, Custodio E (1999) Modelling of non-isothermal multi-component reactive transport in field scale porous media flow systems. J Hydrol 214:144–164

Yeh GT, Tripathi VS (1991) A model for simulating transport of reactive multispecies components: model development and demonstration. Water Resour Res 27:3075–3094

Zhao C, Hobbs BE, Mühlhaus HB (1998) Finite element modelling of temperature gradient driven rock alteration and mineralization in porous rock masses. Comput Methods Appl Mech Eng 165:175–187

Zhao C, Hobbs BE, Mühlhaus HB, Ord A (2001) Finite element modelling of rock alteration and metamorphic process in hydrothermal systems. Commun Numer Methods Eng 17:833–843

Zhao C, Hobbs BE, Mühlhaus HB, Ord A, Lin G (2003) Finite element modeling of three-dimensional steady-state convection and lead/zinc mineralization in fluid-saturated rocks. J Comput Methods Sci Eng 3:73–89

Zhao C, Hobbs BE, Ord A, Peng S, Mühlhaus HB, Liu L (2005) Numerical modeling of chemical effects of magma solidification problems in porous rocks. Int J Num Methods Eng 64:709–728

Zhao C, Hobbs BE, Hornby P, Ord A, Peng S (2006) Numerical modelling of fluids mixing, heat transfer and non-equilibrium redox chemical reactions in fluid-saturated porous rocks. Int J Num Methods Eng 66:1061–1078

Zhao C, Hobbs BE, Hornby P, Ord A, Peng S, Liu L (2008a) Theoretical and numerical analyses of chemical-dissolution front instability in fluid-saturated porous rocks. Int J Numer Anal Methods Geomech 32:1107–1130

Zhao C, Hobbs BE, Ord A, Hornby P, Peng S (2008b) Effect of reactive surface areas associated with different particle shapes on chemical-dissolution front instability in fluid-saturated porous rocks. Transp Porous Media 73:75–94

Zhao C, Hobbs BE, Ord A (2009) Fundamentals of computational geoscience: numerical methods and algorithms. Springer, Berlin

Chapter 5
Effects of Solute Dispersion on Chemical Dissolution-Front Instability in Fluid-Saturated Porous Media

The instability of a chemical dissolution front is an important scientific problem associated with reactive transport processes in fluid-saturated porous media (Chadam et al. 1986, 1988; Ortoleva et al. 1987; Imhoff and Miller 1996; Renard et al. 1998; Imhoff et al. 2003; Chen and Liu 2002; Chen et al. 2009; Zhao et al. 2008a, b, 2010). This problem has a broad scientific and engineering background in the fields of geoscience, mining engineering, petroleum engineering and geo-enviromental engineering. Due to the dissolution of solid minerals, the medium porosity needs to be considered as a major variable for the instability problem of a chemical dissolution front. Since both permeability and solute diffusivity are dependent on the porosity of a porous medium (Bear 1972; Chadam et al. 1986; Nield and Bejan 1992), a chemical dissolution-front instability problem needs to be treated as a fully coupled nonlinear problem between porosity evolution, pore-fluid flow, mass transport and chemical reactions in the fluid-saturated porous medium. Although considerable progress has been made in the study of reactive transport problems (Steefel and Lasage 1990, 1994; Yeh and Tripathi 1991; Raffensperger and Garven 1995; Ormond and Ortoleva 2000; Alt-Epping and Smith 2001; Zhao et al. 1994, 1998, 2001, 2003, 2005, 2006), theoretical consideration of the effects of solute dispersion on chemical dissolution-front instability is still limited.

To understand how solute dispersion affects the instability of a chemical dissolution front, it is important to establish a theoretical criterion that is closely associated with the critical condition, which can be used to assess whether or not a planar chemical dissolution-front becomes unstable in the nonlinear reactive transport system. Toward this end, scientific principles are used for describing porosity evolution, pore-fluid flow, mass transport and chemical reactions in fluid-saturated porous media. Since analytical solutions can be used to investigate the general behavior of the effects of solute dispersion on the instability of a chemical dissolution front during its propagation in the fluid-saturated porous medium, they may produce general conclusions for a broad range of chemical dissolution-front

C. Zhao, *Physical and Chemical Dissolution Front Instability in Porous Media*, Lecture Notes in Earth System Sciences,
DOI: 10.1007/978-3-319-08461-9_5

instability problems. On the contrary, since it is often impossible to use numerical methods to investigate the effects of solute dispersion within the whole parameter space, numerical solutions can only produce some kind of understanding for a particular problem. Therefore, in order to draw general conclusions associated with the effects of solute dispersion on chemical dissolution-front instability in a fluid-saturated porous medium, analytical solutions, if available, are always superior to numerical solutions. On the other hand, analytical solutions provide an important and often unique benchmark solution for verifying numerical methods and algorithms, which are often used to solve the problems of complicated geometries. Since numerical methods are of approximate nature inherently, they must be verified before they are used to solve any new type of scientific and engineering problem. For this reason, it is necessary to derive analytical solutions for the propagation of a planar chemical dissolution-front in a benchmark problem, the geometry of which can be accurately simulated using numerical methods such as the finite element method (Zienkiewicz 1977; Lewis and Schrefler 1998) and the finite difference method (Zhao et al. 1994).

In terms of the theoretical analysis of chemical dissolution-front instability in fluid-saturated porous media when the solute dispersion effect is neglected, a realistic limit case, in which the mineral dissolution ratio approaches zero, is often considered to derive analytical solutions (Chadam et al. 1986; Zhao et al. 2008a, b, 2010). In typical rocks, this ratio may be in a range from 10^{-3} to 10^{-10} (Chadam et al. 1986). The related previous theoretical results have demonstrated that: (1) since fresh pore-fluid flow in the upstream direction of a chemical dissolution front is the main driving force of the chemical dissolution system, the growth rate of the perturbed chemical dissolution-front under supercritical states is strongly dependent on the fresh pore-fluid flow that can be further represented by the pressure gradient of the pore-fluid. (2) Although porosity has a step change at a chemical dissolution front, both the pore-fluid pressure and the solute concentration vary continuously at the chemical dissolution front. By following similar procedures to those used by Chadam et al. (1986), Imhoff and Miller (1996) conducted a theoretical analysis for understanding the dissolution of the nonaqueous phase liquid (NAPL) trapped at residual saturation within the pore space of a porous medium. In their theoretical analysis, the porosity that is used as an independent variable by Chadam et al. (1986) is replaced by the NAPL saturation, while the scalar of diffusivity is replaced by the second-order dispersion tensor. However, their analytical results derived in the limit case (Imhoff and Miller 1996) are questionable because they are independent of the main driving force of the chemical dissolution system. Thus, it is necessary to rederive analytical solutions, in the limit case of the mineral dissolution ratio approaching zero and when solute dispersion is considered as the second-order tensor, for the instability of chemical dissolution fronts in fluid-saturated porous media.

5.1 Mathematical Model for Chemical Dissolution-Front Instability Problems in Fluid-Saturated Porous Media Including Solute Dispersion Effects

The mathematical model of a chemical dissolution-front instability problem in a fluid-saturated porous medium including solute dispersion effects can be established by considering both the related chemical reaction kinetics and the mass conservation of the pore-fluid and dissolved minerals (i.e. solutes hereafter) in the fluid-saturated porous medium. For a fluid-saturated porous medium, Darcy's law is often used to describe pore-fluid flow, while Fick's law and the related chemical reaction kinetics (Steefel and Lasaga 1994) are commonly used to describe reactive mass transport phenomena. If the pore-fluid (i.e. water in this investigation) and solid phase are incompressible, the governing equations of the coupled nonlinear problem between porosity evolution, pore-fluid flow and reactive single-solute transport in the fluid-saturated (rigid) porous medium including solute dispersion effects can be expressed as follows:

$$\frac{\partial \phi}{\partial t} - \nabla \cdot [\psi(\phi) \nabla p] = 0, \tag{5.1}$$

$$\begin{aligned} &\frac{\partial}{\partial t}(\phi C) - \nabla \cdot [\phi \mathbf{D}(\phi) \cdot \nabla C + C\psi(\phi) \nabla p] \\ &+ \rho_s k_{Echemical} \frac{\bar{A}_p}{\bar{V}_p}(\phi_f - \phi)(C - C_{eq}) = 0, \end{aligned} \tag{5.2}$$

$$\frac{\partial \phi}{\partial t} + k_{Echemical} \frac{\bar{A}_p}{\bar{V}_p}(\phi_f - \phi)(C - C_{eq}) = 0, \tag{5.3}$$

$$\psi(\phi) = \frac{k(\phi)}{\mu}, \tag{5.4}$$

where p and C are the pore-fluid pressure and the concentration of the solute; C_{eq} is the equilibrium concentration of the solute; μ is the dynamic viscosity of the pore-fluid; ϕ is the porosity of the porous medium; ϕ_f is the final (i.e. maximum) porosity of the porous medium after the completion of dissolvable mineral dissolution; $\mathbf{D}(\phi)$ is the general dispersion tensor of the solute; $k(\phi)$ is the permeability of the porous medium; $\bar{V}_p$ is the average volume of the dissolvable grain; $\bar{A}_p$ is the average surface area of the dissolvable grain; $k_{Echemical}$ is the comprehensive rate constant of the chemical reaction (as defined in Eq. (2.15) previously); ρ_s is the molar density (i.e. moles per volume) of the dissolvable grains.

Note that since the solute equilibrium concentration of a dissolvable mineral is much smaller than the molar density of the dissolvable mineral in a mineral dissolution system, the mineral dissolution ratio, which is defined as the ratio of the solute equilibrium concentration to the molar density of the dissolvable mineral (Zhao et al. 2010), may be in a range of 10^{-3}–10^{-10} for typical rocks (Chadam et al. 1986). This means that compared with the pore-fluid flow process, the mineral dissolution is a much slower process so that the porosity variation term due to the mineral dissolution can be several orders of magnitude smaller than the pore-fluid velocity variation term in the mass conservation equation of the pore-fluid. In such a case, the porosity variation term due to the mineral dissolution can be neglected in the mass conservation equation of the pore-fluid (i.e. Eq. (5.1)). However, since the mass variation term associated with the porosity change due to the mineral dissolution may have the similar order of magnitude as other mass transport terms in the mass conservation equation of a solute, it needs to be considered in the mass conservation equation of the solute (i.e. Eq. (5.2)). This recognition can be further confirmed later through deriving the dimensionless governing equations of the mineral dissolution system.

As the general dispersion tensor is of the second-order for a two-dimensional problem, it is usually defined by considering solute diffusion, longitudinal and transversal dispersion in a fluid-saturated porous medium (Scheidegger 1961; Bear 1972; Holzbecher 1998).

$$\mathbf{D}(\phi) = (D_{ij}) = \left((D(\phi) + \alpha_T \bar{u}_A)\delta_{ij} + (\alpha_L - \alpha_T)\frac{u_i u_j}{\bar{u}_A} \right), \tag{5.5}$$

where $D(\phi)$ is the diffusivity of the solute; α_T and α_L are the transversal and longitudinal dispersivities of the solute; $\bar{u}_A$ is the absolute magnitude of the mean averaged linear velocity; u_i is the averaged linear velocity of the pore-fluid in the x_i direction; and δ_{ij} ($i = 1$, 2 and $j = 1$, 2 for a two-dimensional problem) is the Kronecker delta.

Based on Darcy's law, the averaged linear velocity vector of the pore-fluid can be expressed as follows:

$$\vec{u}_{linear} = \begin{Bmatrix} u_1 \\ u_2 \end{Bmatrix} = \frac{\vec{u}}{\phi} = -\frac{k(\phi)}{\phi\mu}\nabla p, \tag{5.6}$$

where $\vec{u}$ is the Darcy velocity vector of the pore-fluid; μ is the dynamic viscosity of the pore-fluid; ϕ is the porosity of the porous medium; and $k(\phi)$ is the permeability of the porous medium.

If the porous medium is comprised of sphere particles, then the shape factor which is defined as the ratio of the average volume to the average surface area (i.e. $\bar{V}_p/\bar{A}_p$) of the soluble grain can be expressed as (Zhao et al. 2008b):

$$\frac{\bar{V}_p}{\bar{A}_p} = \left(\sqrt[3]{\frac{1}{36\pi}}\right)\sqrt[3]{\frac{(\phi_f - \phi)}{D_p}} = \alpha_{sphere}\sqrt[3]{\frac{(\phi_f - \phi)}{D_p}}, \tag{5.7}$$

where $\alpha_{sphere} = \sqrt[3]{1/(36\pi)}$ is the shape coefficient of the spherical grains; D_p is the density of the soluble grains, which is defined as the number of the soluble gains per unit medium volume.

Since the solute diffusivity is considered as a function of porosity in the chemical dissolution-front instability problem, a common phenomenological relation can be used for describing this function (Bear 1972; Chadam et al. 1986).

$$D(\phi) = D_0 \phi^q \quad \left(\frac{3}{2} \le q \le \frac{5}{2}\right), \tag{5.8}$$

where D_0 is the chemical species diffusivity in pure water.

To consider the permeability change caused by a change in porosity, the Carman-Kozeny law can be used to express the relationship between permeability and porosity (Scheidegger 1974; Nield and Bejan 1992) as follows:

$$k(\phi) = \frac{k_0 (1-\phi_0)^2 \phi^3}{\phi_0^3 (1-\phi)^2}, \tag{5.9}$$

where ϕ_0 and k_0 are the initial reference porosity and permeability of the porous medium, respectively.

Substituting Eq. (5.7) into Eqs. (5.2) and (5.3) yields the following equations:

$$\begin{aligned} &\frac{\partial}{\partial t}(\phi C) - \nabla \cdot [\phi \mathbf{D}(\phi) \cdot \nabla C + C\psi(\phi) \nabla p] \\ &+ \rho_s k_{Echemical} \frac{\sqrt[3]{D_p}}{\alpha_{sphere}} (\phi_f - \phi)^{\frac{2}{3}} (C - C_{eq}) = 0, \end{aligned} \tag{5.10}$$

$$\frac{\partial \phi}{\partial t} + k_{Echemical} \frac{\sqrt[3]{D_p}}{\alpha_{sphere}} (\phi_f - \phi)^{\frac{2}{3}} (C - C_{eq}) = 0, \tag{5.11}$$

To close the definition of the problem, boundary conditions need to be provided. If the solute is initially in an equilibrium state (i.e. at $t = 0$ and $x \neq -\infty$) within the whole system except for x approaching negative infinity and the fresh pore-fluid (i.e. $C = 0$) is injected at the location of x approaching negative infinity, the following boundary conditions can be used:

$$\lim_{x \to -\infty} C = 0, \quad \lim_{x \to -\infty} \frac{\partial p}{\partial x} = p'_{fx} \quad \text{(upstream boundary)}, \tag{5.12}$$

$$\lim_{x \to \infty} C = C_{eq}, \quad \lim_{x \to \infty} \frac{\partial p}{\partial x} = p'_{0x} \quad \text{(downstream boundary)}, \tag{5.13}$$

where p'_{fx} is the pore-fluid pressure gradient as x approaches negative infinity in the upstream of the pore-fluid flow; p'_{0x} is the unknown pore-fluid pressure gradient as

x approaches positive infinity in the downstream of the pore-fluid flow. Since p'_{fx} drives the pore-fluid flow continuously along the positive x direction, it has a negative algebraic value (i.e. $p'_{fx} < 0$) in this analysis.

The initial condition for this problem is: $\phi(x, 0) = \phi_0$ expect at the negative infinity, where $\lim_{\bar{x} \to -\infty} \phi(x, 0) = \phi_f$. Note that ϕ_0 is the initial porosity of the porous medium.

5.2 Theoretical Consideration of the Effects of Solute Dispersion on Chemical Dissolution-Front Instability in Two-Dimensional Fluid-Saturated Porous Media

A typical theoretical problem involving the propagation of a planar chemical dissolution-front in a full plane, which comprises a fluid-saturated homogeneous porous medium, is considered in this theoretical investigation. For this problem, the pore-fluid flow is only in the x direction when the planar chemical dissolution-front is in a stable state, while it remains dominant in the x direction even when the planar chemical dissolution-front is at the early stages of an unstable state. In the former case, the pore-fluid velocity is exactly equal to zero in the y direction, whereas in the latter case, the pore-fluid velocity in the y direction is relatively small to that in the x direction so that it can be neglected, from the theoretical analysis point of view.

5.2.1 Derivation of Dimensionless Governing Equations for the Theoretical Problem

If the x and y directions are used to replace x_1 and x_2 symbolically, then the general dispersion tensor expressed in Eq. (5.5) can be simplified for this theoretical problem as follows:

$$\mathbf{D}(\phi) = D(\phi)\mathbf{I} + \mathbf{D}_{dispersion}(\phi) = D(\phi)\mathbf{I} + \begin{bmatrix} \alpha_L|u_x| & 0 \\ 0 & \alpha_T|u_x| \end{bmatrix}, \tag{5.14}$$

where $\mathbf{I}$ is a unit second-order tensor (i.e. a 2 by 2 unit matrix); u_x is the averaged linear velocity of the pore-fluid in the x direction; $\mathbf{D}_{dispersion}(\phi)$ is the second-order tensor due to pure dispersion only.

Equation (5.14) clearly indicates that the main difference between this investigation and previous theoretical studies is that the second-order tensor due to pure dispersion is considered here, but it was neglected in the previous theoretical studies (Chadam et al. 1986; Zhao et al. 2008a, b).

Considering Eqs. (5.4), (5.6) and (5.14) simultaneously yields the following expression:

$$\mathbf{D}_{dispersion}(\phi) = \begin{bmatrix} \frac{\alpha_L \psi(\phi)}{\phi}\left|\frac{\partial p}{\partial x}\right| & 0 \\ 0 & \frac{\alpha_T \psi(\phi)}{\phi}\left|\frac{\partial p}{\partial x}\right| \end{bmatrix}. \tag{5.15}$$

This enables Eq. (5.10) to be written in the following form:

$$\begin{aligned} &\frac{\partial}{\partial t}(\phi C) - \nabla \cdot \{\phi[D(\phi)\mathbf{I} + \mathbf{D}_{dispersion}(\phi)] \cdot \nabla C + C\psi(\phi)\nabla p\} \\ &+ \rho_s k_{Echemical} \frac{\sqrt[3]{D_p}}{\alpha_{sphere}} (\phi_f - \phi)^{\frac{2}{3}} (C - C_{eq}) = 0. \end{aligned} \tag{5.16}$$

To translate the dimensional governing equations of the problem (i.e. Eqs. (5.1), (5.11) and (5.16)) into dimensionless ones, the following dimensionless quantities are used in the theoretical analysis:

$$\begin{aligned} &\bar{x} = \frac{x}{L^*}, \quad \bar{y} = \frac{y}{L^*}, \quad \bar{C} = \frac{C}{C_{eq}}, \quad \bar{p} = \frac{p}{p^*}, \\ &\bar{\alpha}_L = \frac{\alpha_L}{L^*}, \quad \bar{\alpha}_T = \frac{\alpha_T}{L^*}, \quad \tau = \frac{t}{t^*}\varepsilon, \quad \varepsilon = \frac{C_{eq}}{\rho_s} << 1, \end{aligned} \tag{5.17}$$

where τ is a slow dimensionless time to describe the slowness of the chemical dissolution that takes place in the chemical dissolution system; ε is the mineral dissolution ratio of the chemical dissolution system. Other characteristic parameters used in Eq. (5.17) can be expressed as follows:

$$\begin{aligned} &t^* = \frac{\alpha_{sphere}}{k_{Echemical} C_{eq} \sqrt[3]{D_p}}, \quad L^* = \sqrt{\phi_f D(\phi_f) t^*}, \quad p^* = \frac{\phi_f D(\phi_f)}{\psi(\phi_f)}, \\ &D^*(\phi) = \frac{\phi D(\phi)}{\phi_f D(\phi_f)}, \quad \psi^*(\phi) = \frac{\psi(\phi)}{\psi(\phi_f)}. \end{aligned} \tag{5.18}$$

Inserting Eqs. (5.17) and (5.18) into Eqs. (5.1), (5.11) and (5.16) yields the following dimensionless equations:

$$\varepsilon \frac{\partial \phi}{\partial \tau} - \nabla \cdot [\psi^*(\phi) \nabla \bar{p}] = 0, \tag{5.19}$$

$$\varepsilon \frac{\partial}{\partial \tau}(\phi \bar{C}) - \nabla \cdot \left\{ \left[D^*(\phi)\mathbf{I} + \mathbf{D}^*_{dispersion}(\phi) \right] \cdot \nabla \bar{C} + \bar{C}\psi^*(\phi)\nabla \bar{p} \right\} - \frac{\partial \phi}{\partial \tau} = 0, \tag{5.20}$$

$$\varepsilon\frac{\partial\phi}{\partial\tau}+(\phi_f-\phi)^{\frac{2}{3}}(\bar{C}-1)=0, \tag{5.21}$$

where $\mathbf{D}^*_{dispersion}(\phi)$ is the dimensionless second-order tensor due to pure dispersion.

$$\mathbf{D}^*_{dispersion}(\phi)=\begin{bmatrix}\bar{\alpha}_L\psi^*(\phi)\left|\frac{\partial\bar{p}}{\partial\bar{x}}\right| & 0\\ 0 & \bar{\alpha}_T\psi^*(\phi)\left|\frac{\partial\bar{p}}{\partial\bar{x}}\right|\end{bmatrix}. \tag{5.22}$$

As discussed previously, Eq. (5.19) clearly demonstrates that the porosity variation term due to the mineral dissolution is several orders of magnitude smaller than the pore-fluid velocity variation term so that it can be neglected in the mass conservation equation of the pore-fluid, while Eq. (5.20) indicates that the mass variation term associated with the porosity change due to the mineral dissolution may have the similar order of magnitude as other mass transport terms so that it needs to be considered in the mass conservation equation of the solute.

Using the above dimensionless quantities (as defined in Eqs. (5.17) and (5.18)), the boundary conditions for this theoretical problem can be expressed in the dimensionless form as follows:

$$\lim_{\bar{x}\to\infty}\bar{C}=1,\quad \lim_{\bar{x}\to\infty}\frac{\partial\bar{p}}{\partial\bar{x}}=\bar{p}'_{0x}\quad \text{(downstream boundary)}, \tag{5.23}$$

$$\lim_{\bar{x}\to-\infty}\bar{C}=0,\quad \lim_{\bar{x}\to-\infty}\frac{\partial\bar{p}}{\partial\bar{x}}=\bar{p}'_{fx}\quad \text{(upstream boundary)}. \tag{5.24}$$

Similarly, the initial condition for this problem is: $\phi(\bar{x},0)=\phi_0$ expect at the negative infinity, where $\lim_{\bar{x}\to-\infty}\phi(\bar{x},0)=\phi_f$.

5.2.2 Derivation of Base Solutions for the Dimensionless Governing Equations of the Theoretical Problem in the Limit Case of the Mineral Dissolution Ratio Approaching Zero

In the limit case of the mineral dissolution ratio (i.e. ε) approaching zero, Eqs. (5.19–5.21) degenerate to the following governing equations for the mineral dissolution system:

$$\nabla\cdot[\psi^*(\phi)\nabla\bar{p}]=0, \tag{5.25}$$

$$\nabla \cdot \left\{ \left[D^*(\phi)\mathbf{I} + \mathbf{D}^*_{dispersion}(\phi) \right] \cdot \nabla \bar{C} + \bar{C}\psi^*(\phi)\nabla \bar{p} \right\} - \frac{\partial \phi}{\partial \tau} = 0, \tag{5.26}$$

$$(\phi_f - \phi)^{\frac{2}{3}}(\bar{C} - 1) = 0. \tag{5.27}$$

In this limit case, Eq. (5.25) clearly indicates that the mass conservation equation of the pore-fluid is indeed independent of the porosity change due to the mineral dissolution, but Eq. (5.26) demonstrates that the mass conservation equation of the solute is dependent on the porosity change due to the mineral dissolution. Because of this difference, the interface condition for the mass flux of the pore-fluid is different from that for the mass flux of the solute at the chemical dissolution front. To keep consistence with Eq. (5.25), the porosity change due to the mineral dissolution should be neglected in the process of deriving the interface condition for the mass flux of the pore-fluid. However, to keep consistence with Eq. (5.26), the porosity change due to the mineral dissolution needs to be considered in the process of deriving the interface condition for the mass flux of the solute.

Equation (5.27) requires that $\bar{C} = 1$ in the undissolved region and $\phi = \phi_f$ in the completely-dissolved region of the dissolvable mineral. Thus a planar chemical dissolution-front divides the problem domain into two regions, an upstream region and a downstream region, relative to the propagation location of the planar chemical dissolution-front. Across the planar chemical dissolution-front, the porosity undergoes a step jump from its initial value to its final value because the chemical reaction rate is considered to be finite in this investigation. As a result, this planar chemical dissolution-front propagation problem can be considered as a Stefan moving boundary problem (Chadam et al. 1986; Zhao et al. 2008a). In this limit case, the corresponding dimensionless governing equations of the problem in both the downstream region and the upstream region can be expressed as follows:

$$\bar{C} = 1, \quad \nabla^2 \bar{p} = 0, \quad \phi = \phi_0 \quad \text{(in the downstream region)}, \tag{5.28}$$

$$\nabla \cdot \left\{ \left[\left(\mathbf{I} + \left| \frac{\partial \bar{p}}{\partial \bar{x}} \right| \bar{\mathbf{D}}^*_{dispersion} \right) \cdot \nabla \bar{C} \right] + \bar{C}\nabla \bar{p} \right\} = 0, \quad \nabla^2 \bar{p} = 0, \quad \phi = \phi_f \quad \text{(in the upstream region)}, \tag{5.29}$$

where

$$\mathbf{D}^*_{dispersion} = \begin{bmatrix} \bar{\alpha}_L & 0 \\ 0 & \bar{\alpha}_T \end{bmatrix}. \tag{5.30}$$

Because $\partial \bar{p}/\partial \bar{x} < 0$ for the theoretical problem considered here, we can substitute $|\partial \bar{p}/\partial \bar{x}| = -\partial \bar{p}/\partial \bar{x}$ in Eq. (5.29) and obtain the following equation:

$$\nabla \cdot \left\{ \left[\left(\mathbf{I} + \left| \frac{\partial \bar{p}}{\partial \bar{x}} \right| \bar{\mathbf{D}}^*_{dispersion} \right) \cdot \nabla \bar{C} \right] + \bar{C} \nabla \bar{p} \right\} = 0, \quad \nabla^2 \bar{p} = 0, \quad \phi = \phi_f \tag{5.31}$$

(in the upstream region),

Since the chemical reaction rate is finite, the solute concentration decreases gradually from its equilibrium value at the step porosity front to zero at somewhere in the upstream region. The physical implication for this is that even though the dissolvable mineral is completely dissolved, it takes some time for the fresh pore-fluid flow to dilute the solution in the upstream region. This is the reason why there is not a step change in the solute concentration across the chemical dissolution front. If the planar chemical dissolution-front is denoted by $S(\bar{x}, \tau) = 0$, then the dimensionless pressure, dimensionless solute concentration and the mass fluxes of both the solute and the pore-fluid should be continuous on $S(\bar{x}, \tau) = 0$. This leads to the following interface conditions for this planar chemical dissolution-front propagation problem:

$$\lim_{S \to 0^-} \bar{C} = \lim_{S \to 0^+} \bar{C}, \qquad \lim_{S \to 0^-} \bar{p} = \lim_{S \to 0^+} \bar{p}, \tag{5.32}$$

$$\lim_{S \to 0^-} (1 - \bar{\alpha}_L \bar{p}'_{fx}) \frac{\partial \bar{C}}{\partial \bar{n}} = \bar{v}_{front}(\phi_f - \phi_0), \quad \lim_{S \to 0^-} \frac{\partial \bar{p}}{\partial \bar{n}} = \frac{\psi(\phi_0)}{\psi(\phi_f)} \lim_{S \to 0^+} \frac{\partial \bar{p}}{\partial \bar{n}} \tag{5.33}$$

where $\bar{n} = n/L^*$; n is the normal vector of the propagating planar chemical dissolution-front; $\bar{v}_{front}$ is the dimensionless propagation speed of the planar chemical dissolution-front.

The first expression in Eq. (5.33) describes the mass flux continuity (conservation) of the solute arising from the propagating planar chemical dissolution-front, while the second expression in Eq. (5.33) describes the pore-fluid flow continuity as a result of the propagating planar chemical dissolution-front. Although the second expression can be straightforwardly derived by considering the Darcy velocity continuity of the pore-fluid, it may be helpful to explain how the first expression in Eq. (5.33) is derived. For this purpose, the interface condition due to the mass flux continuity of the solute is firstly derived in a dimensional form below.

As shown in Fig. 5.1, a planar chemical dissolution-front is propagating from position 1 at time t_1 to position 2 at time t_2. The dimensional propagation speed of the planar chemical dissolution-front is v_{front}. On the left side of the planar chemical dissolution-front (i.e. $S \to 0^-$) at t_1, the mass flux of the solute can be expressed as follows:

$$F_{left} = -\lim_{S \to 0^-} \left[\phi_f \mathbf{D}(\phi_f) \cdot \nabla C + C \psi(\phi_f) \nabla p \right]. \tag{5.34}$$

Similarly, on the right side of the planar chemical dissolution-front (i.e. $S \to 0^+$) at t_2, the mass flux of the solute can be expressed in the following form:

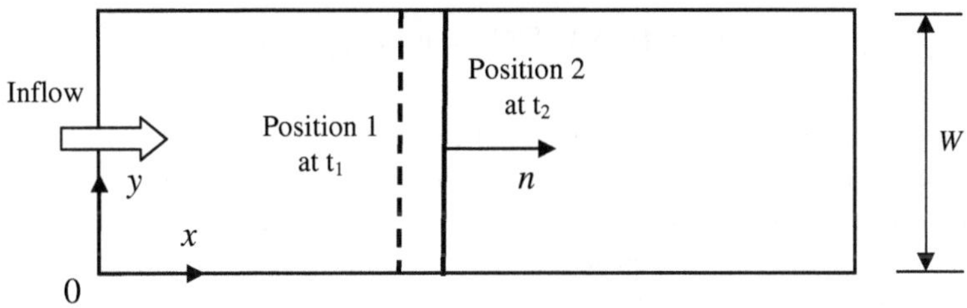

Fig. 5.1 Derivation of the interface condition of a propagating planar chemical dissolution-front in a fluid-saturated porous medium

$$F_{right} = -\lim_{S\to 0^+}[\phi_0 \mathbf{D}(\phi_0)\cdot\nabla C + C\psi(\phi_0)\nabla p]. \tag{5.35}$$

As mentioned previously, the contribution of *the porosity change associated with the chemical dissolution* to the mass flux of the solute needs to be considered. When the planar chemical dissolution-front propagates from position 1 to position 2, the net change in the mass flux of the solute *due to the porosity change associated with the chemical dissolution* can be expressed as follows:

$$F_{net} = v_{front}(\phi_f - \phi_0)\rho_s, \tag{5.36}$$

where ρ_s is the molar density (i.e. moles per volume) of the dissolvable grains.

From the mass conservation condition at the chemical dissolution front, the following expression exists mathematically:

$$F_{right} - F_{left} = F_{net}. \tag{5.37}$$

Substituting Eqs. (5.34)–(5.36) into Eq. (5.37) yields the following equation:

$$-\lim_{S\to 0^+}[\phi_0 \mathbf{D}(\phi_0)\cdot\nabla C + C\psi(\phi_0)\nabla p] + \lim_{S\to 0^-}\left[\phi_f \mathbf{D}(\phi_f)\cdot\nabla C + C\psi(\phi_f)\nabla p\right] = v_{front}(\phi_f - \phi_0)\rho_s. \tag{5.38}$$

As indicated by the second expression in Eq. (5.33), the following expression also exists:

$$\lim_{S\to 0^+}[C\psi(\phi_0)\nabla p] - \lim_{S\to 0^-}\left[C\psi(\phi_f)\nabla p\right]. \tag{5.39}$$

Since the solute concentration is constant in the downstream direction of the chemical dissolution front, the concentration gradient of the solute at $S\to 0^+$ is identical to zero. As a result, Eq. (5.38) can be rewritten as follows:

$$\lim_{S \to 0^-} [\phi_f \mathbf{D}(\phi_f) \cdot \nabla C] = v_{front}(\phi_f - \phi_0)\rho_s. \tag{5.40}$$

Clearly, Eq. (5.40) is the dimensional expression for the mass flux continuity condition of the solute at the chemical dissolution interface between the dissolved and undissolved regions in the fluid-saturated porous medium.

Considering the one-dimensional nature of the planar chemical dissolution-front propagation under subcritical conditions, Eq. (5.40) can be further reduced to the following form:

$$\lim_{S \to 0^-} \left[\phi_f D(\phi_f) - \alpha_L \psi(\phi_f) \frac{\partial p}{\partial n} \right] \frac{\partial C}{\partial n} = v_{front}(\phi_f - \phi_0)\rho_s. \tag{5.41}$$

Based on the dimensionless quantities defined in Eqs. (5.17) and (5.18), the dimensionless propagation speed of the planar chemical dissolution-front can be express as follows:

$$\bar{v}_{front} = \frac{v_{front}}{v^*_{front}}, \quad v^*_{front} = \frac{\varepsilon L^*}{t^*}. \tag{5.42}$$

If the related dimensionless quantities are substituted into Eq. (5.41), then we can obtain the first expression in Eq. (5.33).

Now let us explain how the second expression in Eq. (5.33) is derived. Since the contribution of *the porosity change associated with the chemical dissolution* to the mass flux of the pore-fluid should be neglected during the porosity jump at the interface (as demonstrated by Eq. (5.25)), there is not any pore-fluid production/consumption during this jump. Thus the pore-fluid flow continuity condition requires that the outflow flux of the pore-fluid on the right side of the interface (at t_2) be equal to the inflow flux of the pore-fluid on the left side of the interface (at t_1). Although the averaged linear velocity is different at the left and right sides of the interface, the Darcy velocity is exactly the same because it is equal to the product of the averaged linear velocity and porosity. According to Darcy's law, the inflow flux of the pore-fluid on the left side of the interface (at t_1) for the problem of unit thickness in the direction particular to the xy plane (shown in Fig. 5.1) can be expressed as follows:

$$Inflow = -\lim_{S \to 0^-} \frac{k(\phi_f)}{\mu} \frac{\partial p}{\partial n} W = -\lim_{S \to 0^-} \frac{k(\phi_f) p^*}{\mu L^*} \frac{\partial p}{\partial \bar{n}} W = -\lim_{S \to 0^-} \frac{\psi(\phi_f) p^*}{L^*} \frac{\partial \bar{p}}{\partial \bar{n}} W, \tag{5.43}$$

where W is the width of the interface (as shown in Fig. 5.1).

Similarly, the outflow flux of the pore-fluid on the right side of the interface (at t_2) is:

$$Outflow = -\lim_{S\to 0^+} \frac{k(\phi_0)}{\mu}\frac{\partial p}{\partial n}W = -\lim_{S\to 0^+} \frac{k(\phi_0)p^*}{\mu L^*}\frac{\partial \bar{p}}{\partial \bar{n}}W = -\lim_{S\to 0^+} \frac{\psi(\phi_0)p^*}{L^*}\frac{\partial \bar{p}}{\partial \bar{n}}W \quad (5.44)$$

If the equality condition between the inflow and outflow fluxes of the pore-fluid is considered, then the second expression in Eq. (5.33) can be obtained.

Both expressions in Eq. (5.33) were also derived independently for the chemical dissolution-front propagation problem in a fluid-saturated porous medium by using the method of matched asymptotic expansions in a purely mathematical manner (Chadam et al. 1986). Nevertheless, Eq. (5.33) is only valid for the limit case when the mineral dissolution ratio (i.e. ε) approaches zero. For a general case where the contribution of *the porosity change associated with the chemical dissolution* to the mass flux of the pore-fluid cannot be neglected during the porosity jump in the interface, the second expression of Eq. (5.33) is invalid so that a new expression needs to be derived, but this is beyond the scope of the theoretical analysis carried out in this investigation.

When the planar chemical dissolution-front is in a stable state, base solutions for this theoretical problem can be derived from Eqs. (5.28) to (5.31) with the related boundary and interface conditions (i.e. Eqs. (5.23), (5.24), (5.32) and (5.33)) as follows:

$$\bar{C}(\xi) = 1, \quad \bar{p}(\xi) = \bar{p}'_{0x}\xi + \bar{p}_{C1}, \quad \phi = \phi_0 \quad \text{(in the downstream region)}, \quad (5.45)$$

$$\bar{C}(\xi) = \exp\left(-\frac{\bar{p}'_{fx}}{1-\bar{\alpha}_L\bar{p}'_{fx}}\xi\right), \quad \bar{p}(\xi) = \bar{p}'_{fx}\xi + \bar{p}_{C2}, \quad \phi = \phi_f \quad \text{(in the upstream region)}, \quad (5.46)$$

where $\bar{p}_{C1}$ and $\bar{p}_{C2}$ are two constants to be determined. For example, p_{C1} can be determined by setting the dimensionless pressure $\bar{p}(\xi)$ to be a constant at a prescribed location of the downstream region, while $\bar{p}_{C2}$ can be determined using the pressure continuity condition at the interface between the upstream and the downstream regions. Other parameters are defined below (Zhao et al. 2008a, b):

$$\xi = \bar{x} - \bar{v}_{front}\tau, \quad \bar{p}'_{0x} = \frac{\psi(\phi_f)}{\psi(\phi_0)}\bar{p}'_{fx} \quad (5.47)$$

Considering Eqs. (5.33) and (5.46) yields the following expression for the dimensionless propagation speed of the planar chemical dissolution-front:

$$\bar{v}_{front} = -\frac{\bar{p}'_{fx}}{\phi_f - \phi_0}. \quad (5.48)$$

Equation (5.48) clearly indicates that, compared with the previous studies without including the solute dispersion effect (Zhao et al. 2008a, b), the consideration of solution dispersion does not affect the dimensionless propagation speed of a planar chemical dissolution-front for this theoretical problem.

5.2.3 Derivation of the Critical Condition for Unstable Chemical Dissolution-Fronts in Two-Dimensional Fluid-Saturated Porous Media Including Solute Dispersion Effects

To theoretically determine the critical condition under which a planar chemical dissolution-front in the fluid-saturated porous medium can become unstable, a small time-dependent perturbation is added on the planar chemical dissolution-front. As a result, the total solutions of the system are equal to the sum of the base and perturbed solutions as follows:

$$S(\xi, \bar{y}, \tau) = \xi - \delta \exp(\bar{\omega}\tau) \cos(\overline{my}), \tag{5.49}$$

$$\bar{p}_{total}(\xi, \bar{y}, \tau) = \bar{p}(\xi, \tau) + \delta \hat{p}(\xi) \exp(\bar{\omega}\tau) \cos(\overline{my}), \tag{5.50}$$

$$\bar{C}_{total}(\xi, \bar{y}, \tau) = \bar{C}(\xi, \tau) + \delta \hat{C}(\xi) \exp(\bar{\omega}\tau) \cos(\overline{my}), \tag{5.51}$$

where $\bar{\omega}$ is the dimensionless growth rate of the perturbation; $\bar{m}$ is the dimensionless wavenumber of the perturbation; δ is the amplitude of the perturbation and $\delta << 1$ by the definition of the conventional linear stability analysis.

Since $S(\xi, \bar{y}, \tau)$ is a function of coordinates ξ and $\bar{y}$, the following derivatives exist mathematically (Chadam et al. 1986; Zhao et al. 2008b):

$$\left(\frac{\partial}{\partial \xi}\right)_{\xi} = \frac{\partial S}{\partial \xi}\frac{\partial}{\partial S} = \left(\frac{\partial}{\partial \xi}\right)_{S}, \quad \left(\frac{\partial}{\partial \bar{y}}\right)_{\xi} = \frac{\partial S}{\partial \bar{y}}\frac{\partial}{\partial S} + \left(\frac{\partial}{\partial \bar{y}}\right)_{S} = \frac{\partial S}{\partial \bar{y}}\left(\frac{\partial}{\partial \xi}\right)_{S} + \left(\frac{\partial}{\partial \bar{y}}\right)_{S}, \tag{5.52}$$

$$\left(\frac{\partial^2}{\partial \xi^2}\right)_{\xi} = \left(\frac{\partial^2}{\partial \xi^2}\right)_{S}, \quad \left(\frac{\partial^2}{\partial \bar{y}^2}\right)_{\xi} = \frac{\partial^2 S}{\partial \bar{y}^2}\frac{\partial}{\partial \xi} + \left(\frac{\partial S}{\partial \bar{y}}\right)^2 \frac{\partial^2}{\partial \xi^2} + 2\frac{\partial S}{\partial \bar{y}}\frac{\partial^2}{\partial \xi \partial \bar{y}} + \left(\frac{\partial^2}{\partial \bar{y}^2}\right)_{S}. \tag{5.53}$$

Based on the fact that the total solutions (that are expressed in Eqs. (5.50) and (5.51)) must satisfy the governing equations (that are expressed in Eqs. (5.28) and (5.31)), we can derive the first-order perturbation equations of the chemical dissolution system. As a result, the critical condition, which is used to examine whether or not a planar chemical dissolution-front becomes unstable in the fluid-saturated porous medium, can be determined mathematically. With consideration of Eq. (5.53), the first-order perturbation equations of the chemical dissolution system can be expressed as

$$\hat{C} = 0, \quad \frac{\partial^2 \hat{p}}{\partial \xi^2} - \bar{m}^2 \hat{p} + \bar{m}^2 \bar{p}'_{0x} = 0 \quad \text{(in the downstream region)}, \tag{5.54}$$

$$\left(1-\bar{\alpha}_L\bar{p}'_{fx}\right)\frac{\partial^2\hat{C}}{\partial\xi^2}+\bar{p}'_{fx}\frac{\partial\hat{C}}{\partial\xi}-\left(1-\bar{\alpha}_T\bar{p}'_{fx}\right)\bar{m}^2\hat{C}-\frac{\left(1-\bar{\alpha}_T\bar{p}'_{fx}\right)\bar{m}^2\bar{p}'_{fx}}{1-\bar{\alpha}_L\bar{p}'_{fx}}\exp\left(-\frac{\bar{p}'_{fx}}{1-\bar{\alpha}_L\bar{p}'_{fx}}\xi\right)$$
$$-\frac{\bar{p}'_{fx}}{1-\bar{\alpha}_L\bar{p}'_{fx}}\left(1+\frac{\bar{\alpha}_L\bar{p}'_{fx}}{1-\bar{\alpha}_L\bar{p}'_{fx}}\right)\exp\left(-\frac{\bar{p}'_{fx}}{1-\bar{\alpha}_L\bar{p}'_{fx}}\xi\right)\frac{\partial\hat{p}}{\partial\xi}+\frac{\bar{\alpha}_L\bar{p}'_{fx}}{1-\bar{\alpha}_L\bar{p}'_{fx}}\exp\left(-\frac{\bar{p}'_{fx}}{1-\bar{\alpha}_L\bar{p}'_{fx}}\xi\right)\frac{\partial^2\hat{p}}{\partial\xi^2}=0,$$
$$\frac{\partial^2\hat{p}}{\partial\xi^2}-\bar{m}^2\hat{p}+\bar{m}^2\bar{p}'_{fx}=0 \quad \text{(in the upstream region)},$$
$$(5.55)$$

Since the dispersion tensor is a function of perturbed dimensionless pore-fluid pressure (i.e. $\hat{p}(\xi)$), the first-order perturbation equations when solute dispersion is considered as the second-order tensor are different from those when it is considered as the scalar of a constant (representing the solute diffusivity only) (Chadam et al. 1986). Obviously, consideration of the dependence of the dispersion tensor on the perturbed dimensionless pore-fluid pressure results in much complicated perturbation equations, so that special mathematical treatments must be used to derive the corresponding analytical solutions.

The corresponding boundary conditions for the first-order perturbation problem are:

$$\hat{C}=0, \quad \lim_{x\to\infty}\frac{\partial\hat{p}}{\partial\xi}=0 \quad \text{(downstream boundary)}, \tag{5.56}$$

$$\lim_{x\to-\infty}\hat{C}=0, \quad \lim_{x\to-\infty}\frac{\partial\hat{p}}{\partial\xi}=0 \quad \text{(upstream boundary)}. \tag{5.57}$$

Similarly, the interface conditions for this first-order perturbation problem can be expressed as follows:

$$\hat{C}=0, \quad \lim_{S\to 0^-}\hat{p}=\lim_{S\to 0^+}\hat{p}, \tag{5.58}$$

$$\lim_{S\to 0^-}\left[\left(1-\bar{\alpha}_I\bar{p}'_{fx}\right)\frac{\partial\hat{C}}{\partial n}+\frac{\bar{\alpha}_L\bar{p}'_{fx}}{1-\alpha_L p'_{fx}}\frac{\partial\hat{p}}{\partial n}\right]=\bar{\omega}(\phi_f-\phi_0),$$
$$\lim_{S\to 0^-}\frac{\partial\hat{p}}{\partial\bar{n}}=\frac{\psi(\phi_0)}{\psi(\phi_f)}\lim_{S\to 0^+}\frac{\partial\hat{p}}{\partial\bar{n}}. \tag{5.59}$$

Solving Eqs. (5.54) and (5.55) with the boundary and interface conditions (i.e. Eqs. (5.56)–(5.59)) yields the following analytical solutions:

$$\hat{C}=0, \quad \hat{p}(\xi)=\bar{p}'_{0x}\left[1-\frac{1}{1+\beta}\frac{\beta}{}\exp(-|\bar{m}|\xi)\right] \quad \text{(in the downstream region)},$$
$$(5.60)$$

$$\hat{C}(\xi) = -\frac{\bar{p}'_{fx}}{1-\bar{\alpha}_L\bar{p}'_{fx}}\left\{\exp\left(-\frac{\bar{p}'_{fx}}{1-\bar{\alpha}_L\bar{p}'_{fx}}\xi\right) - \frac{(1+\beta)+\left(\frac{1}{1-\bar{\alpha}_L\bar{p}'_{fx}}-\bar{\alpha}_L|\bar{m}|\right)(1-\beta)}{1+\beta}\exp(\sigma\xi)\right\}$$
$$-\frac{\bar{p}'_{fx}}{1-\bar{\alpha}_L\bar{p}'_{fx}}\left\{\left(\frac{1}{1-\bar{\alpha}_L\bar{p}'_{fx}}-\bar{\alpha}_L|\bar{m}|\right)\frac{1-\beta}{1+\beta}\exp\left[\left(|\bar{m}|-\frac{\bar{p}'_{fx}}{1-\bar{\alpha}_L\bar{p}'_{fx}}\right)\xi\right]\right\},$$
$$\hat{p}(\xi) = \bar{p}'_{fx}\left[1+\frac{1-\beta}{1+\beta}\exp(|\bar{m}|\xi)\right] \quad \text{(in the upstream region)}, \tag{5.61}$$

where

$$\beta = \frac{\psi(\phi_0)}{\psi(\phi_f)} = \frac{k(\phi_0)}{k(\phi_f)}, \tag{5.62}$$

$$\sigma = \frac{\sqrt{\left(\frac{\bar{p}'_{fx}}{1-\bar{\alpha}_L\bar{p}'_{fx}}\right)^2+\frac{4\bar{m}^2\left(1-\bar{\alpha}_T\bar{p}'_{fx}\right)}{1-\bar{\alpha}_L\bar{p}'_{fx}}}-\frac{\bar{p}'_{fx}}{1-\bar{\alpha}_L\bar{p}'_{fx}}}{2}. \tag{5.63}$$

Inserting Eq. (5.61) into Eq. (5.59) yields the following equation for the dimensionless growth rate of the small perturbation:

$$\bar{\omega} = \frac{\bar{p}'_{fx}}{(1+\beta)(\phi_f-\phi_0)}\left\{\frac{(1+\beta)\bar{p}'_{fx}}{1-\bar{\alpha}_L\bar{p}'_{fx}} + \left[(1+\beta)+\left(\frac{1}{1-\bar{\alpha}_L\bar{p}'_{fx}}-\bar{\alpha}_L|\bar{m}|\right)(1-\beta)\right]\sigma\right\}$$
$$-\frac{\bar{p}'_{fx}}{(1+\beta)(\phi_f-\phi_0)}\left[\left(|\bar{m}|-\frac{\bar{p}'_{fx}}{1-\bar{\alpha}_L\bar{p}'_{fx}}-\bar{\alpha}_L\bar{m}^2\left(1-\bar{\alpha}_L\bar{p}'_{fx}\right)\right)\frac{(1-\beta)}{1-\bar{\alpha}_L\bar{p}'_{fx}}\right]. \tag{5.64}$$

If $\bar{\alpha}_L = 0$ and $\bar{\alpha}_T = 0$, Eq. (5.64) can degenerate to the previous form obtained when the dispersion effect is neglected (Zhao et al. 2008a, b). Since $\bar{p}'_{fx} < 0$ when the pore-fluid flow is along the positive x direction, it is useful to define that $Zh = -\bar{p}'_{fx}$ in the theoretical analysis.

Letting $\bar{\omega} = 0$ and $|\bar{m}| = 1$ yields the following equation for the critical Zhao number of the theoretical system.

$$\left[(1+\beta)+\left(\frac{1}{1+\bar{\alpha}_L Zh_{critical}}-\bar{\alpha}_L\right)(1-\beta)\right]\sigma_1 - \frac{(1+\beta)Zh_{critical}}{1+\bar{\alpha}_L Zh_{critical}}$$
$$-\left[\left(1+\frac{Zh_{critical}}{1+\bar{\alpha}_L Zh_{critical}}-\bar{\alpha}_L(1+\bar{\alpha}_L Zh_{critical})\right)\frac{(1-\beta)}{1+\bar{\alpha}_L Zh_{critical}}\right] = 0 \tag{5.65}$$

where

$$\sigma_1 = \frac{\sqrt{\left(\frac{Zh_{critical}}{1+\bar{\alpha}_L Zh_{critical}}\right)^2 + \frac{4(1+\bar{\alpha}_T Zh_{critical})}{1+\bar{\alpha}_L Zh_{critical}}} + \frac{Zh_{critical}}{1+\bar{\alpha}_L Zh_{critical}}}{2}. \tag{5.66}$$

Physically, $\bar{\omega} = 0$ means that the dimensionless growth rate of the perturbation is zero so that the perturbation does not grow in the system, while $|\bar{m}| = 1$ means that the dimensional wavelength (i.e. λ) of the perturbation has the same order of magnitude as the intrinsic characteristic length (i.e. L^* in Eq. (5.18)) of the chemical dissolution system because $|\bar{m}| = 2\pi L^*/\lambda$ (Zhao et al. 2008a).

Since $(1 + \bar{\alpha}_L Zh_{critical}) > 0$, Eq. (5.65) can be rewritten into the following form:

$$\begin{aligned} &\{(1+\beta)(1+\bar{\alpha}_L Zh_{critical}) + [1-\bar{\alpha}_L(1+\bar{\alpha}_L Zh_{critical})](1-\beta)\}\sigma_1 - (1+\beta)Zh_{critical} \\ &- \left(1 + \frac{Zh_{critical}}{1+\bar{\alpha}_L Zh_{critical}} - \bar{\alpha}_L(1+\bar{\alpha}_L Zh_{critical})\right)(1-\beta) = 0. \end{aligned} \tag{5.67}$$

For the case of $\bar{\alpha}_L = \bar{\alpha}_T$, Eq. (5.67) can be further simplified so that an explicit theoretical formula for the critical Zhao number can be obtained. For this case, Eq. (5.67) can be expressed as follows:

$$\begin{aligned} &\{(1+\bar{\alpha}_L Zh_{critical}) + [1-\bar{\alpha}_L(1+\bar{\alpha}_L Zh_{critical})]R\}\sigma_1 - Zh_{critical} \\ &- \left(1 + \frac{Zh_{critical}}{1+\bar{\alpha}_L Zh_{critical}} - \bar{\alpha}_L(1+\bar{\alpha}_L Zh_{critical})\right)R = 0. \end{aligned} \tag{5.68}$$

where

$$R = \frac{1-\beta}{1+\beta}. \tag{5.69}$$

Substituting Eq. (5.66) into Eq. (5.68) yields the following equation:

$$\begin{aligned} &\left\{2R[1-\bar{\alpha}_L(1+\bar{\alpha}_L Zh_{critical})] + \left[(1+\bar{\alpha}_L R) + \frac{R}{1+\bar{\alpha}_L Zh_{critical}}\right] Zh_{critical}\right\}^2 \\ &- \{(1+\bar{\alpha}_L Zh_{critical}) + [1-\bar{\alpha}_L(1+\bar{\alpha}_L Zh_{critical})]R\}^2 \left[\left(\frac{Zh_{critical}}{1+\bar{\alpha}_L Zh_{critical}}\right)^2 + 4\right] = 0. \end{aligned} \tag{5.70}$$

After some mathematical manipulations, Eq. (5.70) can be rewritten as follows:

$$\begin{aligned}&\left(\frac{R^2\bar{\alpha}_L}{1+\bar{\alpha}_L Zh_{critical}}+2\bar{\alpha}_L^3R-\bar{\alpha}_L^2R+\bar{\alpha}_LR-\bar{\alpha}_L^3R^2-\bar{\alpha}_L^2\right)Zh_{critical}^2\\&+\left[\frac{R^2}{1+\bar{\alpha}_L Zh_{critical}}+R-3\bar{\alpha}_LR+4\bar{\alpha}_L^2R-\bar{\alpha}_L^2R^2-2\bar{\alpha}_L\right]Zh_{critical}\\&+(2\bar{\alpha}_LR-2R-1)=0.\end{aligned} \tag{5.71}$$

Finally, the characteristic equation for the critical Zhao number in the case of $\bar{\alpha}_L=\bar{\alpha}_T$ is expressed as the following standard third-power equation:

$$AZh_{critical}^3+BZh_{critical}^2+CZh_{critical}+D=0, \tag{5.72}$$

where

$$A=-\bar{\alpha}_L^3R+2\bar{\alpha}_L^4R+\bar{\alpha}_L^2R-\bar{\alpha}_L^4R^2-\bar{\alpha}_L^3, \tag{5.73}$$

$$B=2\bar{\alpha}_LR-4\bar{\alpha}_L^2R+6\bar{\alpha}_L^3R-2\bar{\alpha}_L^3R^2+\bar{\alpha}_LR^2-3\bar{\alpha}_L^2, \tag{5.74}$$

$$C=-5\bar{\alpha}_LR+6\bar{\alpha}_L^2R-\bar{\alpha}_L^2R^2-3\bar{\alpha}_L+R^2+R, \tag{5.75}$$

$$D=2\bar{\alpha}_LR-2R-1. \tag{5.76}$$

It is obvious that if $\bar{\alpha}_L=0$, then two constants A and B are equal to zero, so that Eq. (5.72) degenerates to the first-power equation with the following solution:

$$Zh_{critical}=\frac{(3-\beta)(1+\beta)}{2(1-\beta)}. \tag{5.77}$$

This solution is identical to the previous theoretical results when the solute dispersion effect is neglected in the chemical dissolution system (Zhao et al. 2008a, b).

Based on the previous study, the dimensionless Zhao number is defined as follows (Zhao et al. 2009):

$$Zh=-\frac{k(\phi_f)p'_{fx}}{\mu\sqrt{\phi_f D(\phi_f)}}\sqrt{\frac{\alpha_{sphere}}{k_{Echemical}C_{eq}\sqrt[3]{D_p}}}. \tag{5.78}$$

If $Zh<Zh_{critical}$, then the chemical dissolution system is subcritical, while if $Zh>Zh_{critical}$, then the chemical dissolution system is supercritical. When $Zh=Zh_{critical}$, the chemical dissolution system is in a critical state. By using this criterion, the instability of chemical dissolution fronts in fluid-saturated porous media can be assessed.

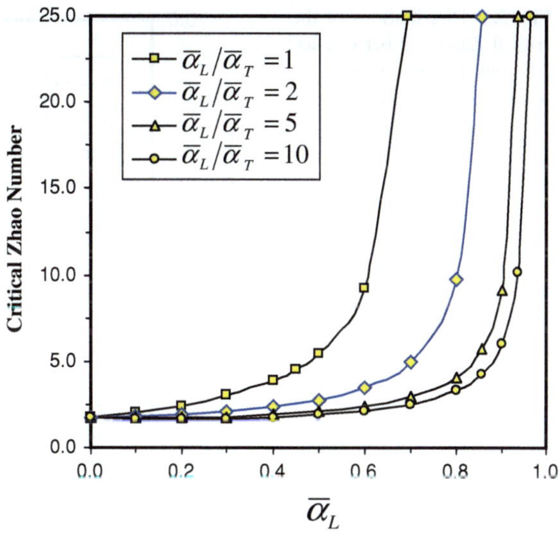

Fig. 5.2 Distributions of the critical Zhao number due to different dispersion ratios ($\phi_0 = 0.1,\ \phi_f = 0.2$)

5.2.4 Theoretical Understanding of the Effects of Solute Dispersion on Chemical Dissolution-Front Instability in Two-Dimensional Fluid-Saturated Porous Media

To theoretically understand how solution dispersion affects the chemical dissolution-front instability in a fluid-saturated porous medium, the critical Zhao numbers associated with different dimensionless solute-dispersion coefficients are evaluated from the above characteristic equation (i.e. Eq. (5.67)). Figure 5.2 shows the distributions of the critical Zhao number due to different solute dispersion ratios. These results are obtained when $\phi_0 = 0.1$ and $\phi_f = 0.2$. In this figure, the solute dispersion ratio (as represented by $\bar{\alpha}_L/\bar{\alpha}_T$) is defined as the ratio of the dimensionless longitudinal dispersivity (as represented by $\bar{\alpha}_L$) to the dimensionless transverse dispersivity (as represented by $\bar{\alpha}_T$). It is obvious that with an increase in the dimensionless longitudinal dispersivity, $\bar{\alpha}_L$, the critical Zhao number can increase significantly, especially for the smaller solute-dispersion ratios such as $\bar{\alpha}_L/\bar{\alpha}_T = 1$ and $\bar{\alpha}_L/\bar{\alpha}_T = 2$. For a given dimensionless longitudinal dispersivity, the critical Zhao number also increases significantly with an increase in the solute dispersion ratio, $\bar{\alpha}_L/\bar{\alpha}_T$. This indicates that an increase in either the longitudinal dispersivity or the dispersion ratio stabilizes the chemical dissolution front so that it becomes more difficult for a planar chemical dissolution-front to evolve into different morphologies.

The effect of the final porosity value on chemical dissolution front instability can be also investigated using the present theoretical results. For this purpose, the solute dispersion ratio is set to be unity (i.e. $\bar{\alpha}_L/\bar{\alpha}_T = 1$). Consequently, Eq. (5.72) can be used to calculate the corresponding critical Zhao numbers. Figure 5.3

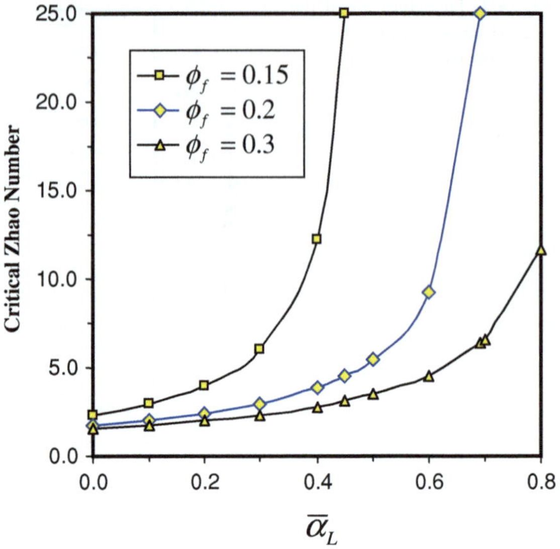

Fig. 5.3 Distributions of the critical Zhao number due to different final porosities ($\phi_0 = 0.1,\ \bar{\alpha}_L = \bar{\alpha}_T$)

shows the distributions of the critical Zhao numbers due to different final porosities. To produce these results, the initial porosity is set to be 0.1 (i.e. $\phi_0 = 0.1$), while three different final porosities, namely $\phi_f = 0.15$, 0.2 and 0.3, are used in the corresponding calculations. Clearly, with an increase in the final porosity value, the critical Zhao number decreases dramatically. This implies that an increase in the final porosity value destabilizes the chemical dissolution front so that it becomes easier for a planar chemical dissolution-front to evolve into different morphologies.

5.3 Application of the Present Theoretical Solutions

The ultimate goal of developing numerical methods is to solve complicated and complex problems in nature. Due to the approximate nature of a numerical method, it is desirable to validate it using existing data obtained from either field measurements or laboratory experiments. However, since such data are usually lacking or very limited for mineralization related instability problems (Chen et al. 2009), it is common practice, as the first step, to verify a numerical method using analytical solutions for benchmark problems. Since the main purpose of this investigation is to deal with theoretical aspects of the effects of solute dispersion on chemical dissolution-front instability problems, generic models can be used to illustrate the possible applications of the present theoretical solutions. Nevertheless, it is useful, in the future, to conduct some field measurements and laboratory experiments for the chemical dissolution-front instability problems in fluid-saturated porous rocks

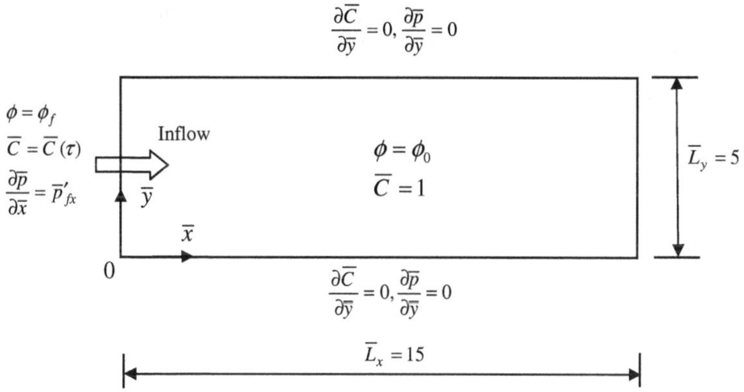

Fig. 5.4 Geometry and boundary conditions of the chemical dissolution problem in a fluid-saturated porous medium

so that the present theoretical solutions can be used to assess chemical dissolution-front instability in real geological and geochemical systems.

As the first step of application, the present theoretical solutions expressed by Eqs. (5.45) and (5.46) in the case of subcritical Zhao numbers are used as benchmark solutions for verifying numerical methods that can be employed to simulate the complicated morphological evolution process of a planar chemical dissolution-front in the case of the chemical dissolution system becoming supercritical. For this purpose, Eqs. (5.19)–(5.21) are solved using a combination of the finite element and finite difference methods. This means that the finite element method is used to discretize the geometrical shape of the problem domain, while the finite difference method is used to discretize the dimensionless time. Since the governing equations of the problem (described by Eqs. (5.19)–(5.21)) when solute dispersion is considered as the second-order tensor are similar to those when the solute dispersion effect is neglected, the numerical method and algorithm used in the previous study (Zhao et al. 2008b) can be employed here to solve the current problem with an appropriate modification to the solute dispersion term. The main purpose of this section is to demonstrate how the present theoretical solutions with consideration of solute dispersion effects can be used as benchmark solutions for verifying the proposed numerical procedure which is a combination of the finite element and finite difference methods (Zhao et al. 2008a). After the proposed numerical procedure is verified, it is used to simulate the complicated morphological evolution process of a planar chemical dissolution-front in the case of the chemical dissolution system becoming supercritical.

Figure 5.4 shows the geometry and boundary conditions of the proposed benchmark problem, for which the dimensionless pore-fluid pressure gradient (i.e. $\bar{p}'_{fx} = -1$) is applied to the left boundary, implying that the corresponding Zhao number of the system is equal to unity. As a result of applying this dimensionless pore-fluid pressure gradient to the left boundary, there is a horizontal throughflow from the left to the right of the computational model. Since the computational domain of the benchmark problem is of finite size, a time-dependent-dimensionless-concentration boundary condition (i.e. $\bar{C}(\tau) = \exp[(\bar{p}'_{fx}\bar{v}_{front}\tau)\Big/(1 - \bar{\alpha}_L\bar{p}'_{fx})])$ needs to be applied to the left boundary, so that numerical solutions can be compared with analytical solutions expressed by Eqs. (5.45) and (5.46). The dimensionless height and width of the computational model are 5 and 15, respectively. Except for the left boundary, the initial porosity of the porous medium is 0.1, while the initial dimensionless concentration is unity within the computational domain. The final porosity with a value of 0.2 is applied to the left boundary of the computational domain. As the value of the dimensionless pore-fluid pressure can be chosen arbitrarily, it is assumed to be 100 (i.e. $\bar{p}(\bar{L}_x, \tau) = 100$) at the right boundary of the computational domain. To consider whether or not a small perturbation affects the morphology of a planar chemical dissolution-front during its propagation in the fluid-saturated porous medium, the initial porosity field is randomly perturbed by a small amount of 1 % of the originally-input initial porosity (i.e. $\phi_0 = 0.1$) before running the computational model. The permeability of the porous medium is calculated using the Carman-Kozeny formula, which has the power of 3 in the power law. The diffusivity of the solute is calculated using the power law, which has the power of 2. Both the top and the bottom boundaries are assumed to be impermeable for the solute and pore-fluid. The mineral dissolution ratio is assumed to be 0.001, while the dimensionless time-step length is set to be 0.001 in the computation. For the purpose of considering the effect of solute dispersion, the dimensionless longitudinal dispersivity is 0.2, while the dimensionless transversal dispersivity is 0.02 in the computational model. Using the above parameters, the critical Zhao number of the system is approximately equal to 1.68, as obtained previously in the theoretical analysis. Since the Zhao number of the system is smaller than its critical value, the chemical dissolution system considered here is in a subcritical state, implying that a planar chemical dissolution-front remains planar during its propagation within the chemical dissolution system. The theoretical value for the dimensionless propagation speed of the chemical dissolution front is equal to 10, which is determined using Eq. (5.48). To appropriately simulate the propagation of the chemical dissolution front, the whole computational domain is simulated by 29601 four-node square elements of 30000 nodal points in total.

Figures 5.5 and 5.6 show the comparison of the numerical results with analytical solutions for the porosity and dimensionless concentration in the computational domain at three different time instants, namely $\tau = 0.45$, 0.75 and 10.5, respectively. In these two figures, thick lines are used to represent the numerical results, while thin lines are used to represent the analytical solutions. Although a small perturbation is applied to the initial porosity field before running the computational model, the planar chemical dissolution-front remains planar throughout the whole numerical simulation process because the Zhao number of the chemical dissolution system under consideration is subcritical. Obviously, the numerical results for both the porosity and the dimensionless concentration agree well with

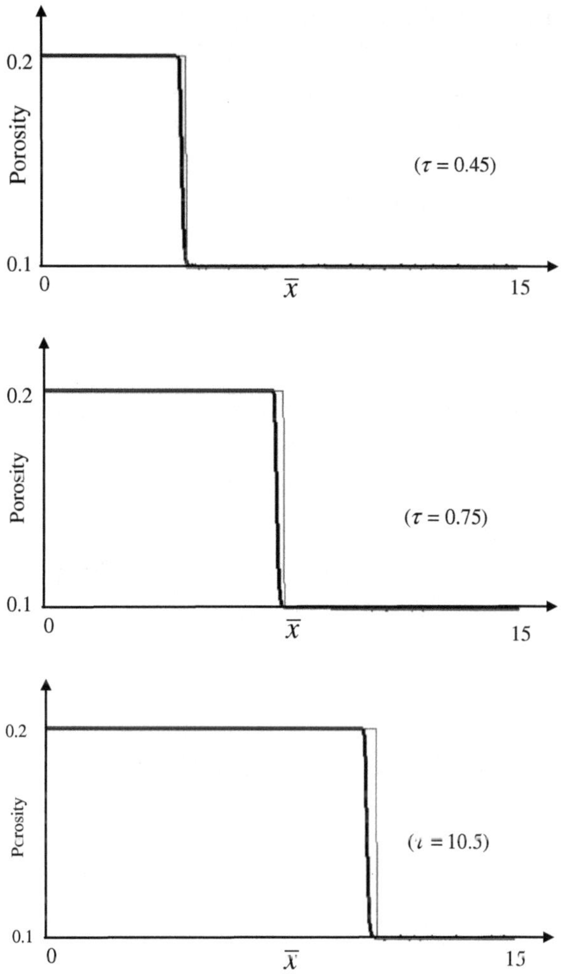

Fig. 5.5 Comparison of the numerical solutions with analytical results at different time instants (Porosity): *thick lines* represent the numerical solutions, while *thin lines* represent the analytical results

the analytical solutions, especially at the early stages of the computational simulation. This fact indicates that the proposed numerical procedure, which comprises a combination of the finite element and finite difference methods, is capable of simulating the planar dissolution-front propagation within the fluid-saturated porous medium. Since a finite value for the mineral dissolution ratio is used in the computational model, there is a discrepancy between the computational model used to produce the numerical results and the theoretical model used to derive the analytical solutions. As mentioned previously, the analytical solutions are derived under the assumption that the mineral dissolution ratio is an infinitesimal quantity, so that the chemical dissolution problem in the fluid-saturated porous medium can be treated as a steady-state boundary value problem, from the mathematical point of view. Due to this discrepancy, both the sharpness and the propagation speed of

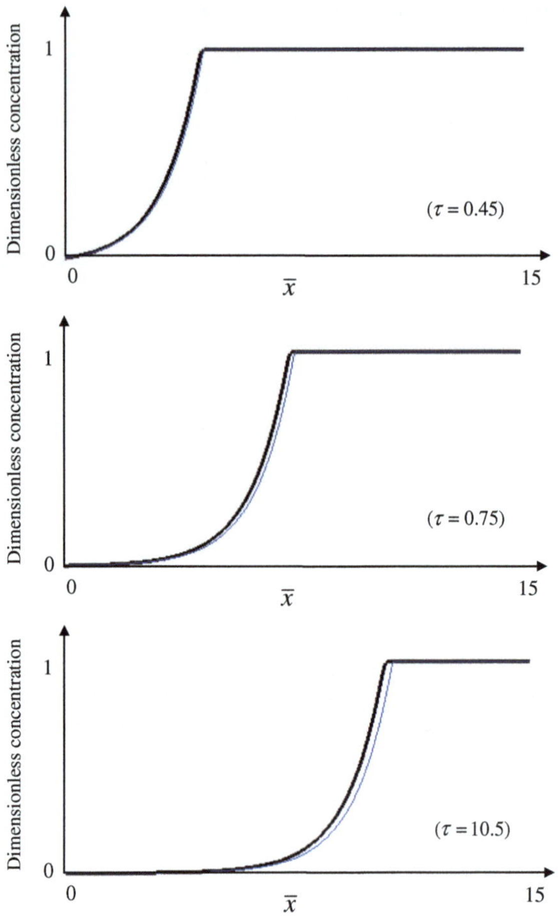

Fig. 5.6 Comparison of the numerical solutions with analytical results at different time instants (Dimensionless concentration): *thick lines* represent the numerical solutions, while *thin lines* represent the analytical results

the chemical dissolution front obtained from the computational model differ slightly from those obtained from the theoretical analysis. To reduce this discrepancy, the value of the mineral dissolution ratio should be used as small as possible in the computational model.

For the purpose of investigating how a planar chemical dissolution-front can evolve into different morphologies when the chemical dissolution system becomes supercritical, the proposed numerical procedure is used to simulate the morphological evolution of the planar chemical dissolution-front in the fluid-saturated porous medium with a supercritical Zhao number. For this purpose, a dimensionless pore-fluid pressure gradient (i.e. $\bar{p}'_{fx} = -10$), which corresponds to a Zhao number of 10, is applied to the left boundary of the computational domain. In this case, the dimensionless propagation speed of the corresponding planar chemical dissolution-front is equal to 100, so that the chemical dissolution front propagates faster than it does when the Zhao number is subcritical, just as what is considered in the previous benchmark problem. Despite this change, the mineral dissolution

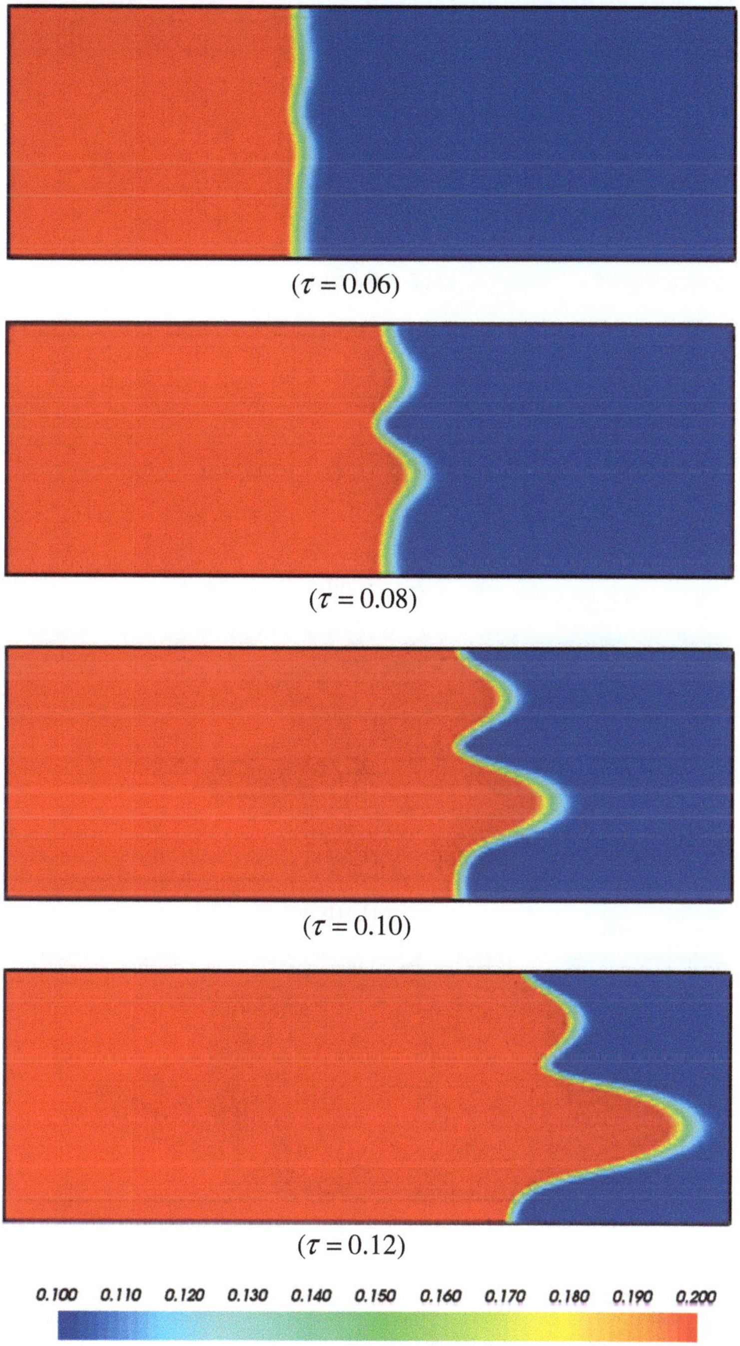

Fig. 5.7 Porosity distributions due to the morphological evolution of the chemical dissolution front in the fluid-saturated porous medium

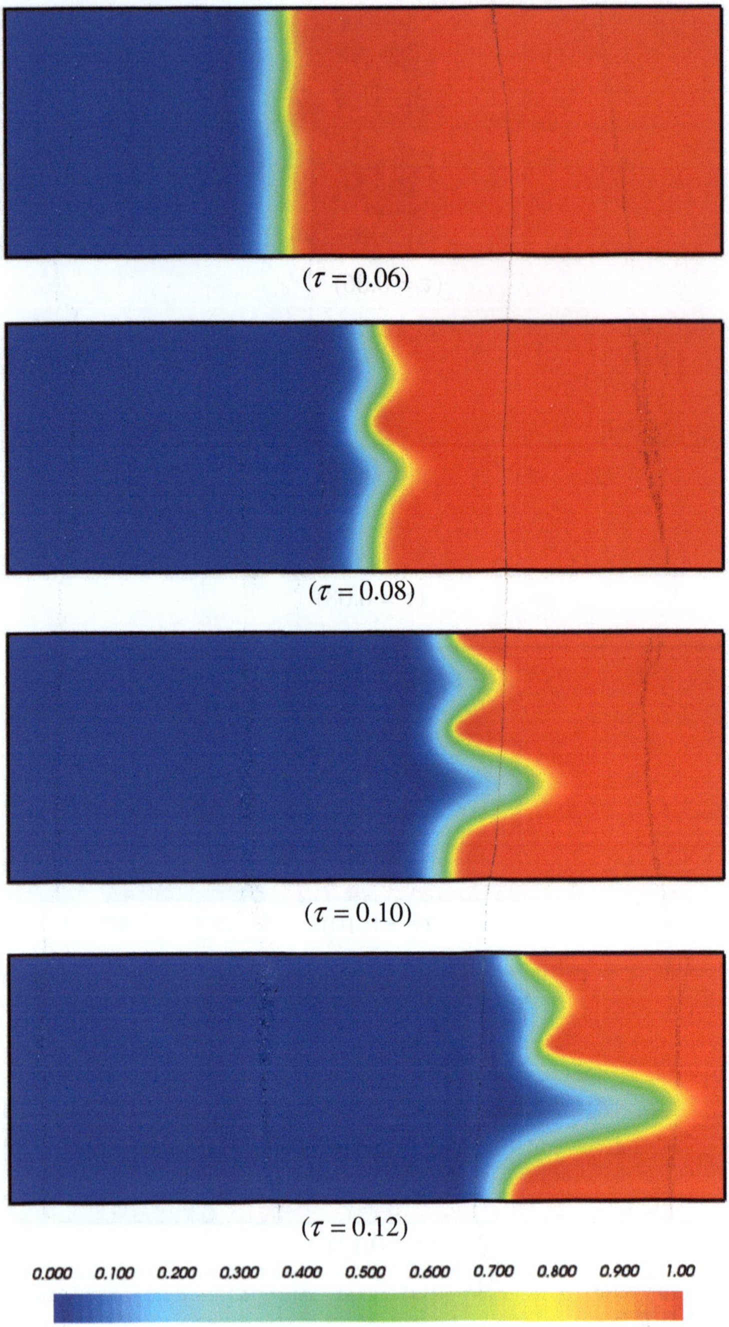

Fig. 5.8 Dimensionless concentration distributions due to the morphological evolution of the chemical dissolution front in the fluid-saturated porous medium

ratio remains 0.001, while the dimensionless time-step length is assumed to be 0.0002 in the computational model. The values of other parameters are exactly the same as those used in the above benchmark problem. Since the Zhao number of the chemical dissolution system is greater than its corresponding critical value (i.e. $Zh_{critical} = 1.68$ for this application example), the chemical dissolution system considered here is supercritical so that a planar chemical dissolution-front should evolve into a complicated morphology during its propagation in the computational model. The same small perturbation as used in the benchmark problem is applied to the initial porosity field of the computational domain before running the corresponding computational model.

Figure 5.7 shows porosity distributions arising from the morphological evolution of the chemical dissolution front in the fluid-saturated porous medium, while Fig. 5.8 shows dimensionless concentration distributions arising from the morphological evolution of the chemical dissolution front within the computational domain. It is noted that since the small perturbation grows with time for a supercritical system, the initial planar chemical dissolution-front evolves gradually into an irregular one after the dimensionless time is greater than 0.07. With the increase of the dimensionless time, the amplitude of the resulting irregular chemical dissolution-front increases significantly, indicating that the chemical dissolution front is morphologically unstable during its propagation within the computational domain. Although both the porosity and the dimensionless concentration have similar propagation front, the distribution of their maximum values along the chemical dissolution front is different. The peak value of the porosity is in good correspondence with the trough value of the dimensionless concentration as a result of the chemical dissolution of the dissolvable mineral in the solid matrix of the porous medium.

In summary, the present theoretical results clearly demonstrated that: (1) the propagation speed of a planar chemical dissolution-front in the case of considering solute dispersion effects is exactly the same as that when solute dispersion effects are neglected. This indicates that solute dispersion does not affect the propagation speed of the planar chemical dissolution-front in a fluid-saturated porous medium. (2) The consideration of solute dispersion can cause a significant increase in the critical Zhao number, which is used to judge whether or not a planar chemical dissolution-front may become unstable in the fluid-saturated porous medium. This means that the consideration of solute dispersion can stabilize a planar chemical dissolution-front because an increase in the critical Zhao number reduces the likelihood of the planar chemical dissolution-front instability in a fluid-saturated porous medium. (3) For both a given solute dispersion ratio and a given longitudinal dispersivity, an increase in the final porosity value destabilizes the chemical dissolution front so that it becomes easier for a planar chemical dissolution-front to evolve into different morphologies. In addition, the present results can be used as benchmark solutions for verifying numerical methods employed to simulate the detailed morphological evolution processes of chemical dissolution fronts in two-dimensional fluid-saturated porous media.

References

Alt-Epping P, Smith L (2001) Computing geochemical mass transfer and water/rock ratios in submarine hydrothermal systems: implications for estimating the vigour of convection. Geofluids 1:163–181

Bear J (1972) Dynamics of fluids in porous media. American Elsevier Publishing Company, New York

Chadam J, Hoff D, Merino E, Ortoleva P, Sen A (1986) Reactive infiltration instabilities. IMA J Appl Math 36:207–221

Chadam J, Ortoleva P, Sen A (1988) A weekly nonlinear stability analysis of the reactive infiltration interface. IMA J Appl Math 48:1362–1378

Chen JS, Liu CW (2002) Numerical simulation of the evolution of aquifer porosity and species concentrations during reactive transport. Comput Geosci 28:485–499

Chen JS, Liu CW, Lai GX, Ni CF (2009) Effects of mechanical dispersion on the morphological evolution of a chemical dissolution front in a fluid-saturated porous medium. J Hydrol 373:96–102

Holzbecher EO (1998) Modeling density-driven flow in porous media. Springer, Berlin

Imhoff PT, Miller CT (1996) Dissolution fingering during the solubilization of nonaqueous phase liquids in saturated porous media: 1. Model predictions. Water Resour Res 32:1919–1928

Imhoff PT, Farthing MW, Miller CT (2003) Modeling NAPL dissolution fingering with upscaled mass transfer rate coefficients. Adv Water Resour 26:1097–1111

Lewis RW, Schrefler BA (1998) The finite element method in the static and dynamic deformation and consolidation of porous media. Wiley, New York

Nield DA, Bejan A (1992) Convection in porous media. Springer, New York

Ormond A, Ortoleva P (2000) Numerical modeling of reaction-induced cavities in a porous rock. J Geophys Res 105:16737–16747

Ortoleva P, Chadam J, Merino E, Sen A (1987) Geochemical self-organization II: The reactive-infiltration instability. Am J Sci 287:1008–1040

Raffensperger JP, Garven G (1995) The formation of unconformity-type uranium ore deposits: Coupled hydrochemical modelling. Am J Sci 295:639–696

Renard F, Gratier JP, Ortoleva P, Brosse E, Bazin B (1998) Self-organization during reactive fluid flow in a porous medium. Geophys Res Lett 25:385–388

Scheidegger AE (1961) General theory of dispersion in porous media. J Geophys Res 66:3273–3278

Scheidegger AE (1974) The physics of flow through porous media. University of Toronto Press, Toronto

Steefel CI, Lasaga AC (1990) Evolution of dissolution patterns: permeability change due to coupled flow and reaction. In: Melchior DC, Basset RL (eds) Chemical modeling in Aqueous systems II. American chemistry society symposium series, vol 416. pp 213–225

Steefel CI, Lasaga AC (1994) A coupled model for transport of multiple chemical species and kinetic precipitation/dissolution reactions with application to reactive flow in single phase hydrothermal systems. Am J Sci 294:529–592

Yeh GT, Tripathi VS (1991) A model for simulating transport of reactive multispecies components: Model development and demonstration. Water Resour Res 27:3075–3094

Zhao C, Xu TP, Valliappan S (1994) Numerical modelling of mass transport problems in porous media: A review. Comput Struct 53:849–860

Zhao C, Hobbs BE, Mühlhaus HB (1998) Finite element modelling of temperature gradient driven rock alteration and mineralization in porous rock masses. Comput Methods Appl Mech Eng 165:175–187

Zhao C, Hobbs BE, Mühlhaus HB, Ord A (2001) Finite element modelling of rock alteration and metamorphic process in hydrothermal systems. Commun Numer Methods Eng 17:833–843

Zhao C, Hobbs BE, Mühlhaus HB, Ord A, Lin G (2003) Finite element modeling of three-dimensional steady-state convection and lead/zinc mineralization in fluid-saturated rocks. J Comput Methods Sci Eng 3:73–89

Zhao C, Hobbs BE, Ord A, Peng S, Mühlhaus HB, Liu L (2005) Numerical modeling of chemical effects of magma solidification problems in porous rocks. Int J Numer Meth Eng 64:709–728

Zhao C, Hobbs BE, Hornby P, Ord A, Peng S (2006) Numerical modelling of fluids mixing, heat transfer and non-equilibrium redox chemical reactions in fluid-saturated porous rocks. Int J Numer Meth Eng 66:1061–1078

Zhao C, Hobbs BE, Hornby P, Ord A, Peng S, Liu L (2008a) Theoretical and numerical analyses of chemical-dissolution front instability in fluid-saturated porous rocks. Int J Numer Anal Meth Geomech 32:1107–1130

Zhao C, Hobbs BE, Ord A, Hornby P, Peng S (2008b) Effect of reactive surface areas associated with different particle shapes on chemical-dissolution front instability in fluid-saturated porous rocks. Transp Porous Media 73:75–94

Zhao C, Hobbs BE, Ord A (2009) Fundamentals of computational geoscience: numerical methods and algorithms. Springer, Berlin

Zhao C, Hobbs BE, Ord A, Peng S (2010) Effects of mineral dissolution ratios on chemical-dissolution front instability in fluid-saturated porous media. Transp Porous Media 82:317–335

Zienkiewicz OC (1977) The finite element method. McGraw-Hill, London

Chapter 6
Effects of Medium Permeability Anisotropy on Chemical Dissolution-Front Instability in Fluid-Saturated Porous Media

In the previous studies, mathematical analyses were conducted to establish a theoretical criterion, which is used to judge whether or not a planar chemical dissolution-front can become unstable during its propagation in the fluid-saturated porous medium (Chadam et al. 1986, 1988; Zhao et al. 2008a, 2009). The established theoretical criterion consists of a comparison between the Zhao number and the critical Zhao number of the chemical dissolution system under consideration (Zhao et al. 2008a, 2009). The Zhao number is a dimensionless number that can be used to represent the three major controlling mechanisms, namely pore-fluid flow, mass transport and chemical reactions, simultaneously taking place in a chemical dissolution system, while the critical Zhao number is a special value of the Zhao number, at which the chemical dissolution system becomes neutrally unstable. Thus, if the Zhao number is smaller than the corresponding critical Zhao number, then the chemical dissolution system is in a stable state. If the Zhao number is greater than the corresponding critical Zhao number, then the chemical dissolution system is in an unstable state. Due to this key role played by the critical Zhao number, it is natural to turn the study of different factors, including mineral dissolution ratios, solute (or mechanical) dispersion, medium/pore-fluid compressibility and reactive surface areas of particles, on the chemical dissolution-front instability into the study of these factors on the critical Zhao number of the chemical dissolution system (Zhao et al. 2008b, 2010a, b, 2012a, b). Generally, if a particular factor can result in an increase in the critical Zhao number of a chemical dissolution system, then this factor can stabilize the chemical dissolution system so that it reduces the likelihood of the chemical dissolution-front instability in the chemical dissolution system. On the contrary, if a particular factor can result in a decrease in the critical Zhao number of a chemical dissolution system, then this factor can destabilize the chemical dissolution system so that it increases the likelihood of the chemical dissolution-front instability in the chemical dissolution system.

Despite of the fact that mathematical analyses were conducted to investigate how several factors, such as mineral dissolution ratios, solute (or mechanical) dispersion,

C. Zhao, *Physical and Chemical Dissolution Front Instability in Porous Media*, Lecture Notes in Earth System Sciences,
DOI: 10.1007/978-3-319-08461-9_6

medium/pore-fluid compressibility and reactive surface areas of particles, can affect the instability of chemical dissolution fronts in fluid-saturated porous media, it is very difficult, if not impossible, to mathematically describe the detailed morphological evolution processes of chemical dissolution fronts in a supercritical chemical dissolution system, in which the Zhao number of the chemical dissolution system is greater than the corresponding critical Zhao number. As a complementary way, computational simulation methods, such as a combination of the finite element method and finite difference method (Zienkiewicz 1977; Lewis and Schrefler 1998; Zhao et al. 2004, 2009), are commonly used to simulate the detailed morphological evolution processes of chemical dissolution fronts in the supercritical chemical dissolution system (Ortoleva et al. 1987; Renard et al. 1998; Ormond and Ortoleva 2000; Chen and Liu 2002, 2004; Chen et al. 2009; Zhao et al. 2008a, 2009). Although considerable progress has been made in the study of chemical dissolution-front instability problems (Chadam et al. 1986, 1988; Zhao et al. 2008a, 2009, 2010b, 2012a), theoretical consideration of the effect of medium permeability anisotropy on the chemical dissolution-front instability is only available very recently (Zhao et al. 2013).

Medium permeability anisotropic phenomena may be encountered in some natural systems. For example, in sedimentary basins, the vertical permeability of the porous rock can be smaller than the horizontal one as a result of gravity compaction in the vertical direction. More importantly, from the theoretical point of view, a multiply layered porous medium, in which each layer is viewed as an isotropic material, can be approximately treated as an equivalent anisotropic porous medium. With a horizontally layered porous medium taken as an example, the horizontal permeability of the equivalent anisotropic porous medium can be determined by the pore-fluid flow equity condition, which requires that the total pore-fluid flow of the equivalent anisotropic porous medium in the horizontal direction be equal to the sum of the horizontal pore-fluid flow in every layer of the layered porous medium (Freeze and Cherry 1979). In an isothermal case, in which the dynamic viscosity of the pore-fluid can be treated as a constant, the horizontal permeability of the equivalent anisotropic porous medium can be evaluated using the following formula:

$$k_{H-equivalent} = \frac{\sum_{i=1}^{N} k_i h_i}{H}, \tag{6.1}$$

where k_i and h_i are the horizontal permeability and thickness of layer i; H is the total thickness of the layered porous medium and N is the total number of layers in the layered porous medium.

Similarly, the vertical permeability of the equivalent anisotropic porous medium can be determined by considering the pore-fluid pressure difference equity condition between the equivalent anisotropic porous medium and the layered porous medium as follows:

$$k_{V-equivalent} = \frac{H}{\sum_{i=1}^{N} \frac{h_i}{k_i}}. \tag{6.2}$$

This indicates that since the layered porous medium can be approximately treated as an anisotropic porous medium, the study of the chemical dissolution-front instability problem in an anisotropic porous medium is not only of theoretical significance, but also of certain practical application backgrounds.

6.1 Mathematical Governing Equations for Chemical Dissolution-Front Instability Problems in Fluid-Saturated Porous Media Including Medium Anisotropy Effects

From the previous studies (Chadam et al. 1986; Zhao et al. 2008a, 2009), a chemical dissolution-front instability problem can be treated as a coupled non-linear problem between porosity evolution, pore-fluid flow and reactive (single-solute) transport in the fluid-saturated (rigid) porous medium. As a result, the governing equations and related expressions for the chemical dissolution-front instability problem in a two-dimensional fluid-saturated porous medium including medium anisotropy effects can be expressed as follows:

$$\frac{\partial \phi}{\partial t} - \nabla \cdot [\boldsymbol{\psi}(\phi) \cdot \nabla p] = 0, \tag{6.3}$$

$$\frac{\partial}{\partial t}(\phi C) - \nabla \cdot [\phi \mathbf{D}(\phi) \cdot \nabla C + C\boldsymbol{\psi}(\phi) \cdot \nabla p] + \rho_s k_{Echemical} \frac{\bar{A}_p}{\bar{V}_p}(\phi_f - \phi)(C - C_{eq}) = 0, \tag{6.4}$$

$$\frac{\partial \phi}{\partial t} + k_{Echemical} \frac{\bar{A}_p}{\bar{V}_p}(\phi_f - \phi)(C - C_{eq}) = 0, \tag{6.5}$$

$$\boldsymbol{\psi}(\phi) = \begin{bmatrix} \frac{k_{xx}(\phi)}{\mu} & \frac{k_{xy}(\phi)}{\mu} \\ \frac{k_{yx}(\phi)}{\mu} & \frac{k_{yy}(\phi)}{\mu} \end{bmatrix}, \tag{6.6}$$

$$\mathbf{D}(\phi) = \begin{bmatrix} D(\phi) + \frac{\alpha_L v_x^2 + \alpha_T v_y^2}{\sqrt{v_x^2 + v_y^2}} & (\alpha_L - \alpha_T)\frac{v_x v_y}{\sqrt{v_x^2 + v_y^2}} \\ (\alpha_L - \alpha_T)\frac{v_x v_y}{\sqrt{v_x^2 + v_y^2}} & D(\phi) + \frac{\alpha_L v_y^2 + \alpha_T v_x^2}{\sqrt{v_x^2 + v_y^2}} \end{bmatrix}, \tag{6.7}$$

where p and C are the pore-fluid pressure and the concentration of the solute (i.e. the dissolvable mineral here); C_{eq} is the equilibrium concentration of the solute with units of moles per unit pore-fluid volume; μ is the dynamic viscosity of the pore-fluid; ϕ is the porosity of the porous medium; ϕ_f is the final (i.e. maximum) porosity of the porous medium after the completion of dissolvable mineral dissolution; $\mathbf{D}(\phi)$ is the general dispersion tensor of the solute (Scheidegger 1961;

Bear 1972; Holzbecher 1998); α_T and α_L are the transversal and longitudinal dispersivities of the solute; $D(\phi)$ is the molecular diffusivity of the solute; $k_{xx}(\phi)$, $k_{xy}(\phi)$, $k_{yx}(\phi)$ and $k_{yy}(\phi)$ are the components of the permeability tensor of the porous medium; $\bar{V}_p$ is the average volume of the dissolvable grain; $\bar{A}_p$ is the average surface area of the dissolvable grain; $k_{Echemical}$ is the comprehensive rate constant of the chemical reaction (as defined in Eq. (2.15) previously); ρ_s is the molar density (i.e. moles per unit volume) of the dissolvable grains; v_x and v_y are the averaged linear velocity components of the pore-fluid flow in the x and y directions respectively.

Based on Darcy's law, the averaged linear velocity vector of the pore-fluid can be expressed as follows:

$$\vec{u}_{linear} = \begin{Bmatrix} v_x \\ v_y \end{Bmatrix} = \frac{\vec{u}}{\phi} = -\begin{bmatrix} \frac{k_{xx}(\phi)}{\phi\mu} & \frac{k_{xy}(\phi)}{\phi\mu} \\ \frac{k_{yx}(\phi)}{\phi\mu} & \frac{k_{yy}(\phi)}{\phi\mu} \end{bmatrix} \begin{Bmatrix} \frac{\partial p}{\partial x} \\ \frac{\partial p}{\partial y} \end{Bmatrix}, \tag{6.8}$$

where $\vec{u}_{linear}$ is the averaged linear velocity vector of the pore-fluid; $\vec{u}$ is the Darcy velocity vector of the pore-fluid.

Note that the ratio of the average volume to the average surface area (i.e. $\bar{V}_p/\bar{A}_p$) of the soluble grain can be used to represent the soluble grain shape, so that it is called the shape factor of the soluble grain. If the porous medium is comprised of sphere particles, then the corresponding shape factor of the soluble grain can be expressed as (Zhao et al. 2008b):

$$\frac{\bar{V}_p}{\bar{A}_p} = \left(\sqrt[3]{\frac{1}{36\pi}}\right)\sqrt[3]{\frac{(\phi_f - \phi)}{D_p}} = \alpha_{sph}\sqrt[3]{\frac{(\phi_f - \phi)}{D_p}}, \tag{6.9}$$

where $\alpha_{sph} = \sqrt[3]{1/(36\pi)}$ is the shape coefficient of the spherical grains; D_p is the density of the soluble grains, which is defined as the number of the soluble gains per unit medium volume.

Since the solute diffusivity is considered as a function of porosity in the chemical dissolution-front instability problem, a common phenomenological relation can be used for describing this function (Chadam et al. 1986) as follows:.

$$D(\phi) = D_0\phi^q \quad \left(\frac{3}{2} \le q \le \frac{5}{2}\right), \tag{6.10}$$

where D_0 is the chemical species diffusivity in pure water.

Although $q = 0$ is often used in many applications, without loss of generality, we keep q as a parameter in the following theoretical analysis, so that more general situations could be considered for the solute diffusivity in the fluid-saturated porous medium.

To consider the permeability change caused by a change in porosity, the Carman-Kozeny law can be used to express the relationship between permeability and porosity (Scheidegger 1974; Nield and Bejan 1992) as follows:

$$\begin{aligned} k_{xx}(\phi) &= \frac{k_{xx0}(1-\phi_0)^2\phi^3}{\phi_0^3(1-\phi)^2}, k_{yy}(\phi) = \frac{k_{yy0}(1-\phi_0)^2\phi^3}{\phi_0^3(1-\phi)^2}, \\ k_{xy}(\phi) &= \frac{k_{xy0}(1-\phi_0)^2\phi^3}{\phi_0^3(1-\phi)^2}, k_{yx}(\phi) = \frac{k_{yx0}(1-\phi_0)^2\phi^3}{\phi_0^3(1-\phi)^2}, \end{aligned} \tag{6.11}$$

where ϕ_0 is the initial reference porosity of the porous medium; k_{xx0}, k_{xy0}, k_{yx0} and k_{yy0} are the components of the corresponding initial permeability tensor of the porous medium, respectively.

Substituting Eq. (6.9) into Eqs. (6.4) and (6.5) yields the following equations:

$$\frac{\partial}{\partial t}(\phi C) - \nabla \cdot [\phi \mathbf{D}(\phi) \cdot \nabla C + C\boldsymbol{\psi}(\phi) \cdot \nabla p] + \rho_s k_{Echemical} \frac{\sqrt[3]{D_p}}{\alpha_{sph}} (\phi_f - \phi)^{\frac{2}{3}} (C - C_{eq}) = 0, \tag{6.12}$$

$$\frac{\partial \phi}{\partial t} + k_{Echemical} \frac{\sqrt[3]{D_p}}{\alpha_{sph}} (\phi_f - \phi)^{\frac{2}{3}} (C - C_{eq}) = 0. \tag{6.13}$$

To close the definition of the problem, boundary conditions need to be provided. If the solute is initially in an equilibrium state (i.e. at $t = 0$ and $x \neq -\infty$) within the whole system except for x approaching negative infinity and the fresh pore-fluid (i.e. $C = 0$) is injected at the location of x approaching negative infinity, the following boundary conditions can be obtained:

$$\lim_{x \to -\infty} C(x, t) = 0, \quad \lim_{x \to -\infty} \frac{\partial p(x, t)}{\partial x} = p'_{fx} \quad \text{(upstream boundary)}, \tag{6.14}$$

$$\lim_{x \to \infty} C(x, t) = C_{eq}, \quad \lim_{x \to \infty} \frac{\partial p(x, t)}{\partial x} = p'_{0x} \quad \text{(downstream boundary)}, \tag{6.15}$$

where p'_{fx} is the pore-fluid pressure gradient as x approaches negative infinity in the upstream of the pore-fluid flow; p'_{0x} is the unknown pore-fluid pressure gradient as x approaches positive infinity in the downstream of the pore-fluid flow. Since p'_{fx} drives the pore-fluid flow continuously along the positive x direction, it has a negative algebraic value (i.e. $p'_{fx} < 0$) in this study. This means that for the sake of facilitating the forthcoming mathematical deductions, we assume the porous medium to be an infinite porous medium in the theoretical analysis.

The initial condition for this theoretical problem is: $\phi(x, 0) = \phi_0$ expect at the negative infinity, where $\lim_{\bar{x} \to -\infty} \phi(x, 0) = \phi_f$. Note that ϕ_0 is the initial porosity of the porous medium.

6.2 Theoretical Analyses of Medium Permeability Anisotropy Effects

For a two-dimensional porous medium, the medium anisotropy can generally lead to a second-order permeability tensor of non-zero off-diagonal components in an arbitrary rectangular coordinate system. However, it has been demonstrated that through using the appropriate rectangular coordinate system transformation, it is possible to enable all the non-zero off-diagonal components of the second-order permeability tensor to be zero, resulting in a diagonal second-order permeability tensor. For this diagonal second-order permeability tensor, it has two principal permeability directions that are parallel to a specific rectangular coordinate system. This means that if the rectangular coordinate system is chosen in such a way that its two coordinate axes (i.e. the x and y axes) are parallel to the two principal permeability directions of a second-order permeability tensor, all the non-zero off-diagonal components of the second-order permeability tensor can be equal to zero. For this reason, Eq. (6.6) can be rewritten for an orthotropic porous medium as follows:

$$\boldsymbol{\psi}(\phi) = \begin{bmatrix} \psi_{xx}(\phi) & 0 \\ 0 & \psi_{yy}(\phi) \end{bmatrix} = \begin{bmatrix} \frac{k_{xx}(\phi)}{\mu} & 0 \\ 0 & \frac{k_{yy}(\phi)}{\mu} \end{bmatrix}. \tag{6.16}$$

To facilitate the mathematical deductions and analyses, a typical theoretical problem involving the propagation of a planar chemical dissolution-front in a full plane, which comprises a fluid-saturated orthotropic and homogeneous porous medium, is considered in this theoretical investigation. For this problem, the pore-fluid flow is only in the x direction when the planar chemical dissolution-front is in a stable state, while it remains dominant in the x direction even when the planar chemical dissolution-front is at the early stages of an unstable state. In the former case, the pore-fluid velocity is exactly equal to zero in the y direction, whereas in the latter case, the pore-fluid velocity in the y direction is relatively small to that in the x direction so that it can be neglected, from the theoretical analysis point of view (Zhao et al. 2010b). Consequently, the second-order dispersion tensor for the orthotropic porous medium can be expressed in the following form:

$$\mathbf{D}(\phi) = D(\phi)\mathbf{I} + \mathbf{D}_{disper}(\phi) = D(\phi)\mathbf{I} + \begin{bmatrix} \alpha_L|v_x| & 0 \\ 0 & \alpha_T|v_x| \end{bmatrix}, \tag{6.17}$$

where $\mathbf{I}$ is a unit second-order tensor (i.e. a 2 by 2 unit matrix); v_x is the averaged linear velocity of the pore-fluid in the x direction; $\mathbf{D}_{disper}(\phi)$ is the second-order tensor due to pure dispersion only.

Note that physically, Eqs. (6.16) and (6.17) can be used to represent the behaviour of an orthotropic porous material, such as a layered porous medium. The layered sedimentary rock (e.g. shale) is a typical example of the orthotropic porous material.

Considering Eqs. (6.8) and (6.17) simultaneously yields the following expression:

$$\mathbf{D}_{disper}(\phi) = \begin{bmatrix} \frac{\alpha_L k_{xx}(\phi)}{\phi\mu}\left|\frac{\partial p}{\partial x}\right| & 0 \\ 0 & \frac{\alpha_T k_{xx}(\phi)}{\phi\mu}\left|\frac{\partial p}{\partial x}\right| \end{bmatrix}. \tag{6.18}$$

6.2.1 Derivation of Dimensionless Governing Equations

Substituting Eq. (6.17) into Eq. (6.12) yields the following equation:

$$\frac{\partial}{\partial t}(\phi C) - \nabla \cdot \{\phi[D(\phi)\mathbf{I} + \mathbf{D}_{disper}(\phi)] \cdot \nabla C + C\psi(\phi) \cdot \nabla p\} + \rho_s k_{Echemical} \frac{\sqrt[3]{D_p}}{\alpha_{sph}} (\phi_f - \phi)^{\frac{2}{3}} (C - C_{eq}) = 0. \tag{6.19}$$

To translate the dimensional governing equations of the problem (i.e. Eqs. (6.3), (6.13) and (6.19)) into dimensionless ones, the following dimensionless quantities are used in the theoretical analysis:

$$\bar{x} = \frac{x}{L^*}, \quad \bar{y} = \frac{y}{L^*}, \quad \bar{C} = \frac{C}{C_{eq}}, \quad \bar{p} = \frac{p}{p^*}, \tag{6.20}$$

$$\bar{\alpha}_L = \frac{\alpha_L}{L^*}, \quad \bar{\alpha}_T = \frac{\alpha_T}{L^*}, \quad \tau = \frac{t}{t^*}\varepsilon, \quad \varepsilon = \frac{C_{eq}}{\rho_s} << 1, \tag{6.21}$$

where ε is the mineral dissolution ratio of the chemical dissolution system; τ is a dimensionless time that is defined to be directly proportional to the mineral dissolution ratio and therefore, is generally much less than one due to the relative slowness of the chemical reaction. For example, in typical rocks, the mineral dissolution ratio that is used to describe the relative slowness of the chemical reaction may range from 10^{-3} to 10^{-10} (Chadam et al. 1986). Other characteristic parameters used in Eqs. (6.20) and (6.21) can be expressed as follows:

$$t^* = \frac{\alpha_{sph}}{k_{Echemical} C_{eq} \sqrt[3]{D_p}}, \quad L^* = \sqrt{\phi_f D(\phi_f) t^*}, \quad p^* = \frac{\phi_f D(\phi_f)}{\psi_{xx}(\phi_f)}, \tag{6.22}$$

$$D^*(\phi) = \frac{\phi D(\phi)}{\phi_f D(\phi_f)}, \quad \psi^*(\phi) = \frac{\psi(\phi)}{\psi_{xx}(\phi_f)}. \tag{6.23}$$

Inserting Eqs. (6.20)–(6.23) into Eqs. (6.3), (6.13) and (6.19) yields the following dimensionless equations:

$$\varepsilon\frac{\partial\phi}{\partial\tau} - \nabla\cdot[\boldsymbol{\psi}^*(\phi)\cdot\nabla\bar{p}] = 0, \tag{6.24}$$

$$\varepsilon\frac{\partial}{\partial\tau}(\phi\bar{C}) - \nabla\cdot\left\{\left[D^*(\phi)\mathbf{I} + \mathbf{D}^*_{disper}(\phi)\right]\cdot\nabla\bar{C} + \bar{C}\boldsymbol{\psi}^*(\phi)\cdot\nabla\bar{p}\right\} - \frac{\partial\phi}{\partial\tau} = 0, \tag{6.25}$$

$$\varepsilon\frac{\partial\phi}{\partial\tau} + (\phi_f - \phi)^{\frac{2}{3}}(\bar{C} - 1) = 0, \tag{6.26}$$

where $\mathbf{D}^*_{disper}(\phi)$ is the dimensionless second-order tensor due to pure dispersion; $\boldsymbol{\psi}^*(\phi)$ is the dimensionless second-order tensor of the orthotropic porous medium. They can be expressed as follows:

$$\mathbf{D}^*_{disper}(\phi) = \begin{bmatrix} \bar{\alpha}_L\psi^*_{xx}(\phi)\left|\frac{\partial\bar{p}}{\partial\bar{x}}\right| & 0 \\ 0 & \bar{\alpha}_T\psi^*_{xx}(\phi)\left|\frac{\partial\bar{p}}{\partial\bar{x}}\right| \end{bmatrix}, \tag{6.27}$$

$$\boldsymbol{\psi}^*(\phi) = \begin{bmatrix} \psi^*_{xx}(\phi) & 0 \\ 0 & \psi^*_{yy}(\phi) \end{bmatrix} = \begin{bmatrix} \frac{k_{xx}(\phi)}{k_{xx}(\phi_f)} & 0 \\ 0 & \frac{k_{yy}(\phi)}{k_{xx}(\phi_f)} \end{bmatrix}. \tag{6.28}$$

Using the above dimensionless quantities (as defined in Eqs. (6.20) and (6.21)), the boundary conditions for this theoretical problem can be expressed in the following dimensionless form:

$$\lim_{\bar{x}\to\infty}\bar{C}(\bar{x},\tau) = 1, \quad \lim_{\bar{x}\to\infty}\frac{\partial\bar{p}(\bar{x},\tau)}{\partial\bar{x}} = \bar{p}'_{0x} \quad \text{(downstream boundary)}, \tag{6.29}$$

$$\lim_{\bar{x}\to-\infty}\bar{C}(\bar{x},\tau) = 0, \quad \lim_{\bar{x}\to-\infty}\frac{\partial\bar{p}(\bar{x},\tau)}{\partial\bar{x}} = \bar{p}'_{fx} \quad \text{(upstream boundary)}. \tag{6.30}$$

In this case, the initial condition for this theoretical problem is: $\phi(\bar{x}, 0) = \phi_0$ expect at the negative infinity, where $\lim_{\bar{x}\to-\infty}\phi(\bar{x}, 0) = \phi_f$.

6.2.2 Derivation of Base Solutions for the Dimensionless Governing Equations of the Theoretical Problem Including the Consideration of Medium Orthotropic Effects (in the Case of $\varepsilon \to 0$)

In the limit case of the mineral dissolution ratio (i.e. ε) approaching zero, Eqs. (6.24)–(6.26) degenerate to the following governing equations for the chemical dissolution system:

$$\nabla \cdot [\psi^*(\phi) \cdot \nabla \bar{p}] = 0, \tag{6.31}$$

$$\nabla \cdot \left\{ \left[D^*(\phi)\mathbf{I} + \mathbf{D}^*_{disper}(\phi) \right] \cdot \nabla \bar{C} + \bar{C}\psi^*(\phi) \cdot \nabla \bar{p} \right\} - \frac{\partial \phi}{\partial \tau} = 0, \tag{6.32}$$

$$(\phi_f - \phi)^{\frac{2}{3}}(\bar{C} - 1) = 0. \tag{6.33}$$

From the mathematical point of view, Eq. (6.33) requires that $\bar{C} = 1$ in the undissolved region and $\phi = \phi_f$ in the completely dissolved region of the dissolvable mineral. This means that a planar chemical dissolution-front divides the problem domain into two regions: an upstream region and a downstream region, relative to the propagation location of the planar chemical dissolution-front. Across the planar chemical dissolution-front, the porosity undergoes a step jump from its initial value into its final value. As a result, the corresponding dimensionless governing equations of the problem in both the downstream region and the upstream region can be expressed as follows:

$$\bar{C} = 1, \quad \nabla \cdot [\psi^*(\phi) \cdot \nabla \bar{p}] = 0, \quad \phi = \phi_0 \quad \text{(in the downstream region)}, \tag{6.34}$$

$$\nabla \cdot \left\{ \left[\left(\mathbf{I} - \frac{\partial \bar{p}}{\partial \bar{x}} \bar{\mathbf{D}}^*_{disper} \right) \cdot \nabla \bar{C} \right] + \bar{C}\psi^*(\phi) \cdot \nabla \bar{p} \right\} = 0, \quad \nabla \cdot [\psi^*(\phi) \cdot \nabla \bar{p}] = 0, \quad \phi = \phi_f \quad \text{(in the upstream region)}, \tag{6.35}$$

where

$$\bar{\mathbf{D}}^*_{disper} = \begin{bmatrix} \bar{\alpha}_L & 0 \\ 0 & \bar{\alpha}_T \end{bmatrix}, \quad \psi^*(\phi) = \begin{bmatrix} 1 & 0 \\ 0 & \lambda \end{bmatrix} = \begin{bmatrix} 1 & 0 \\ 0 & \frac{k_{yy}(\phi_f)}{k_{xx}(\phi_f)} \end{bmatrix}. \tag{6.36}$$

It needs to be pointed out that since $\partial \bar{p}/\partial \bar{x} < 0$ for the theoretical problem considered in this study, the expression of $|\partial \bar{p}/\partial \bar{x}| = -\partial \bar{p}/\partial \bar{x}$ is used to derive Eq. (6.35). In addition, the expression of $\lambda = k_{yy}(\phi_0)/k_{xx}(\phi_0) = k_{yy}(\phi_f)/k_{xx}(\phi_f)$ is used in the process of deriving Eq. (6.34).

If the planar chemical dissolution-front is denoted by $S(\bar{x}, \tau) = 0$, then the dimensionless pore-fluid pressure, dimensionless solute concentration and the mass fluxes of both the solute and the pore-fluid should be continuous on $S(\bar{x}, \tau) = 0$. This leads to the following interface conditions for this planar chemical dissolution-front propagation problem (Chadam et al. 1986; Zhao et al. 2008a, 2009):

$$\lim_{S \to 0^-} \bar{C} = \lim_{S \to 0^+} \bar{C}, \quad \lim_{S \to 0^-} p = \lim_{S \to 0^+} p, \tag{6.37}$$

$$\lim_{S\to 0^-}(1-\bar{\alpha}_L\bar{p}'_{fx})\frac{\partial \bar{C}}{\partial \bar{n}}=\bar{v}_{front}(\phi_f-\phi_0),\quad \lim_{S\to 0^-}\frac{\partial \bar{p}}{\partial \bar{n}}=\frac{\psi_{xx}(\phi_0)}{\psi_{xx}(\phi_f)}\lim_{S\to 0^+}\frac{\partial \bar{p}}{\partial \bar{n}}, \tag{6.38}$$

where $\bar{n}=n/L^*$; n is the normal vector of the propagating planar chemical dissolution-front; $\bar{v}_{front}$ is the dimensionless propagation speed of the planar chemical dissolution-front.

When the planar chemical dissolution-front is in a stable state, it remains planar shape during propagation in the fluid-saturated porous medium, indicating that the pore-fluid flow is independent of the $\bar{y}$ coordinate in the theoretical problem under consideration (Zhao et al. 2008a, 2009). In this case, the corresponding base solutions for this theoretical problem can be derived from Eqs. (6.34) and (6.35) with the related boundary and interface conditions (i.e. Eqs. (6.29), (6.30), (6.37) and (6.38)) as follows:

$$\bar{C}(\xi)=1,\quad \bar{p}(\xi)=\bar{p}'_{0x}\xi+\bar{p}_{C1},\quad \phi=\phi_0\quad \text{(in the downstream region)}, \tag{6.39}$$

$$\bar{C}(\xi)=\exp\left(-\frac{\bar{p}'_{fx}}{1-\bar{\alpha}_L\bar{p}'_{fx}}\xi\right),\quad \bar{p}(\xi)=\bar{p}'_{fx}\xi+\bar{p}_{C2},\quad \phi=\phi_f \quad \text{(in the upstream region)}, \tag{6.40}$$

where $\bar{p}_{C1}$ and $\bar{p}_{C2}$ are two constants to be determined. Other parameters are defined below (Zhao et al. 2008a, b):

$$\xi=\bar{x}-\bar{v}_{front}\tau,\quad \bar{p}'_{0x}=\frac{\psi_{xx}(\phi_f)}{\psi_{xx}(\phi_0)}\bar{p}'_{fx}. \tag{6.41}$$

Considering Eqs. (6.38) and (6.40) yields the following expression for the dimensionless propagation speed of the planar chemical dissolution-front:

$$\bar{v}_{front}=-\frac{\bar{p}'_{fx}}{\phi_f-\phi_0}. \tag{6.42}$$

Equation (6.42) clearly indicates that, compared with the previous studies without including the medium permeability anisotropy effect (Zhao et al. 2008a, b), the consideration of medium permeability anisotropy does not affect the dimensionless propagation speed of a planar chemical dissolution-front for this theoretical problem.

6.2.3 Derivation of the Critical Condition for Unstable Chemical Dissolution-Fronts in Two-Dimensional Fluid-Saturated Porous Media Including Medium Orthotropic Effects (in the Case of $\varepsilon \to 0$)

In the limit case of the mineral dissolution ratio (i.e. ε) approaching zero, the linear stability analysis can be conducted to theoretically determine the critical condition, which is used to judge whether a planar chemical dissolution-front in the fluid-saturated porous medium can become unstable (Zhao et al. 2008a, 2009). Toward this end, a small time-dependent perturbation is added on the planar chemical dissolution-front, so that the total solutions of the system are equal to the sum of the base and perturbed solutions as follows:

$$S(\xi, \bar{y}, \tau) = \xi - \delta \exp(\bar{\omega}\tau)\cos(\overline{my}), \tag{6.43}$$

$$\bar{p}_{total}(\xi, \bar{y}, \tau) = \bar{p}(\xi, \tau) + \delta\hat{p}(\xi)\exp(\bar{\omega}\tau)\cos(\overline{my}), \tag{6.44}$$

$$\bar{C}_{total}(\xi, \bar{y}, \tau) = \bar{C}(\xi, \tau) + \delta\hat{C}(\xi)\exp(\bar{\omega}\tau)\cos(\overline{my}), \tag{6.45}$$

where $\bar{\omega}$ is the dimensionless growth rate of the perturbation; $\bar{m}$ is the dimensionless wavenumber of the perturbation; δ is the amplitude of the perturbation and $\delta << 1$ by the definition of the conventional linear stability analysis.

Since $S(\xi, \bar{y}, \tau)$ is a function of coordinates ξ and $\bar{y}$, the following derivatives exist mathematically (Chadam et al. 1986; Zhao et al. 2008a, 2009):

$$\left(\frac{\partial}{\partial \xi}\right)_{\xi} = \frac{\partial S}{\partial \xi}\frac{\partial}{\partial S} = \left(\frac{\partial}{\partial \xi}\right)_{S}, \quad \left(\frac{\partial}{\partial \bar{y}}\right)_{\xi} = \frac{\partial S}{\partial \bar{y}}\frac{\partial}{\partial S} + \left(\frac{\partial}{\partial \bar{y}}\right)_{S} = \frac{\partial S}{\partial \bar{y}}\left(\frac{\partial}{\partial \xi}\right)_{S} + \left(\frac{\partial}{\partial \bar{y}}\right)_{S}, \tag{6.46}$$

$$\left(\frac{\partial^2}{\partial \xi^2}\right)_{\xi} = \left(\frac{\partial^2}{\partial \xi^2}\right)_{S}, \quad \left(\frac{\partial^2}{\partial \bar{y}^2}\right)_{\xi} = \frac{\partial^2 S}{\partial \bar{y}^2}\frac{\partial}{\partial \xi} + \left(\frac{\partial S}{\partial \bar{y}}\right)^2\frac{\partial^2}{\partial \xi^2} + 2\frac{\partial S}{\partial \bar{y}}\frac{\partial^2}{\partial \xi \partial \bar{y}} + \left(\frac{\partial^2}{\partial \bar{y}^2}\right)_{S}. \tag{6.47}$$

Based on the fact that the total solutions must satisfy the governing equations, the first-order perturbation equations of the chemical dissolution system under consideration can be derived and expressed as

$$\hat{C} = 0, \quad \frac{\partial^2 \hat{p}}{\partial \xi^2} - \lambda\bar{m}^2\hat{p} + \lambda\bar{m}^2\bar{p}'_{0x} = 0 \quad \text{(in the downstream region)}, \tag{6.48}$$

$$\left(1-\bar{\alpha}_L\bar{p}'_{fx}\right)\frac{\partial^2\hat{C}}{\partial\xi^2}+\bar{p}'_{fx}\frac{\partial\hat{C}}{\partial\xi}-\left(1-\bar{\alpha}_T\bar{p}'_{fx}\right)\bar{m}^2\hat{C}-\frac{\left(1-\bar{\alpha}_T\bar{p}'_{fx}\right)\bar{m}^2\bar{p}'_{fx}}{1-\bar{\alpha}_L\bar{p}'_{fx}}\exp\left(-\frac{\bar{p}'_{fx}}{1-\bar{\alpha}_L\bar{p}'_{fx}}\xi\right)$$
$$-\frac{\bar{p}'_{fx}}{1-\bar{\alpha}_L\bar{p}'_{fx}}\left(1+\frac{\bar{\alpha}_L\bar{p}'_{fx}}{1-\bar{\alpha}_L\bar{p}'_{fx}}\right)\exp\left(-\frac{\bar{p}'_{fx}}{1-\bar{\alpha}_L\bar{p}'_{fx}}\xi\right)\frac{\partial\hat{p}}{\partial\xi}+\frac{\bar{\alpha}_L\bar{p}'_{fx}}{1-\bar{\alpha}_L\bar{p}'_{fx}}\exp\left(-\frac{\bar{p}'_{fx}}{1-\bar{\alpha}_L\bar{p}'_{fx}}\xi\right)\frac{\partial^2\hat{p}}{\partial\xi^2}=0$$

$$\frac{\partial^2\hat{p}}{\partial\xi^2}-\lambda\bar{m}^2\hat{p}+\lambda\bar{m}^2\bar{p}'_{fx}=0 \quad \text{(in the upstream region)} \tag{6.49}$$

In the process of deriving the first-order perturbation equation for the chemical dissolution-front instability problem, the following expression has been considered:

$$\begin{aligned}
&\nabla\cdot\left[\left(\mathbf{I}-\frac{\partial\bar{p}_{total}}{\partial\bar{x}}\bar{\mathbf{D}}^*_{disper}\right)\cdot\nabla\bar{C}_{total}\right]\\
&=\frac{\partial}{\partial\bar{x}}\left[\left(1-\bar{\alpha}_L\frac{\partial\bar{p}_{total}}{\partial\bar{x}}\right)\frac{\partial\bar{C}_{total}}{\partial\bar{x}}\right]+\frac{\partial}{\partial\bar{y}}\left[\left(1-\bar{\alpha}_T\frac{\partial\bar{p}_{total}}{\partial\bar{x}}\right)\frac{\partial\bar{C}_{total}}{\partial\bar{y}}\right]\\
&=\left[\left(1-\bar{\alpha}_L\frac{\partial\bar{p}_{total}}{\partial\bar{x}}\right)\frac{\partial^2\bar{C}_{total}}{\partial\bar{x}^2}-\bar{\alpha}_L\frac{\partial^2\bar{p}_{total}}{\partial\bar{x}^2}\frac{\partial\bar{C}_{total}}{\partial\bar{x}}\right]\\
&+\left[\left(1-\bar{\alpha}_T\frac{\partial\bar{p}_{total}}{\partial\bar{x}}\right)\frac{\partial^2\bar{C}_{total}}{\partial\bar{y}^2}-\bar{\alpha}_T\frac{\partial^2\bar{p}_{total}}{\partial\bar{x}\partial\bar{y}}\frac{\partial\bar{C}_{total}}{\partial\bar{y}}\right].
\end{aligned}\tag{6.50}$$

From the first expression of Eq. (6.41), the following expression is used to derive the first-order perturbation equations:

$$\frac{\partial}{\partial\xi}=\frac{\partial}{\partial\bar{x}}.\tag{6.51}$$

Substituting Eqs. (6.44) and (6.45) into the first term on the right-hand side of Eq. (6.50) and considering $\partial^2\bar{p}/(\partial\bar{x}^2)=0$ simultaneously yield the following equation:

$$\begin{aligned}
&\left(1-\bar{\alpha}_L\frac{\partial\bar{p}_{total}}{\partial\bar{x}}\right)\frac{\partial^2\bar{C}_{total}}{\partial\bar{x}^2}-\bar{\alpha}_L\frac{\partial^2\bar{p}_{total}}{\partial\bar{x}^2}\frac{\partial\bar{C}_{total}}{\partial\bar{x}}\\
&=\left(1-\bar{\alpha}_L\frac{\partial\bar{p}}{\partial\bar{x}}\right)\frac{\partial^2\bar{C}}{\partial\bar{x}^2}+\left(1-\bar{\alpha}_L\frac{\partial\bar{p}}{\partial\bar{x}}\right)\frac{\partial^2\hat{C}}{\partial\bar{x}^2}\delta\exp(\bar{\omega}\,\tau)\cos(\overline{my})\\
&-\bar{\alpha}_L\frac{\partial\hat{p}}{\partial\bar{x}}\frac{\partial^2\bar{C}}{\partial\bar{x}^2}\delta\exp(\bar{\omega}\,\tau)\cos(\overline{my})-\bar{\alpha}_L\frac{\partial\hat{p}}{\partial\bar{x}}\frac{\partial^2\hat{C}}{\partial\bar{x}^2}[\delta\exp(\bar{\omega}\,\tau)\cos(\overline{my})]^2\\
&-\bar{\alpha}_L\frac{\partial^2\hat{p}}{\partial\bar{x}^2}\frac{\partial\bar{C}}{\partial\bar{x}}\delta\exp(\bar{\omega}\,\tau)\cos(\overline{my})-\bar{\alpha}_L\frac{\partial^2\hat{p}}{\partial\bar{x}^2}\frac{\partial\hat{C}}{\partial\bar{x}}[\delta\exp(\bar{\omega}\,\tau)\cos(\overline{my})]^2.
\end{aligned}\tag{6.52}$$

If the second-order terms due to the perturbation are neglected, then Eq. (6.52) can be rewritten in the following form:

$$\left(1-\bar{\alpha}_L\frac{\partial\bar{p}_{atotal}}{\partial\bar{x}}\right)\frac{\partial^2\bar{C}_{total}}{\partial\bar{x}^2}-\bar{\alpha}_L\frac{\partial^2\bar{p}_{total}}{\partial\bar{x}^2}\frac{\partial\bar{C}_{total}}{\partial\bar{x}}$$
$$=\left(1-\bar{\alpha}_L\frac{\partial\bar{p}}{\partial\bar{x}}\right)\frac{\partial^2\bar{C}}{\partial\bar{x}^2}+\left(1-\bar{\alpha}_L\frac{\partial\bar{p}_{ab}}{\partial\bar{x}}\right)\frac{\partial^2\hat{C}}{\partial\bar{x}^2}\delta\exp(\bar{\omega}\,\tau)\cos(\overline{my}) \tag{6.53}$$
$$-\bar{\alpha}_L\frac{\partial\hat{p}}{\partial\bar{x}}\frac{\partial^2\bar{C}}{\partial\bar{x}^2}\delta\exp(\bar{\omega}\,\tau)\cos(\bar{m}y)-\bar{\alpha}_L\frac{\partial^2\hat{p}}{\partial\bar{x}^2}\frac{\partial\bar{C}}{\partial\bar{x}}\delta\exp(\bar{\omega}\,\tau)\cos(\overline{my}).$$

Equation (6.53) clearly indicates that in the process of deriving the first-order perturbation equation for the chemical dissolution-front instability problem when dispersion is considered as the second-order tensor, the following three first-order perturbation terms need to be considered:

$$\left(1-\bar{\alpha}_L\frac{\partial\bar{p}}{\partial\bar{x}}\right)\frac{\partial^2\hat{C}}{\partial\bar{x}^2}\delta\exp(\bar{\omega}\,\tau)\cos(\overline{my}),\quad -\bar{\alpha}_L\frac{\partial\hat{p}}{\partial\bar{x}}\frac{\partial^2\bar{C}}{\partial\bar{x}^2}\delta\exp(\bar{\omega}\,\tau)\cos(\bar{m}y),$$
$$-\bar{\alpha}_L\frac{\partial^2\hat{p}}{\partial\bar{x}^2}\frac{\partial\bar{C}}{\partial\bar{x}}\delta\exp(\bar{\omega}\,\tau)\cos(\overline{my}). \tag{6.54}$$

Similarly, if Eqs. (6.44) and (6.45) are substituted into the second term on the right-hand side of Eq. (6.50) with consideration of $\partial^2\bar{p}/(\partial\bar{x}\partial\bar{y})=0$ and $\partial\bar{C}/\partial\bar{y}=0$, then the following equation can be obtained when the second-order terms are neglected:

$$\left(1-\bar{\alpha}_T\frac{\partial\bar{p}_{total}}{\partial\bar{x}}\right)\frac{\partial^2\bar{C}_{total}}{\partial\bar{y}^2}-\bar{\alpha}_T\frac{\partial^2\bar{p}_{total}}{\partial\bar{x}\partial\bar{y}}\frac{\partial\bar{C}_{total}}{\partial\bar{y}}$$
$$=\left(1-\bar{\alpha}_T\frac{\partial\bar{p}}{\partial\bar{x}}\right)\frac{\partial^2\bar{C}}{\partial\bar{y}^2}+\left(1-\bar{\alpha}_T\frac{\partial\bar{p}}{\partial\bar{x}}\right)\left(\frac{\partial^2\left[\hat{C}\cos(\overline{my})\right]}{\partial\bar{y}^2}\right)_{\xi}\delta\exp(\bar{\omega}\,\tau)$$
$$=\left(1-\bar{\alpha}_T\frac{\partial\bar{p}}{\partial\xi}\right)\left[m^2\frac{\partial\bar{C}}{\partial\xi}\delta\cos(\overline{my})\exp(\bar{\omega}\,\tau)-\hat{C}m^2\delta\cos(my)\exp(\bar{\omega}\,\tau)\right]. \tag{6.55}$$

In this situation, Eq. (6.47) is used to derive the right-hand side of Eq. (6.55).

The corresponding boundary conditions for the first-order perturbation problem are:

$$\hat{C}=0,\quad \lim_{x\to\infty}\frac{\partial\hat{p}}{\partial\xi}=0\quad\text{(downstream boundary)}, \tag{6.56}$$

$$\lim_{x\to-\infty}\hat{C}=0,\quad \lim_{x\to-\infty}\frac{\partial\hat{p}}{\partial\xi}=0\quad\text{(upstream boundary)}. \tag{6.57}$$

Similarly, the interface conditions for this first-order perturbation problem can be expressed as follows:

$$\hat{C} = 0, \quad \lim_{S \to 0^-} \hat{p} = \lim_{S \to 0^+} \hat{p}, \tag{6.58}$$

$$\begin{aligned} &\lim_{S \to 0^-} \left[\left(1 - \bar{\alpha}_L \bar{p}'_{fx}\right) \frac{\partial \hat{C}}{\partial \bar{n}} + \frac{\bar{\alpha}_L \bar{p}'_{fx}}{1 - \bar{\alpha}_L \bar{p}'_{fx}} \frac{\partial \hat{p}}{\partial \bar{n}} \right] = \bar{\omega}(\phi_f - \phi_0), \\ &\lim_{S \to 0^-} \frac{\partial \hat{p}}{\partial \bar{n}} = \frac{\psi_{xx}(\phi_0)}{\psi_{xx}(\phi_f)} \lim_{S \to 0^+} \frac{\partial \hat{p}}{\partial \bar{n}}. \end{aligned} \tag{6.59}$$

Solving Eqs. (6.48) and (6.49) with the boundary and interface conditions yields the following analytical solutions:

$$\hat{C} = 0, \quad \hat{p}(\xi) = \bar{p}'_{0x} \left[1 - \frac{1 - \beta}{1 + \beta} \exp\left(-\sqrt{\lambda} |\bar{m}| \xi \right) \right] \quad \text{(in the downstream region)}, \tag{6.60}$$

$$\begin{aligned} \hat{C}(\xi) = &\frac{-\bar{p}'_{fx}}{1 - \bar{\alpha}_L \bar{p}'_{fx}} \left\{ \exp\left(\frac{-\bar{p}'_{fx}}{1 - \bar{\alpha}_L \bar{p}'_{fx}} \xi \right) - \frac{(1+\beta) + \left(\frac{1}{1 - \bar{\alpha}_L \bar{p}'_{fx}} - \bar{\alpha}_L |\bar{m}| \sqrt{\lambda} \right)(1 - \beta)}{1 + \beta} \exp(\sigma \xi) \right\} \\ &- \frac{\bar{p}'_{fx}}{1 - \bar{\alpha}_L \bar{p}'_{fx}} \left\{ \left(\frac{1}{1 - \bar{\alpha}_L \bar{p}'_{fx}} - \bar{\alpha}_L |\bar{m}| \sqrt{\lambda} \right) \frac{1 - \beta}{1 + \beta} \exp\left[\left(|\bar{m}| \sqrt{\lambda} - \frac{\bar{p}'_{fx}}{1 - \bar{\alpha}_L \bar{p}'_{fx}} \right) \xi \right] \right\}, \\ \hat{p}(\xi) = &\bar{p}'_{fx} \left[1 + \frac{1 - \beta}{1 + \beta} \exp(\sqrt{\lambda} |\bar{m}| \xi) \right] \quad \text{(in the upstream region)}, \end{aligned} \tag{6.61}$$

where

$$\beta = \frac{\psi_{xx}(\phi_0)}{\psi_{xx}(\phi_f)} = \frac{k_{xx}(\phi_0)}{k_{xx}(\phi_f)}, \tag{6.62}$$

$$\sigma = \frac{\sqrt{\left(\frac{\bar{p}'_{fx}}{1 - \bar{\alpha}_L \bar{p}'_{fx}} \right)^2 + \frac{4 \bar{m}^2 \left(1 - \bar{\alpha}_T \bar{p}'_{fx}\right)}{1 - \bar{\alpha}_L \bar{p}'_{fx}}} - \frac{\bar{p}'_{fx}}{1 - \bar{\alpha}_L \bar{p}'_{fx}}}{2}. \tag{6.63}$$

Inserting Eq. (6.61) into Eq. (6.59) yields the following equation for the dimensionless growth rate of the small perturbation:

$$\begin{aligned} \bar{\omega} = &\frac{\bar{p}'_{fx}}{(1+\beta)(\phi_f - \phi_0)} \left\{ \frac{(1+\beta)\bar{p}'_{fx}}{1 - \bar{\alpha}_L \bar{p}'_{fx}} + \left[(1+\beta) + \left(\frac{1}{1 - \bar{\alpha}_L \bar{p}'_{fx}} - \bar{\alpha}_L |\bar{m}| \sqrt{\lambda} \right)(1 - \beta) \right] \sigma \right\} \\ &- \frac{\bar{p}'_{fx}}{(1+\beta)(\phi_f - \phi_0)} \left[\left(|\bar{m}| \sqrt{\lambda} - \frac{\bar{p}'_{fx}}{1 - \bar{\alpha}_L \bar{p}'_{fx}} - \bar{\alpha}_L \bar{m}^2 \lambda \left(1 - \bar{\alpha}_L \bar{p}'_{fx}\right) \right) \frac{(1 - \beta)}{1 - \bar{\alpha}_L \bar{p}'_{fx}} \right]. \end{aligned} \tag{6.64}$$

If $\bar{\alpha}_L = 0$ and $\bar{\alpha}_T = 0$, Eq. (6.64) can degenerate to the previous form obtained when the medium permeability anisotropic effect is neglected (Zhao et al. 2008a, 2009). Since $\bar{p}'_{fx} < 0$ when the pore-fluid flow is along the positive x direction, it is useful to define that $Zh = -\bar{p}'_{fx}$ in the theoretical analysis, where Zh is known as the Zhao number of the chemical dissolution system (Zhao et al. 2009, 2010a, b).

Letting $\bar{\omega} = 0$ and $|\bar{m}| = 1$ yields the following equation for the critical Zhao number (i.e. $Zh_{critical}$) of the chemical dissolution system under the theoretical consideration.

$$\left[(1+\beta) + \left(\frac{1}{1+\bar{\alpha}_L Zh_{critical}} - \bar{\alpha}_L\sqrt{\lambda}\right)(1-\beta)\right]\sigma_1 - \frac{(1+\beta)Zh_{critical}}{1+\bar{\alpha}_L Zh_{critical}} - \left[\left(\sqrt{\lambda} + \frac{Zh_{critical}}{1+\bar{\alpha}_L Zh_{critical}} - \bar{\alpha}_L\lambda(1+\bar{\alpha}_L Zh_{critical})\right)\frac{(1-\beta)}{1+\bar{\alpha}_L Zh_{critical}}\right] = 0, \tag{6.65}$$

where

$$\sigma_1 = \frac{\sqrt{\left(\frac{Zh_{critical}}{1+\bar{\alpha}_L Zh_{critical}}\right)^2 + \frac{4(1+\bar{\alpha}_T Zh_{critical})}{1+\bar{\alpha}_L Zh_{critical}}} + \frac{Zh_{critical}}{1+\bar{\alpha}_L Zh_{critical}}}{2}. \tag{6.66}$$

Since $(1+\bar{\alpha}_L Zh_{critical}) > 0$, Eq. (6.65) can be rewritten into the following form:

$$\left\{(1+\beta)(1+\bar{\alpha}_L Zh_{critical}) + \left[1 - \bar{\alpha}_L(1+\bar{\alpha}_L Zh_{critical})\sqrt{\lambda}\right](1-\beta)\right\}\sigma_1 - (1+\beta)Zh_{critical} - \left(\sqrt{\lambda} + \frac{Zh_{critical}}{1+\bar{\alpha}_L Zh_{critical}} - \bar{\alpha}_L\lambda(1+\bar{\alpha}_L Zh_{critical})\right)(1-\beta) = 0. \tag{6.67}$$

Although Eq. (6.67) can be solved using the conventional searching method, it is desirable to explore its explicit solution for the critical Zhao number in some special cases. For example, in the case of $\bar{\alpha}_L = \bar{\alpha}_T$, Eq. (6.67) can be further simplified as follows:

$$\left\{(1+\bar{\alpha}_L Zh_{critical}) + \left[1 - \bar{\alpha}_L(1+\bar{\alpha}_L Zh_{critical})\sqrt{\lambda}\right]R\right\}\sigma_1 - Zh_{critical} - \left(\sqrt{\lambda} + \frac{Zh_{critical}}{1+\bar{\alpha}_L Zh_{critical}} - \bar{\alpha}_L\lambda(1+\bar{\alpha}_L Zh_{critical})\right)R = 0, \tag{6.68}$$

where

$$R = \frac{1-\beta}{1+\beta}. \tag{6.69}$$

Substituting Eq. (6.66) into Eq. (6.68) yields the following equation:

$$\left\{2R\left[\sqrt{\lambda}-\bar{\alpha}_L\lambda(1+\bar{\alpha}_L Zh_{critical})\right]+\left[\left(1+\bar{\alpha}_L\sqrt{\lambda}R\right)+\frac{R}{1+\bar{\alpha}_L Zh_{critical}}\right]Zh_{critical}\right\}^2$$
$$-\left\{(1+\bar{\alpha}_L Zh_{critical})+\left[1-\bar{\alpha}_L(1+\bar{\alpha}_L Zh_{critical})\sqrt{\lambda}\right]R\right\}^2\left[\left(\frac{Zh_{critical}}{1+\bar{\alpha}_L Zh_{critical}}\right)^2+4\right]=0. \tag{6.70}$$

After some mathematical manipulations, Eq. (6.70) can be rewritten as follows:

$$\begin{aligned}
&\left(\frac{R^2\bar{\alpha}_L\sqrt{\lambda}}{1+\bar{\alpha}_L Zh_{critical}}+2\bar{\alpha}_L^3R\sqrt{\lambda}-\bar{\alpha}_L^2R\lambda\right)Zh_{critical}^2\\
&+\left(\bar{\alpha}_L R\sqrt{\lambda}-\bar{\alpha}_L^3R^2\lambda\sqrt{\lambda}-\bar{\alpha}_L^2+\bar{\alpha}_L^4R^2(\lambda-1)\lambda\right)Zh_{critical}^2\\
&+\left[\frac{R^2\sqrt{\lambda}}{1+\bar{\alpha}_L Zh_{critical}}+R\sqrt{\lambda}-\bar{\alpha}_L R(\lambda+2)+4\bar{\alpha}_L^2R\sqrt{\lambda}\right]Zh_{critical}\\
&+\left[\bar{\alpha}_L^2R^2(2-3\lambda)\sqrt{\lambda}-2\bar{\alpha}_L+2\bar{\alpha}_L^3R^2(\lambda-1)\lambda\right]Zh_{critical}\\
&+\left(2\bar{\alpha}_L R\sqrt{\lambda}-2R-1\right)+R^2(\lambda-1)-2\bar{\alpha}_L R^2(\lambda-1)\sqrt{\lambda}\\
&+\bar{\alpha}_L^2R^2(\lambda-1)\lambda=0.
\end{aligned} \tag{6.71}$$

Finally, the characteristic equation for the critical Zhao number in the case of $\bar{\alpha}_L=\bar{\alpha}_T$ and $\lambda=1$ is expressed as the following standard third-power equation:

$$AZh_{critical}^3+BZh_{critical}^2+CZh_{critical}+D=0, \tag{6.72}$$

where

$$A=-\bar{\alpha}_L^3R+2\bar{\alpha}_L^4R+\bar{\alpha}_L^2R-\bar{\alpha}_L^4R^2-\bar{\alpha}_L^3, \tag{6.73}$$

$$B=2\bar{\alpha}_L R-4\bar{\alpha}_L^2R+6\bar{\alpha}_L^3R-2\bar{\alpha}_L^3R^2+\bar{\alpha}_L R^2-3\bar{\alpha}_L^2, \tag{6.74}$$

$$C=-5\bar{\alpha}_L R+6\bar{\alpha}_l^2R-\bar{\alpha}_l^2R^2-3\bar{\alpha}_L+R^2+R, \tag{6.75}$$

$$D=2\bar{\alpha}_L R-2R-1. \tag{6.76}$$

It is obvious that if $\bar{\alpha}_L = \bar{\alpha}_T = 0$, then Eq. (6.71) can be rewritten in the following form:

$$\left[R^2\sqrt{\lambda} + R\sqrt{\lambda}\right] Zh_{critical} - (2R+1) + R^2(\lambda - 1) = 0. \quad (6.77)$$

Substitution of Eq. (6.69) into Eq. (6.77) yields the following expression:

$$Zh_{critical} = \frac{(1+R)^2 - R^2\lambda}{\sqrt{\lambda}R(1+R)} = \frac{(3-\beta)(1+\beta)}{2(1-\beta)\sqrt{\lambda}} - \frac{(1-\lambda)(1-\beta)}{2\beta\sqrt{\lambda}}. \quad (6.78)$$

For an isotropic porous medium, λ is equal to unity, so that Eq. (6.78) degenerates to the following solution:

$$Zh_{critical} = \frac{(3-\beta)(1+\beta)}{2(1-\beta)}. \quad (6.79)$$

This solution is identical to the previous theoretical results when the medium permeability anisotropic effect is neglected in the chemical dissolution system (Zhao et al. 2008a, 2009), indicating that the above mathematical deduction is correct.

Based on the previous study, the dimensionless Zhao number is defined as follows (Zhao et al. 2009):

$$Zh = -\frac{k_{xx}(\phi_f)p'_{fx}}{\mu\sqrt{\phi_f D(\phi_f)}}\sqrt{\frac{\alpha_{sph}}{k_{Echemical}C_{eq}\sqrt[3]{D_p}}}. \quad (6.80)$$

Note that if $Zh < Zh_{critical}$, then the chemical dissolution system is subcritical, while if $Zh > Zh_{critical}$, then the chemical dissolution system is supercritical. In the case of $Zh = Zh_{critical}$, the chemical dissolution system is in a critical state. This means that by using this criterion, the instability of chemical dissolution fronts in fluid-saturated porous media can be assessed in a theoretical manner.

Although the theoretical analysis is conducted to derive the critical instability criterion of planar chemical dissolution-fronts in a two-dimensional case, the same procedure as used in this study can be employed to derive the critical instability criterion of planar chemical dissolution-fronts in a three-dimensional situation, as already demonstrated in the previous study (Zhao et al. 2008c). Nevertheless, in the three-dimensional case, the related theoretical analysis and mathematical deductions become much more lengthy and tedious.

6.3 Application of the Present Theoretical Solutions

6.3.1 Theoretical Understanding of the Effects of Medium Permeability Anisotropy on Chemical Dissolution-Front Instability in Two-Dimensional Fluid-Saturated Porous Media

The present theoretical solutions for the critical Zhao numbers can be used to investigate the effects of medium permeability anisotropy on the chemical dissolution-front instability in two-dimensional fluid-saturated porous media. Generally, the greater the critical Zhao number of a chemical dissolution system, the more stable the chemical dissolution front in the chemical dissolution system. In the case of $|\bar{m}| = 1$, which means that the dimensional wavelength (i.e. λ_{wave}) of the perturbation has the same order of magnitude as the intrinsic characteristic length (i.e. L^* in Eq. (6.22)) of the chemical dissolution system because $|\bar{m}| = 2\pi L^*/\lambda_{wave}$ (Zhao et al. 2008a, 2009), the characteristic equation of the critical Zhao number (i.e. Eq. (6.65)) indicates that the critical Zhao number of the chemical dissolution system is a function of β (representing the ratio of the initial principal permeability to the finial principal permeability in the pore-fluid flow direction), λ (representing the ratio of the principal permeability in the transversal direction to that in the longitudinal direction that is parallel to the pore-fluid inflow direction), $\bar{\alpha}_L$ (representing the dimensionless longitudinal dispersivity of the solute) and $\bar{\alpha}_T$ (representing the dimensionless transversal dispersivity of the solute). Since β is only dependent on the initial and finial porosities (see Eqs. (6.11) and (6.62)), it can be directly replaced by both the initial porosity (i.e. ϕ_0) and the finial porosity (i.e. ϕ_f) of the chemical dissolution system in the following theoretical investigations.

Figure 6.1 shows the variations of the critical Zhao number with the medium anisotropic permeability factor (represented by λ), which is defined as the ratio of the principal permeability in the transversal direction to that in the longitudinal direction that is parallel to the pore-fluid inflow direction, due to three different final porosities (represented by ϕ_f). These results are obtained when $\phi_0 = 0.1$ and $\bar{\alpha}_L = \bar{\alpha}_T = 0.2$. It is observed that with an increase in the medium anisotropic permeability factor (represented by λ), the critical Zhao number can decrease significantly, especially when the medium anisotropic permeability factor is less than unity. This indicates that if the principal permeability in the transversal direction, which is perpendicular to the inflow direction, is smaller than that in the longitudinal direction, which is parallel to the inflow direction, the growth of the small perturbation is hindered in the transversal direction. As a result, the likelihood of the chemical dissolution-front instability is significantly reduced. Generally, the smaller the medium anisotropic permeability factor, the greater the critical Zhao number, implying that it becomes more difficult for the chemical dissolution-front instability to take place in the chemical dissolution system. On the other hand, for a given medium anisotropic permeability factor, the critical Zhao number

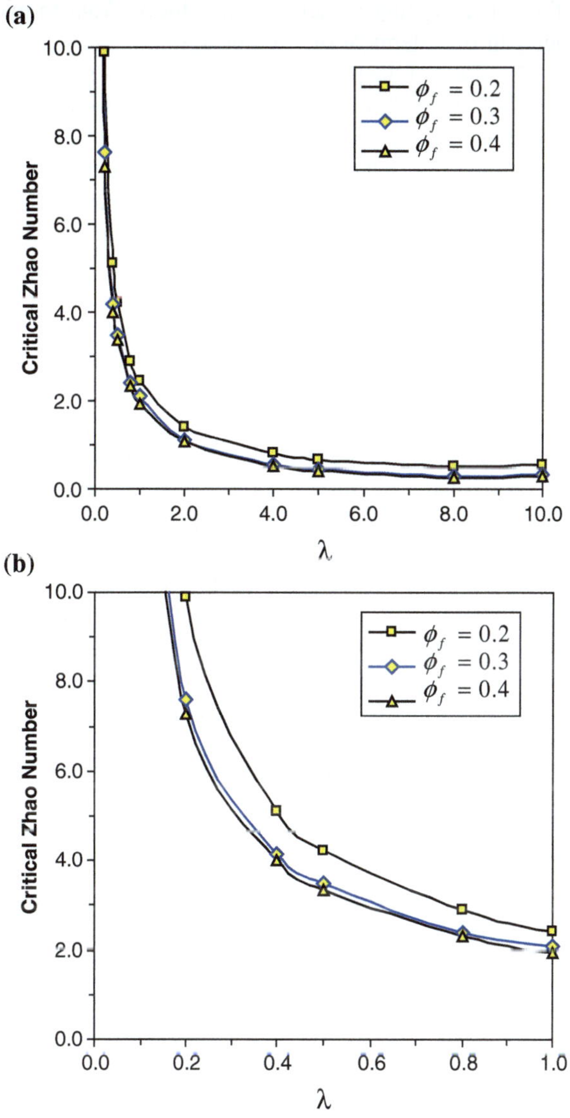

Fig. 6.1 Distributions of the critical Zhao number due to different final porosities ($\phi_0 = 0.1,\ \bar{\alpha}_L = \bar{\alpha}_T = 0.2$): **a** Entire view; **b** Zoom-in view

also decreases with an increase in the final porosity (i.e. ϕ_f) of the chemical dissolution system. For instance, in the case of the medium anisotropic permeability factor being equal to 5.0, the corresponding critical Zhao number of the chemical dissolution system is equal to 0.6697, 0.4485 and 0.4084 when the final porosity (i.e. ϕ_f) of the chemical dissolution system is equal to 0.2, 0.3 and 0.4 respectively. It can be concluded that a decrease in the medium anisotropic permeability factor stabilizes the chemical dissolution front so that it becomes more difficult for a planar chemical dissolution-front to evolve into different morphologies in the chemical dissolution system.

The effect of the medium anisotropic dispersivity factor, which is represented by the ratio of the dimensionless longitudinal dispersivity to the dimensionless transversal dispersivity (i.e. $\bar{\alpha}_L/\bar{\alpha}_T$), on the chemical dissolution instability can be also investigated using the present theoretical results. Toward this end, the initial and final porosities of the chemical dissolution system are assigned to be 0.1 and 0.2 respectively, while the dimensionless longitudinal dispersivity is assigned to be 0.2 (i.e. $\bar{\alpha}_L = 0.2$) during the evaluation of the corresponding critical Zhao numbers using Eq. (6.67), which is derived in the previous section. Figure 6.2 shows the effects of the medium anisotropic dispersivity factor (represented by $\bar{\alpha}_L/\bar{\alpha}_T$) on the distributions of the critical Zhao numbers due to four different medium anisotropic dispersivity factor, namely $\bar{\alpha}_L/\bar{\alpha}_T = 1, 2, 5$ and 10 respectively. Note that with an increase in the value of the medium anisotropic dispersivity factor (represented by $\bar{\alpha}_L/\bar{\alpha}_T$), the value of the critical Zhao number decreases accordingly. For example, in the case of the medium anisotropic permeability factor being equal to 0.5 (i.e. $\lambda = 0.5$), the corresponding critical Zhao number of the chemical dissolution system is equal to 4.233, 3.077, 2.622 and 2.496 when the medium anisotropic dispersivity factor (represented by $\bar{\alpha}_L/\bar{\alpha}_T$) of the chemical dissolution system is equal to 1, 2, 5 and 10 respectively. This implies that an increase in the value of the medium anisotropic dispersivity factor destabilizes the chemical dissolution-front so that it becomes easier for a planar chemical dissolution-front to evolve into different morphologies in the chemical dissolution system.

6.3.2 *Effects of Medium Permeability Anisotropy on the Morphological Evolution of the Chemical Dissolution Front in Two-Dimensional Fluid-Saturated Porous Media*

To investigate how the medium permeability anisotropy can affect the morphological evolution of a planar chemical dissolution-front in a two-dimensional fluid-saturated porous medium, the existing numerical methods that were used in the previous studies (Zhao et al. 2008b, 2009) are modified to include the consideration of the medium permeability anisotropy in the computational simulation, so that Eqs. (6.24)–(6.26) can be solved numerically (Zhao et al. 2013). Figure 6.3 shows the geometry and boundary conditions of the computational domain, for which the dimensionless pore-pressure gradient (i.e. $\bar{p}'_{fx} = -10$) is applied to the left boundary, implying that the corresponding Zhao number of the system is equal to 10. As a result, there is a horizontal throughflow from the left to the right of the computational model. The dimensionless height and width of the computational model are 5 and 15, respectively. Since the computational domain is of finite size,

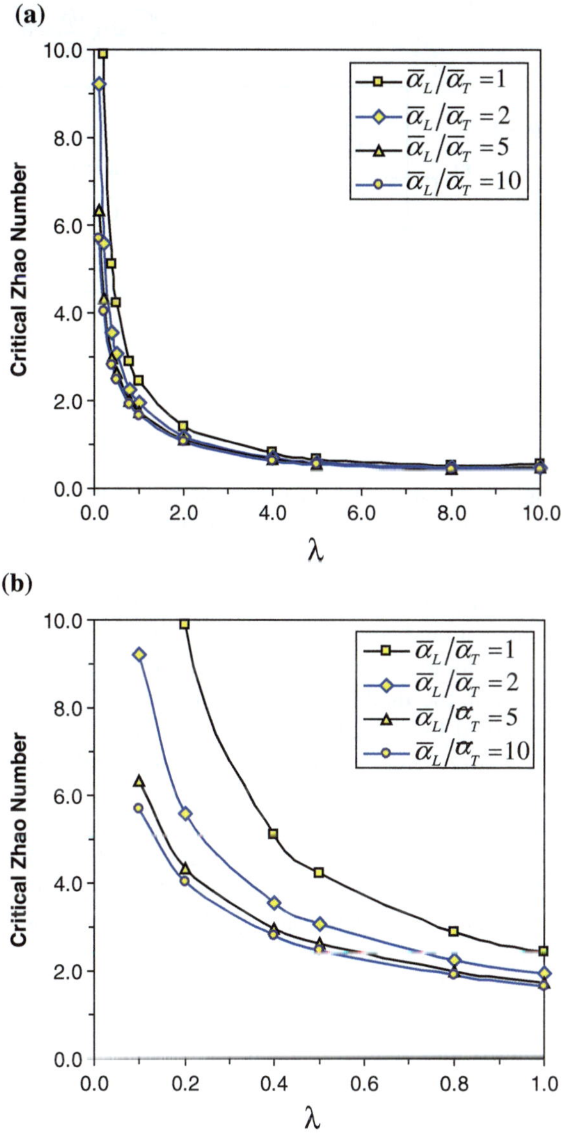

Fig. 6.2 Distributions of the critical Zhao number due to different final porosities ($\phi_0 = 0.1,\ \phi_f = 0.2,\ \bar{\alpha}_L = 0.2$): **a** Entire view; **b** Zoom-in view

a time-dependent dimensionless-concentration boundary condition (i.e. $\bar{C}(\tau) = \exp\left[(\bar{p}'_{fx}\bar{v}_{front}\tau)\Big/(1 - \bar{\alpha}_L\bar{p}'_{fx})\right]$) is applied to the left boundary. Except for the left boundary, the initial porosity of the porous medium is 0.1, while the initial dimensionless concentration is unity within the computational domain. The final porosity with a value of 0.2 is applied to the left boundary of the computational domain. As the value of the dimensionless pore-fluid pressure can be chosen

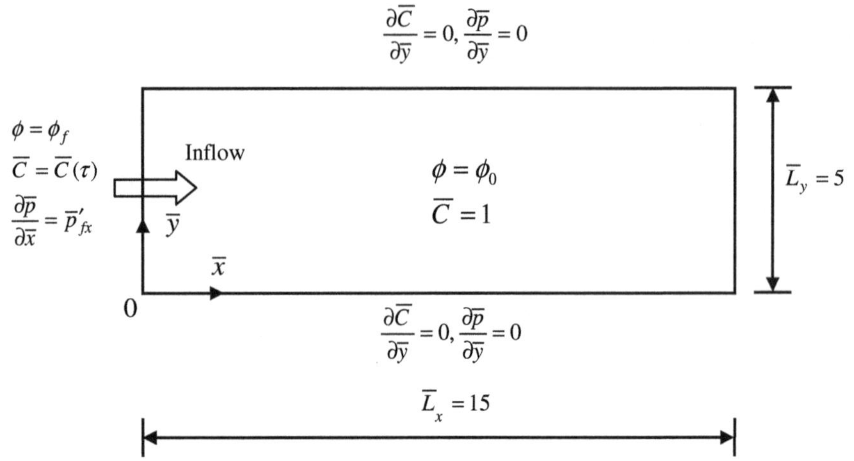

Fig. 6.3 Geometry and boundary conditions of the chemical dissolution problem in a fluid-saturated porous medium including medium anisotropic effects

arbitrarily, it is assumed to be 100 (i.e. $\bar{p}(\bar{L}_x, \tau) = 100$) at the right boundary of the computational domain. To consider whether or not a small perturbation affects the morphological evolution of a planar chemical dissolution-front during its propagation in the fluid-saturated porous medium, the initial porosity field is randomly perturbed by a small amount of 1 % of the originally-input initial porosity (i.e. $\phi_0 = 0.1$) before running the computational model. The dimensionless permeability tensor (involved in Eqs. (6.24) and (6.25)) of the anisotropic porous medium is calculated using the Carman-Kozeny formula, which has the power of three in the power law. The diffusivity of the solute is calculated using the power law, which has the power of two. Both the top and the bottom boundaries are assumed to be impermeable for the solute and pore-fluid. The mineral dissolution ratio (i.e. ε) of the chemical dissolution system is assumed to be 0.001, while the dimensionless time-step length is set to be 0.0002 in the computation. For the purpose of considering the medium dispersivity anisotropic effect, the dimensionless longitudinal dispersivity is 0.2, while the dimensionless transversal dispersivity is 0.02 in the computational model. To appropriately simulate the propagation of the chemical dissolution-front, the whole computational domain is simulated by 29,601 four-node square elements of 30,000 nodal points in total.

Figures 6.4 and 6.5 show the effects of the medium anisotropic permeability ratios (i.e. λ) on the morphological evolution of the chemical dissolution-front in the fluid-saturated porous medium at two different dimensionless time instants, namely $\tau = 40$ and $\tau = 50$, respectively. In these two figures, four different values of the medium anisotropic permeability ratios, namely $\lambda = 0.2$, 1.0, 5.0 and 10.0, have been used to show how the medium anisotropic permeability ratio affects the morphological evolution of the chemical dissolution-front in the fluid-saturated porous medium. Note that on the basis of the above parameters, the critical Zhao

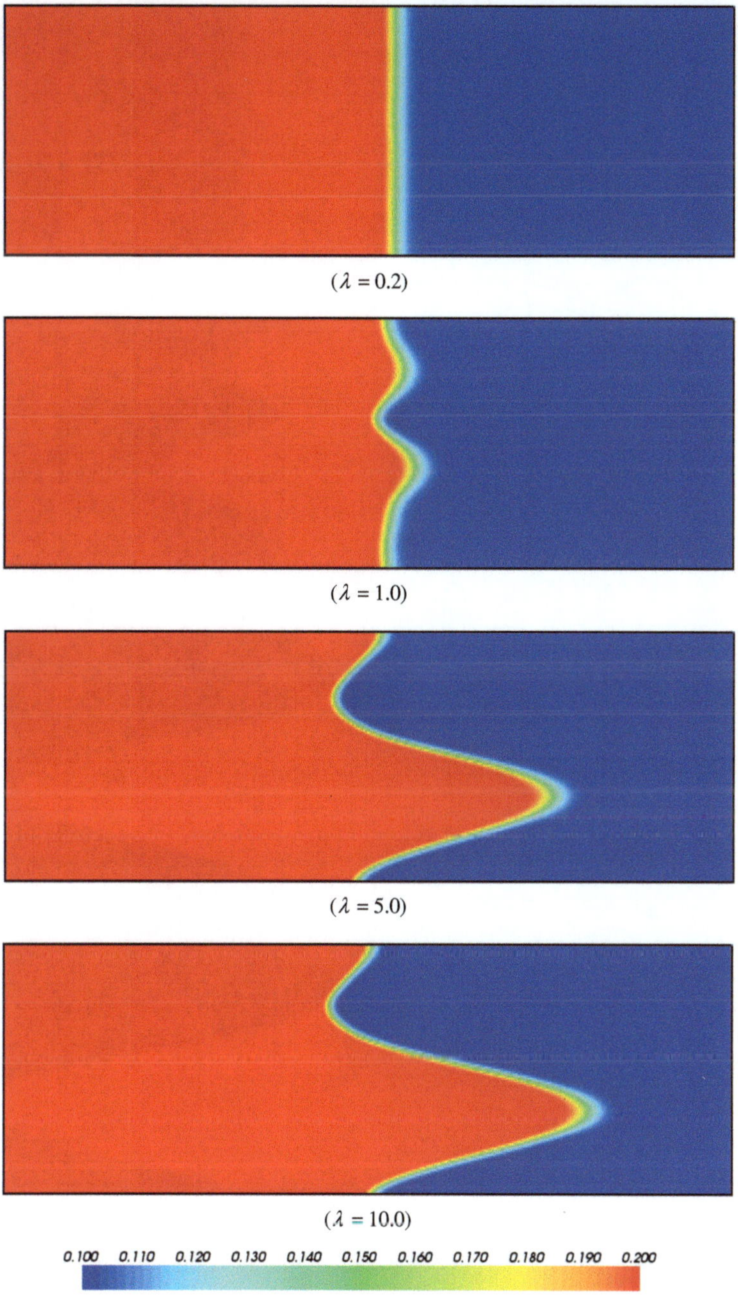

Fig. 6.4 Effects of medium anisotropic permeability ratios on the morphological evolution of the chemical dissolution front in the fluid-saturated porous medium ($\tau = 40$)

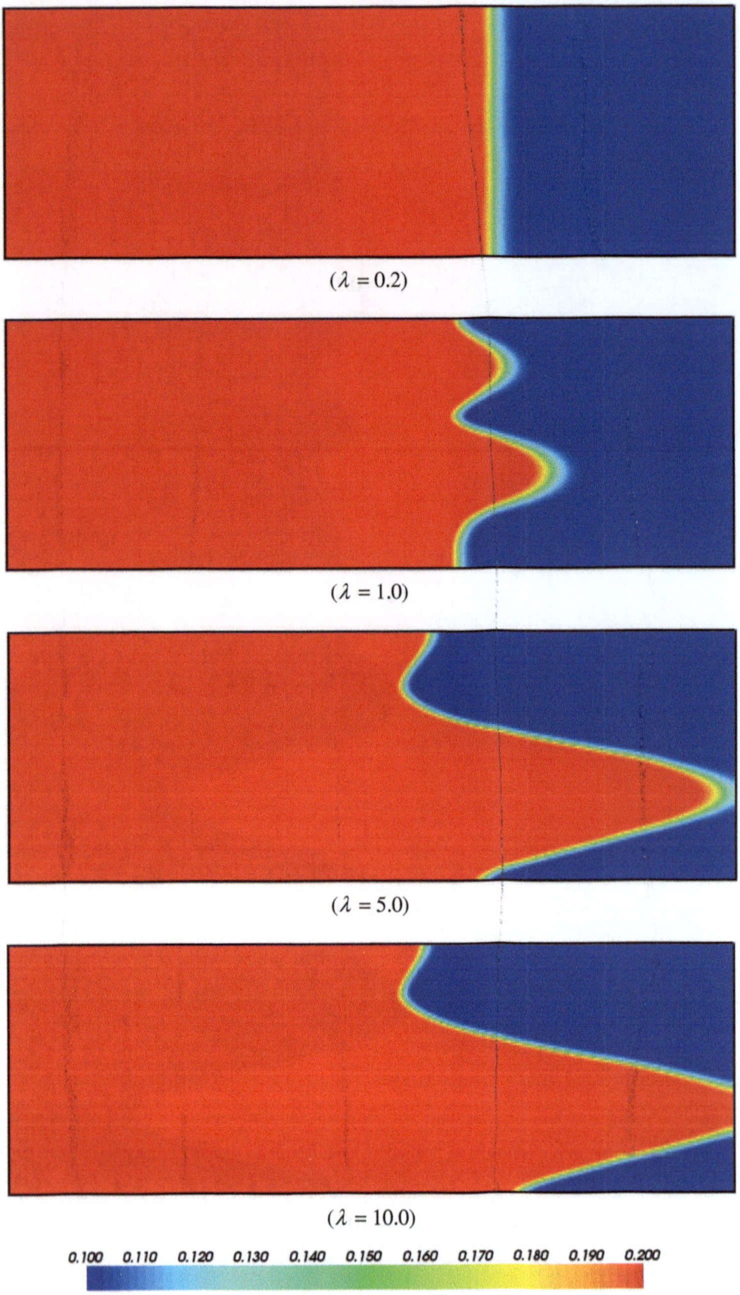

Fig. 6.5 Effects of medium anisotropic permeability ratios on the morphological evolution of the chemical dissolution front in the fluid-saturated porous medium ($\tau = 50$)

number of the chemical dissolution system is approximately equal to 4.030, 1.679, 0.549 and 0.460 when the medium anisotropic permeability ratio (i.e. λ) is equal to 0.2, 1.0, 5.0 and 10.0 respectively. These critical Zhao number can be obtained from the theoretical analysis conducted in the previous section of this study. Since the Zhao number (i.e. $Zh = 10$) of the simulated chemical dissolution system is greater than all the corresponding critical values, the chemical dissolution system considered here is in a supercritical state, implying that a planar chemical dissolution-front can evolve into irregular morphologies during its propagation within the chemical dissolution system. It is observed that the medium anisotropic permeability ratios (i.e. λ) can have significant effects on the morphological evolution of the chemical dissolution-front in the computational simulation models. Generally, when the Zhao number of the chemical dissolution system is greater than its critical value, the greater the medium anisotropic permeability ratios (i.e. λ), the faster the irregular chemical dissolution front grows. This indicates that since the same small perturbation is applied to the computational models involving different medium anisotropic permeability ratios, the dimensionless growth rate of the applied small perturbation in the chemical dissolution system increases with the increase of the medium anisotropic permeability ratio in the fluid-saturated porous medium.

It is interesting to investigate the effects of the medium anisotropic permeability ratios (i.e. λ) on the appearing time of the initial irregular chemical dissolution-front in the fluid-saturated porous medium including the medium permeability anisotropy. Figure 6.6 shows the related simulation results in this aspect. It is noted that the appearing time of the initial irregular chemical dissolution-front is 0.146, 0.06, 0.04 and 0.034 when the medium anisotropic permeability ratio (i.e. λ) is equal to 0.2, 1.0, 5.0 and 10.0 respectively. This means that with the increase of the medium anisotropic permeability ratio, the appearing time of the initial irregular chemical dissolution-front decreases accordingly. The reason for this phenomenon is that as indicated by Eq. (6.43), the amplitude of an irregular chemical dissolution-front is equal to the product of the applied small perturbation (i.e. δ) and the value of the exponential function, $\exp(\bar{\omega}\tau)$, where $\bar{\omega}$ is the dimensionless growth rate of the perturbation and τ is the dimensionless time. For the given values of both the applied small perturbation (i.e. δ) and the amplitude of an irregular chemical dissolution-front, an increase in the dimensionless growth rate can result in a decrease in the appearing (dimensionless) time of the initial irregular chemical dissolution-front. This further demonstrates that the present theoretical results in this chapter can be used to theoretically understand the morphological evolution of the chemical dissolution front in the fluid-saturated porous medium including the medium permeability anisotropic effects.

It needs to be pointed out that since the main purpose of this chapter is to mathematically derive the critical instability criterion of planar chemical dissolution-fronts in a fluid-saturated porous medium, many factors, such as the complicated domain shape, medium heterogeneity, different boundary conditions, corner effects of an irregular domain and so forth, are not considered in the numerical simulation. These factors should be considered in the future research.

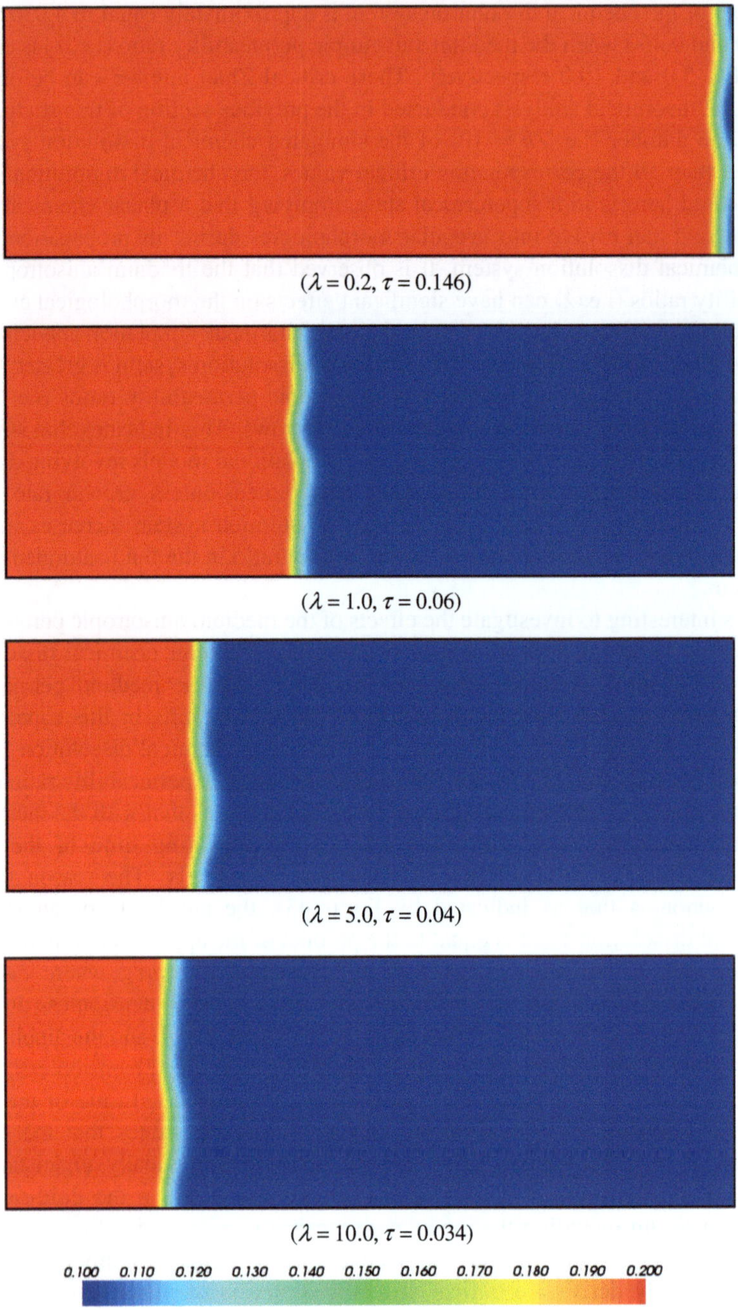

Fig. 6.6 Effects of medium anisotropic permeability ratios on the appearing time of the initial irregular chemical dissolution-front in the fluid-saturated porous medium

In summary, the related theoretical and numerical results demonstrated that: (1) a decrease in the medium anisotropic permeability factor (or ratio), which is defined as the ratio of the principal permeability in the transversal direction to that in the longitudinal direction parallel to the pore-fluid inflow direction, can stabilize the chemical dissolution front so that it becomes more difficult for a planar chemical dissolution-front to evolve into different morphologies in the chemical dissolution system; (2) the medium anisotropic permeability ratio can have significant effects on the morphological evolution of the chemical dissolution front. When the Zhao number of the chemical dissolution system is greater than its critical value, the greater the medium anisotropic permeability ratio, the faster the irregular chemical dissolution-front grows.

References

Bear J (1972) Dynamics of fluids in porous media. American Elsevier Publishing Company, New York

Chadam J, Hoff D, Merino E, Ortoleva P, Sen A (1986) Reactive infiltration instabilities. IMA J Appl Math 36:207–221

Chadam J, Ortoleva P, Sen A (1988) A weekly nonlinear stability analysis of the reactive infiltration interface. IMA J Appl Math 48:1362–1378

Chen JS, Liu CW (2002) Numerical simulation of the evolution of aquifer porosity and species concentrations during reactive transport. Comput Geosci 28:485–499

Chen JS, Liu CW (2004) Interaction of reactive fronts during transport in a homogeneous porous medium with initial small non-uniformity. J Contam Hydrol 72:47–66

Chen JS, Liu CW, Lai GX, Ni CF (2009) Effects of mechanical dispersion on the morphological evolution of a chemical dissolution front in a fluid-saturated porous medium. J Hydrol 373:96–102

Freeze RA, Cherry JA (1979) Groundwater. Prentice-Hall, Englewood Cliffs

Holzbecher EO (1998) Modeling density-driven flow in porous media. Springer, Berlin

Lewis RW, Schrefler BA (1998) The finite element method in the static and dynamic deformation and consolidation of porous media. Wiley, New York

Nield DA, Bejan A (1992) Convection in porous media. Springer, New York

Ormond A, Ortoleva P (2000) Numerical modeling of reaction-induced cavities in a porous rock. J Geophys Res 105:16737–16747

Ortoleva P, Chadam J, Merino E, Sen A (1987) Geochemical self-organization II: the reactive infiltration instability. Am J Sci 287:1008–1040

Renard F, Gratier JP, Ortoleva P, Brosse E, Bazin B (1998) Self-organization during reactive fluid flow in a porous medium. Geophys Res Lett 25:385–388

Scheidegger AE (1961) General theory of dispersion in porous media. J Geophys Res 66:3273–3278

Scheidegger AE (1974) The physics of flow through porous media. University of Toronto Press, Toronto

Zhao C, Hobbs BE, Ord A, Peng S, Mühlhaus HB, Liu L (2004) Theoretical investigation of convective instability in inclined and fluid-saturated three-dimensional fault zones. Tectonophysics 387:47–64

Zhao C, Hobbs BE, Hornby P, Ord A, Peng S, Liu L (2008a) Theoretical and numerical analyses of chemical-dissolution front instability in fluid-saturated porous rocks. Int J Numer Anal Meth Geomech 32:1107–1130

Zhao C, Hobbs BE, Ord A, Hornby P, Peng S (2008b) Effect of reactive surface areas associated with different particle shapes on chemical-dissolution front instability in fluid-saturated porous rocks. Transp Porous Media 73:75–94

Zhao C, Hobbs BE, Ord A, Hornby P, Peng S (2008c) Morphological evolution of three-dimensional chemical dissolution front in fluid-saturated porous media: a numerical simulation approach. Geofluids 8:113–127

Zhao C, Hobbs BE, Ord A (2009) Fundamentals of computational geoscience: numerical methods and algorithms. Springer, Berlin

Zhao C, Hobbs BE, Ord A, Peng S (2010a) Effects of mineral dissolution ratios on chemical-dissolution front instability in fluid-saturated porous media. Transp Porous Media 82:317–335

Zhao C, Hobbs BE, Ord A (2010b) Theoretical analyses of the effects of solute dispersion on chemical-dissolution front instability in fluid-saturated porous rocks. Transp Porous Media 84:629–653

Zhao C, Hobbs BE, Ord A (2012a) Effects of medium and pore-fluid compressibility on chemical-dissolution front instability in fluid-saturated porous media. Int J Numer Anal Meth Geomech 36:1077–1100

Zhao C, Reid LB, Regenauer-Lieb K, Poulet T (2012b) A porosity-gradient replacement approach for computational simulation of chemical-dissolution front propagation in fluid-saturated porous media including pore-fluid compressibility. Comput Geosci 16:735–755

Zhao C, Hobbs BE, Ord A (2013) Effects of medium permeability anisotropy on chemical-dissolution front instability in fluid-saturated porous rocks. Transp Porous Media 99:119–143

Zienkiewicz OC (1977) The finite element method. McGraw-Hill, London

Chapter 7
Effects of Medium and Pore-Fluid Compressibility on Chemical Dissolution-Front Instability in Fluid-Saturated Porous Media

Both medium and pore-fluid compressibility was neglected in the previous study of the classical reactive infiltration instability problem within a fluid-saturated porous medium (Chadam et al. 1986, 1988; Ortoleva et al. 1987). Due to this simplification, the critical condition, which is used to assess whether or not the fully-coupled nonlinear system between porosity, pore-fluid pressure and reactive chemical-species transport within the fluid-saturated porous medium can become unstable, was rigorously derived in a pure mathematical manner. Since a change in the pore-fluid pressure of a porous medium causes a change in the effective stress, which in turn leads to a change in the deformation of the porous medium, a considerable change in the pore-fluid pressure can cause a considerable change in the porosity of the porous medium. On the other hand, a pore-fluid pressure change can also result in a variation of the pore-fluid volume so that it can cause a change in the density of the pore-fluid. In the case of a small change in the porosity of a porous medium, the simplification made in the previous study (Chadam et al. 1986, 1988; Ortoleva et al. 1987) may be reasonable for the porous medium consisting of hard rocks (Steefel and Lasage 1990, 1994; Yeh and Tripathi 1991; Raffensperger and Garven 1995; Ormond and Ortoleva 2000; Chen and Liu 2002; Zhao et al. 1997, 1998, 2002, 2007). However, if the porosity change is remarkable or if the porous medium is comprised of soft materials, the above-mentioned simplification may be questionable, depending on how medium and pore-fluid compressibility affects the chemical dissolution-front instability in the fluid-saturated porous medium. For this reason, it is necessary to investigate the effect of both medium and pore-fluid compressibility on the chemical dissolution-front propagation within a deformable and fluid-saturated porous medium.

Pore-fluid compressibility is usually considered using the equation of state, in which the density of pore-fluid is usually expressed as a function of the pore-fluid pressure, while medium compressibility can be considered in the following two different approaches. In the first approach, the deformation of a porous medium is explicitly considered by the displacement variables (Turcotte and Schubert 1982; Gow et al. 2002; Schaubs and Zhao 2002; Ord et al. 2002), so that the porosity change caused by a pore-fluid pressure change can be considered by computing the volumetric strain change in the porous medium (Zhao et al. 2000). On the contrary,

C. Zhao, *Physical and Chemical Dissolution Front Instability in Porous Media*, Lecture Notes in Earth System Sciences,
DOI: 10.1007/978-3-319-08461-9_7

in the second approach, the porosity change caused by a pore-fluid pressure change is directly evaluated using the medium compressibility coefficient, which is equal to the inverse of the bulk modulus of the porous medium, without considering the medium displacements directly. The main advantage in using the first approach is that the detailed deformation process including a stress-strain constitutive relationship in a porous medium can be considered in the analysis, while the main disadvantage in using the first approach is that due to the use of displacement variables, the total number of the governing equations of the problem increases significantly so that it is very difficult to find analytical solutions for the problem. Since the medium deformation is indirectly considered, the main advantage in using the second approach is that the total number of the partial differential equations of the problem can be reduced so that it is possible to derive analytical solutions for the problem of a simple geometrical domain. Because the major focus of the chemical dissolution-front instability problem is on the detailed process of pore-fluid flow and chemical dissolution-front propagation within the fluid-saturated porous medium, the second approach is more appropriate for conducting the theoretical analysis of the problem.

7.1 Theoretical Considerations of the Chemical Dissolution-Front Instability Problem in Deformable Fluid-Saturated Porous Media

7.1.1 Dimensional and Dimensionless Governing Equations of the Problem

If pore-fluid compressibility in a deformable and fluid-saturated porous medium is considered, the governing equations of the coupled nonlinear problem between porosity, pore-fluid density, pore-fluid flow and reactive multi-chemical-species transport in the deformable and fluid-saturated porous medium can be expressed as follows:

$$\frac{\partial}{\partial t}(\rho_f \phi) + \nabla \cdot (\rho_f \phi \vec{u}_{linear}) = 0, \tag{7.1}$$

$$\vec{u} = \phi \vec{u}_{linear} = -\frac{k(\phi)}{\mu} \nabla p, \tag{7.2}$$

$$\frac{\partial}{\partial t}(\phi C_i) + \nabla \cdot (\phi C_i \vec{u}_{linear}) = \nabla \cdot [\phi D_i(\phi) \nabla C_i] + R_i \quad (i = 1, 2, \ldots, N), \tag{7.3}$$

$$\frac{\partial \phi}{\partial t} = \alpha \frac{\partial p}{\partial t} - D_p R_p, \tag{7.4}$$

$$\frac{\partial \rho_f}{\partial p} = \rho_f \beta, \tag{7.5}$$

where $\vec{u}_{linear}$ is the averaged linear velocity vector within the pore space of the porous medium; $\vec{u}$ is the Darcy velocity vector within the porous medium; p and C_i are the pore-fluid pressure and the concentration (moles per pore-fluid volume) of chemical species i; μ is the dynamic viscosity of the pore-fluid; ϕ is the porosity of the porous medium; $D_i(\phi)$ is the diffusivity of chemical species i; ρ_f is the density of the pore-fluid; N is the total number of all the chemical species to be considered in the system; R_i is the source/sink term of chemical species i due to the dissolution/precipitation of solid minerals within the system; $k(\phi)$ is the permeability of the porous medium; D_p is the density of the soluble grains, which is defined as the number of the soluble gains per unit medium volume; α is the compressibility coefficient of the porous medium; β is the compressibility coefficient of the pore-fluid; R_p is the rate of grain-volume change due to a chemical (precipitation) reaction at the particle level (Zhao et al. 2008).

The first term at the right hand side of Eq. (7.4) represents the contribution of a pore-fluid pressure change rate to the increase of the porosity change rate, while the second term at the right hand side of Eq. (7.4) represents the contribution of a chemical precipitation reaction to the increase of the porosity change rate. Since an increase in pore-fluid pressure causes a decrease in the effective (compressive) stress within the solid matrix of a porous medium, it can result in an increase in the volumetric strain of the solid matrix, indicating an increase in the porosity change rate. This is the reason why the first term at the right hand side of Eq. (7.4) has a positive sign. On the other hand, the particle precipitation due to a chemical precipitation reaction can fill the pore space of the porous medium, so that it results in a decrease in the porosity change rate. This is the reason why the second term at the right hand side of Eq. (7.4) has a negative sign.

To consider the variations of both the diffusivity of each chemical species and the permeability of the porous medium with porosity, the following formulas are commonly used in the theoretical investigation of the chemical dissolution-front instability problem (Bear 1972; Scheidegger 1974; Chadam et al. 1986; Detournay and Cheng 1993).

$$D_i(\phi) = D_{0i}\phi^q \quad \left(\frac{3}{2} \le q \le \frac{5}{2}\right), \tag{7.6}$$

$$k(\phi) = \frac{k_0(1-\phi_0)^2\phi^3}{\phi_0^3(1-\phi)^2}, \tag{7.7}$$

where D_{0i} is the diffusivity of chemical species i in pure water; ϕ_0 and k_0 are the initial reference porosity and permeability of the porous medium, respectively.

The source/sink term of chemical species i due to the dissolution/precipitation of solid minerals within the system can be expressed as follows (Chadam et al. 1986):

$$R_i = -\chi_i k_{chemical} \frac{A_p}{\bar{V}_p}(\phi_f - \phi)\left(\prod_{i=1}^{N} C_i^{\chi_i} - K_{eq}\right), \tag{7.8}$$

where χ_i is the stoichiometric coefficient of the ith chemical species; A_p is the averaged surface area of soluble grains; $k_{chemical}$ and K_{eq} are the rate constant and equilibrium constant of the chemical reaction respectively; $\bar{V}_p$ is the average volume of each individual soluble grain.

If a single reactive chemical species is considered in the chemical system, then the above-mentioned governing equations (i.e. Eqs. (7.1)–(7.5)) can be simplified as follows:

$$\frac{\partial(\phi\rho_f)}{\partial t} - \nabla\cdot[\rho_f\psi(\phi)\nabla p] = 0, \tag{7.9}$$

$$\frac{\partial}{\partial t}(\phi C) - \nabla\cdot[\phi D(\phi)\nabla C + C\psi(\phi)\nabla p] + \rho_s k_{Echemical}\frac{A_p}{\bar{V}_p}(\phi_f - \phi)(C - C_{eq}) = 0, \tag{7.10}$$

$$\frac{\partial\phi}{\partial t} - \alpha\frac{\partial p}{\partial t} + k_{Echemical}\frac{A_p}{\bar{V}_p}(\phi_f - \phi)(C - C_{eq}) = 0, \tag{7.11}$$

$$\frac{\partial\rho_f}{\partial p} = \rho_f\beta, \tag{7.12}$$

$$\psi(\phi) = \frac{k(\phi)}{\mu}, \tag{7.13}$$

where C and C_{eq} are the concentration and equilibrium concentration of the single chemical species; $k_{Echemical}$ is the comprehensive rate constant of the chemical reaction (as defined in Eq. (2.15) previously); ϕ_f is the final (i.e. maximum) porosity of the porous medium after the completion of dissolvable mineral dissolution; ρ_s is the molar density (i.e. moles per unit volume) of the dissolvable grains. Other quantities in Eqs. (7.9)–(7.13) are of the same meanings as those defined in Eqs. (7.1)–(7.5) and (7.8).

Obviously, Eq. (7.12) has the following solution:

$$\rho_f = \rho_{ref}\exp[\beta(p - p_{ref})], \tag{7.14}$$

where ρ_{ref} is the reference density of the pore-fluid at the prescribed reference pressure, p_{ref}.

It is noted that Eq. (7.14) is suitable for a fully-compressible pore-fluid, the volume of which approaches zero as the pressure approaches infinity so that the corresponding density of the pore-fluid approaches infinity. However, for a partially-compressible pore-fluid, there exists a critical compressible pressure, known as the up-limit compressible pressure, above which the compressibility of the pore-fluid is negligible. If the ratio of the minimum limit density at the zero pressure (i.e. $p = 0$) to the critical compressible density at the critical compressible pressure (i.e. $p = p_{critical}$) is defined as $\lambda = \rho_{f\text{minimum}}/\rho_{fcritical}$, then the critical compressible pressure of the pore-fluid can be determined by replacing p_{ref} with $p_{critical}$ in Eq. (7.14).

$$p_{critical} = -\frac{\ln \lambda}{\beta}. \tag{7.15}$$

Thus, for a partially-compressible pore-fluid, Eq. (7.14) needs to be modified as follows:

$$\rho_f = \rho_{fcritical} \exp[\beta(p - p_{critical})] \quad (p \le p_{critical}), \tag{7.16}$$

$$\rho_f = \rho_{fcritical} \quad (p > p_{critical}). \tag{7.17}$$

We consider a special case, in which the solid molar density greatly exceeds the equilibrium concentration of the chemical species (Phillips 1991; Chadam et al. 1986; Zhao et al. 2008, 2010). For this special case, it is useful to define a small parameter, known as the mineral dissolution ratio (Zhao et al. 2010), as follows:

$$\varepsilon = \frac{C_{eq}}{\rho_s} << 1. \tag{7.18}$$

To facilitate the theoretical analysis, the following dimensionless parameters and variables are defined:

$$\bar{x} = \frac{x}{L^*}, \quad \bar{y} = \frac{y}{L^*}, \quad \bar{z} = \frac{z}{L^*}, \quad \bar{C} = \frac{C}{C_{eq}}, \quad \bar{p} = \frac{p}{p^*}, \tag{7.19}$$

$$\tau = \frac{t}{t^*}\varepsilon, \quad \bar{\rho}_f = \frac{\rho_f}{\rho_{ref}}, \quad \bar{\alpha} = \alpha p^*, \quad \bar{\beta} = \beta p^*, \tag{7.20}$$

where τ is a slow dimensionless time to describe the slowness of the chemical dissolution that takes place in the system. Other characteristic parameters used in Eqs. (7.19) and (7.20) can be expressed as follows:

$$t^* = \frac{\bar{V}_p}{k_{Echemical} A_p C_{eq}}, \quad L^* = \sqrt{\psi_f D(\psi_f) t^*}, \quad p^* = \frac{\phi_f D(\phi_f)}{\psi(\phi_f)}, \tag{7.21}$$

$$D^*(\phi) = \frac{\phi D(\phi)}{\phi_f D(\phi_f)}, \quad \psi^*(\phi) = \frac{\psi(\phi)}{\psi(\phi_f)}. \tag{7.22}$$

Using the related dimensionless quantities and variables, Eqs. (7.14) and (7.9)–(7.11) can be rewritten in the following dimensionless form:

$$\bar{\rho}_f = \exp\left[\bar{\beta}(\bar{p} - \bar{p}_{ref})\right], \tag{7.23}$$

$$\varepsilon \frac{\partial(\phi \bar{\rho}_f)}{\partial \tau} - \nabla \cdot [\bar{\rho}_f \psi^*(\phi) \nabla \bar{p}] = 0, \tag{7.24}$$

$$\varepsilon \frac{\partial}{\partial \tau}(\phi \bar{C}) - \nabla \cdot [D^*(\phi) \nabla \bar{C} + \bar{C} \psi^*(\phi) \nabla \bar{p}] - \frac{\partial(\phi - \bar{\alpha}\bar{p})}{\partial \tau} = 0, \tag{7.25}$$

$$\varepsilon \frac{\partial(\phi - \bar{\alpha}\bar{p})}{\partial \tau} + (\phi_f - \phi)(\bar{C} - 1) = 0. \tag{7.26}$$

Note that if $p_{ref} = p^*$, then $\bar{p}_{ref} = 1$ in Eq. (7.23). As a result, Eq. (7.23) can be rewritten as

$$\bar{\rho}_f = \exp\left[\bar{\beta}(\bar{p} - 1)\right]. \tag{7.27}$$

If the geometry of the problem domain is an infinite plane, then the boundary conditions of this special problem can be expressed in the dimensionless form as follows (Zhao et al. 2008):

$$\lim_{\bar{x} \to \infty} \bar{C} = 1, \quad \lim_{\bar{x} \to \infty} \frac{\partial \bar{p}}{\partial \bar{x}} = \bar{p}'_{0x} \quad \text{(downstream boundary)}, \tag{7.28}$$

$$\lim_{\bar{x} \to -\infty} \bar{C} = 0, \quad \lim_{\bar{x} \to -\infty} \frac{\partial \bar{p}}{\partial \bar{x}} = \bar{p}'_{fx} \quad \text{(upstream boundary)}. \tag{7.29}$$

where $\bar{p}'_{fx}$ is the dimensionless pore-fluid pressure gradient as x approaches negative infinity in the upstream of the pore-fluid flow; $\bar{p}'_{0x}$ is the unknown dimensionless pore-fluid pressure gradient as x approaches positive infinity in the downstream of the pore-fluid flow.

In this case, the initial condition for this theoretical problem is: $\phi(\bar{x}, 0) = \phi_0$ expect at the negative infinity, where $\lim_{\bar{x} \to -\infty} \phi(\bar{x}, 0) = \phi_f$. Note that ϕ_0 is the initial porosity of the porous medium.

The propagating chemical dissolution-front divides the problem domain into two regions, an upstream region and a downstream region. Across this propagating chemical dissolution-front, the porosity undergoes a jump from its initial value into its final value. This implies that the governing equations of the problem are different for these two regions. Since there are two kinds of pore-fluid compressibility models, namely a fully-compressible pore-fluid model and a partially-compressible pore-fluid model, it is necessary to derive the analytical solutions using these two kinds of pore-fluid compressibility models in this investigation.

7.1.2 Analytical Solutions Derived from Using a Fully-Compressible Pore-Fluid Model

When both pore-fluid and medium compressibilities are considered simultaneously, it is very difficult, if not impossible, to derive analytical solutions for the problem. However, if they are considered separately, it is possible to derive some analytical solutions when pore-fluid compressibility is only considered. For this purpose, a limit case, in which ε approaches zero, is taken into account. In this limit case, the corresponding governing equations for the dimensionless variables of the problem in both the downstream region and the upstream region can be expressed below.

$$\bar{C} = 1, \quad \nabla^2 \bar{p} + \bar{\beta}(\nabla \bar{p} \cdot \nabla \bar{p}) = 0, \quad \phi = \phi_0 \quad \text{(in the downstream region)}, \tag{7.30}$$

$$\nabla \cdot (\nabla \bar{C} + \bar{C} \nabla \bar{p}) = 0, \quad \nabla^2 p + \bar{\beta}(\nabla p \cdot \nabla p) = 0, \quad \phi = \phi_f \quad \text{(in the upstream region)}, \tag{7.31}$$

If the chemical dissolution front is denoted by $S(\bar{x}, \tau) = 0$, then the following interface conditions exist for this propagating chemical dissolution-front problem.

$$\lim_{S \to 0^-} \bar{C} = \lim_{S \to 0^+} \bar{C}, \quad \lim_{S \to 0^-} \bar{p} = \lim_{S \to 0^+} \bar{p}, \tag{7.32}$$

$$\lim_{S \to 0^-} \frac{\partial \bar{C}}{\partial \bar{n}} = \bar{v}_{front}(\phi_f - \phi_0), \quad \lim_{S \to 0^-} \frac{\partial \bar{p}}{\partial \bar{n}} = \frac{\psi(\phi_0)}{\psi(\phi_f)} \lim_{S \to 0^+} \frac{\partial \bar{p}}{\partial \bar{n}}, \tag{7.33}$$

where $\bar{n} = n/L^*$; n is the normal vector of the propagating chemical dissolution-front.

When a planar dissolution front is under stable conditions, general solutions for the fully-compressible pore fluid model can be derived from Eqs. (7.30) and (7.31) with the related interface and boundary conditions. The resulting general solutions are expressed as follows:

$$\bar{C}(\xi) = 1, \quad \bar{p}(\xi) = \frac{1}{\bar{\beta}} \ln\left(\frac{\psi(\phi_f)}{\psi(\phi_0)} \bar{p}_{C1} \xi + 1 \right) + \bar{p}_{C2}, \quad \phi = \phi_0 \quad \text{(in the downstream region)}, \tag{7.34}$$

$$\bar{C}(\xi) = (\bar{p}_{C1} \xi + 1)^{-\frac{1}{\bar{\beta}}}, \quad \bar{p}(\xi) = \frac{1}{\bar{\beta}} \ln(\bar{p}_{C1} \xi + 1) + \bar{p}_{C2}, \quad \phi = \phi_f \quad \text{(in the upstream region)}, \tag{7.35}$$

where $\bar{p}_{C1}$ and $\bar{p}_{C2}$ are two constants to be determined.

Due to the consideration of the fully-compressible pore-fluid, the corresponding pore-fluid pressure gradient approaches zero, implying that the pore-fluid cannot be further injected on the upstream boundary where ξ approaches negative infinity. This means that the upstream boundary condition in the case of ξ approaching negative infinity cannot be satisfied when the pore-fluid is considered to be fully compressible. For this reason, it is desirable to examine the effect of the partially-compressible pore-fluid model on the chemical dissolution-front propagation in a fluid-saturated porous medium.

7.1.3 Analytical Solutions When the Pore-Fluid in the Upstream Region Is Incompressible

Although the analytical solution derived from the fully-compressible pore-fluid model cannot satisfy the upstream boundary condition in the case of ξ approaching negative infinity, it may provide some useful information about the pore-fluid pressure variation within the vicinity region of the chemical dissolution front. For example, the solution derived from the fully-compressible pore-fluid model indicates that there is a dramatic drop in the pore-fluid pressure within the downstream region. This dramatic drop can be located near the chemical dissolution front, while the amount of this drop can approach negative infinity. Because of this dramatic drop, it may be reasonable to assume that the real pore-fluid in the upstream region has reached a relatively incompressible state and that the pore-fluid pressure at the upstream–downstream interface is equal to the critical compressible pressure. It is noted that when the pore-fluid becomes incompressible, the density of the pore-fluid is kept as a constant in the analysis. As a result, the corresponding governing equations for the dimensionless variables of the problem in both the downstream region and the upstream region can be expressed below.

$$\bar{C} = 1, \quad \nabla^2 \bar{p} + \bar{\beta}(\nabla \bar{p} \cdot \nabla \bar{p}) = 0, \quad \phi = \phi_0 \quad \text{(in the downstream region)}, \tag{7.36}$$

$$\nabla \cdot (\nabla \bar{C} + \bar{C} \nabla \bar{p}) = 0, \quad \nabla^2 \bar{p} = 0, \quad \phi = \phi_f \quad \text{(in the upstream region)}. \tag{7.37}$$

In this case, Eqs. (7.16) and (7.17) can be rewritten in the dimensionless form as follows:

$$\bar{\rho}_f = \bar{\rho}_{fcritical} \exp\left[\bar{\beta}(\bar{p} - \bar{p}_{critical})\right] \quad (\bar{p} \leq \bar{p}_{critical}), \tag{7.38}$$

$$\bar{\rho}_f = \bar{\rho}_{fcritical} \quad (\bar{p} > \bar{p}_{critical}). \tag{7.39}$$

Similarly, for a planar chemical dissolution-front propagation in the fluid-saturated porous medium under stable conditions, the base solutions for this problem can be derived from Eqs. (7.36) and (7.37) with the related boundary and interface conditions mentioned previously. The resulting base solutions are expressed as follows:

$$\bar{C}(\xi) = 1, \quad \bar{p}(\xi) = \frac{1}{\bar{\beta}}\ln(E_1\xi + 1) + E_2, \quad \phi = \phi_0 \quad \text{(in the downstream region)}, \tag{7.40}$$

$$\bar{C}(\xi) = \exp(-\bar{p}'_{fx}\xi), \quad \bar{p}(\xi) = \bar{p}'_{fx}\xi + \bar{p}_{critical}, \quad \phi = \phi_f \quad \text{(in the upstream region)}, \tag{7.41}$$

where E_1 and E_2 are the following two constants:

$$E_1 = \frac{\psi(\phi_f)}{\psi(\phi_0)}\bar{\beta}\bar{p}'_{fx}, \quad E_2 = \bar{p}_{critical}. \tag{7.42}$$

Other parameters can be defined below (Zhao et al. 2008).

$$\xi = \bar{x} - \bar{v}_{front}\tau, \quad \bar{v}_{front} = -\frac{\bar{p}'_{fx}}{\phi_f - \phi_0}. \tag{7.43}$$

Mathematically, the stability analysis of the base solutions for the problem can be carried out using the linear stability theory (Nield and Bejan 1992), so that the instability of the planar chemical dissolution-front can be investigated (Chadam et al. 1986; Zhao et al. 2008). For this purpose, a small time-dependent perturbation is added on the planar chemical dissolution-front. Consequently, the total solution, which is equal to the sum of the base solution and the perturbed solution of the system, can be expressed in a two-dimensional case as follows:

$$S(\xi, \bar{y}, \tau) = \xi - \delta\exp(\bar{\omega}\tau)\cos(\overline{my}), \tag{7.44}$$

$$\bar{p}_{total}(\xi, \bar{y}, \tau) = \bar{p}(\xi, \tau) + \delta\hat{p}(\xi)\exp(\bar{\omega}\tau)\cos(\overline{my}), \tag{7.45}$$

$$\bar{C}_{total}(\xi, \bar{y}, \tau) = \bar{C}(\xi, \tau) + \delta\hat{C}(\xi)\exp(\bar{\omega}\tau)\cos(\overline{my}), \tag{7.46}$$

where $\bar{\omega}$ is the dimensionless growth rate of the perturbation; $\bar{m}$ is the dimensionless wavenumber of the perturbation; δ is the dimensionless amplitude of the perturbation and $\delta \ll 1$ by the definition of a linear stability analysis.

Since $S(\xi, \bar{y}, \tau)$ is a function of coordinates ξ and $\bar{y}$, the following derivatives exist mathematically (Zhao et al. 2008).

$$\left(\frac{\partial}{\partial\xi}\right)_{\xi} = \frac{\partial S}{\partial\xi}\frac{\partial}{\partial S} = \left(\frac{\partial}{\partial\xi}\right)_S, \quad \left(\frac{\partial}{\partial\bar{y}}\right)_{\xi} = \frac{\partial S}{\partial\bar{y}}\frac{\partial}{\partial S} + \left(\frac{\partial}{\partial\bar{y}}\right)_S = \frac{\partial S}{\partial\bar{y}}\left(\frac{\partial}{\partial\xi}\right)_S + \left(\frac{\partial}{\partial\bar{y}}\right)_S, \tag{7.47}$$

$$\left(\frac{\partial^2}{\partial \xi^2}\right)_\xi = \left(\frac{\partial^2}{\partial \xi^2}\right)_S, \quad \left(\frac{\partial^2}{\partial \bar{y}^2}\right)_\xi = \frac{\partial^2 S}{\partial \bar{y}^2}\frac{\partial}{\partial \xi} + \left(\frac{\partial S}{\partial \bar{y}}\right)^2 \frac{\partial^2}{\partial \xi^2} + 2\frac{\partial S}{\partial \bar{y}}\frac{\partial^2}{\partial \xi \partial \bar{y}} + \left(\frac{\partial^2}{\partial \bar{y}^2}\right)_S. \tag{7.48}$$

The total solutions expressed in Eqs. (7.45) and (7.46) must satisfy the governing equations that are expressed in Eqs. (7.36) and (7.37). With consideration of Eqs. (7.47) and (7.48), the first-order perturbation equations of this chemical dissolution system can be expressed as

$$\begin{aligned} &\hat{C} = 0, \\ &\frac{\partial^2 \hat{p}}{\partial \xi^2} - \bar{m}^2 \hat{p} + \bar{m}^2 \frac{\partial \bar{p}}{\partial \xi} + 2\bar{\beta}\frac{\partial \bar{p}}{\partial \xi}\frac{\partial \hat{p}}{\partial \xi} = 0 \quad \text{(in the downstream region)}, \end{aligned} \tag{7.49}$$

$$\begin{aligned} &\frac{\partial^2 \hat{C}}{\partial \xi^2} + \bar{p}'_{fx}\frac{\partial \hat{C}}{\partial \xi} - \bar{m}^2 \hat{C} - \bar{m}^2 \bar{p}'_{fx}\exp(-\bar{p}'_{fx}\xi) - \bar{p}'_{fx}\exp(-\bar{p}'_{fx}\xi)\frac{\partial \hat{p}}{\partial \xi} = 0, \\ &\frac{\partial^2 \hat{p}}{\partial \xi^2} - \bar{m}^2 \hat{p} + \bar{m}^2 \bar{p}'_{fx} = 0 \qquad \text{(in the upstream region)}. \end{aligned} \tag{7.50}$$

Note that in the process of deriving Eq. (7.49), the following mathematical equality is used:

$$\begin{aligned} \nabla \cdot \nabla &= \left(\frac{\partial}{\partial \xi}\right)^2 + \left[\frac{\partial S}{\partial \bar{y}}\left(\frac{\partial}{\partial \xi}\right) + \frac{\partial}{\partial \bar{y}}\right]^2 \\ &= \left(\frac{\partial}{\partial \xi}\right)^2 + \left(\frac{\partial}{\partial \xi}\right)^2\left(\frac{\partial S}{\partial \bar{y}}\right)^2 + 2\frac{\partial S}{\partial \bar{y}}\left(\frac{\partial}{\partial \xi}\right)\left(\frac{\partial}{\partial \bar{y}}\right) + \left(\frac{\partial}{\partial \bar{y}}\right)^2. \end{aligned} \tag{7.51}$$

Since $\partial \bar{p}/\partial \bar{y} = 0$, only the first term on the right hand side of Eq. (7.51) has a contribution to the first-order perturbation equation (i.e. Eq. (7.49)) of this system.

The corresponding boundary conditions of the first-order perturbation problem arc:

$$\hat{C} = 0, \quad \lim_{x \to \infty} \frac{\partial \hat{p}}{\partial \xi} = 0 \quad \text{(downstream boundary)}, \tag{7.52}$$

$$\lim_{x \to -\infty} \hat{C} = 0, \quad \lim_{x \to -\infty} \frac{\partial \hat{p}}{\partial \xi} = 0, \quad \text{(upstream boundary)}. \tag{7.53}$$

Similarly, the interface conditions for this first-order perturbation problem can be expressed as follows:

$$\hat{C} = 0, \quad \lim_{S \to 0^-} \hat{p} = \lim_{S \to 0^+} \hat{p}, \tag{7.54}$$

$$\lim_{S\to 0}\frac{\partial \hat{C}}{\partial \bar{n}} = \bar{\omega}(\phi_f - \phi_0), \quad \lim_{S\to 0^-}\frac{\partial \hat{p}}{\partial \bar{n}} = \frac{\psi(\phi_0)}{\psi(\phi_f)}\lim_{S\to 0^+}\frac{\partial \hat{p}}{\partial \bar{n}}. \tag{7.55}$$

If the pore-fluid is incompressible (i.e. $\bar{\beta} = 0$), Eq. (7.49) can be straightforwardly degenerated into that derived in the previous study (Zhao et al. 2008), in which the compressibility of the pore-fluid is neglected. Although the compressibility of pore-fluid can affect the perturbation solution in the downstream region, it does not affect the critical condition, under which the planar chemical dissolution-front becomes unstable. This can be demonstrated below.

Solving Eq. (7.50) with the corresponding boundary and interface conditions yields the following analytical results:

$$\begin{aligned}
\hat{C}(\xi) &= -\bar{p}'_{fx}\left\{\exp(-\bar{p}'_{fx}\xi) - \frac{2}{1+\chi}\exp(\sigma\xi) + \frac{1-\chi}{1+\chi}\exp\left[\left(|\bar{m}| - \bar{p}'_{fx}\right)\xi\right]\right\}, \\
\hat{p}(\xi) &= \bar{p}'_{fx}\left[1 + \frac{1-\chi}{1+\chi}\exp(|\bar{m}|\xi)\right] \quad \text{(in the upstream region)},
\end{aligned} \tag{7.56}$$

where

$$\chi = \frac{\psi(\phi_0)}{\psi(\phi_f)} = \frac{k(\phi_0)}{k(\phi_f)}, \quad \sigma = \frac{\sqrt{(\bar{p}'_{fx})^2 + 4m^2} - \bar{p}'_{fx}}{2}. \tag{7.57}$$

Substituting Eq. (7.56) into Eq. (7.55) yields the following equation for the dimensionless growth rate of the small perturbation.

$$\bar{\omega}(\bar{m}) = \frac{-\bar{p}'_{fx}}{(1+\chi)(\phi_f - \phi_0)}\left[-\bar{p}'_{fx} - \sqrt{(\bar{p}'_{fx})^2 + 4\bar{m}^2} + (1-\chi)|\bar{m}|\right]. \tag{7.58}$$

Equation (7.58) clearly indicates that the planar chemical dissolution-front within the reactive transport system is stable to short wavelength (i.e. the large dimensionless wavenumber $\bar{m}$) perturbations but it is unstable to long wavelength (i.e. the small dimensionless wavenumber $\bar{m}$) perturbations.

Letting $\bar{\omega}(\bar{m}) = 0$ and $\bar{m} = 1$ yields the following critical condition, which can be used to examine whether or not the planar chemical dissolution-front within the reactive transport system becomes unstable.

$$Zh_{critical} = -\bar{p}'_{fx}\Big|_{critical} = \frac{(3-\chi)(1+\chi)}{2(1-\chi)}, \tag{7.59}$$

where $Zh_{critical}$ is the critical Zhao number, which is equal to the negative value of the dimensionless pore-fluid pressure gradient, $\bar{p}'_{fx}\Big|_{critical}$, in the far upstream direction as x approaching negative infinity. Note that the Zhao number is a dimensionless number and has the same definition as that used in the previous study (Zhao et al. 2009).

$$Zh = -\bar{p}'_{fx} = \frac{p'_{fx}L^*}{p^*} = -\frac{k(\phi_f)p'_{fx}}{\mu\sqrt{\phi_f D(\phi_f)}}\sqrt{\frac{\bar{V}_p}{k_{Echemical}A_p C_{eq}}}, \tag{7.60}$$

where Zh is the Zhao number of the chemical dissolution system in a fluid-saturated porous medium.

7.2 Effects of Pore-Fluid Compressibility on the Evolution of a Chemical Dissolution Front in Subcritical Chemical Dissolution Systems

Since the main difference between this study and the previous ones is to consider pore-fluid compressibility, the numerical procedure proposed in the previous study can be used to simulate the morphological evolution of a chemical dissolution front after some appropriate modifications are made. However, it is still desirable to investigate the effect of pore-fluid compressibility on the base solution (i.e. trivial solution) of a chemical dissolution system. Note that the base solution is the only solution for a subcritical chemical-dissolution system, while it is a solution before the chemical dissolution front becomes unstable for either a critical or a supercritical chemical-dissolution system. For this purpose, a conceptual model (shown in Fig. 7.1) is constructed to derive the analytical solution. Because the geometry of this conceptual model can be exactly simulated using the conventional numerical methods such as the finite element and finite difference methods (Zienkiewicz 1977; Zhao et al. 1994; Lewis and Schrefler 1998), the analytical solution derived for this conceptual model can be also used to serve as a benchmark solution for verifying numerical methods. In this regard, the conceptual model under consideration can be viewed as a benchmark problem.

7.2.1 Derivation of Analytical Solutions for the Benchmark Problem in the Subcritical Chemical Dissolution System of a Finite Domain

Although the general analytical solution derived from the fully-compressible pore-fluid model cannot satisfy the boundary condition at the (negative) infinity of an unbounded domain, it can satisfy the boundary condition at the upstream boundary of a finite domain. This means that the analytical solution for the conceptual model (shown in Fig. 7.1) can be derived from the related analytical solution that is derived from the fully-compressible pore-fluid model as follows:

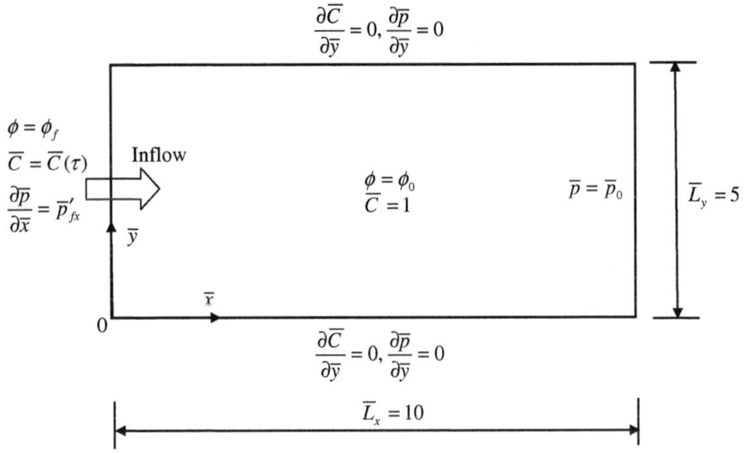

Fig. 7.1 Conceptual model of the chemical dissolution problem in a fluid-saturated porous medium

$$\bar{C}(\bar{x},\tau)=1,\quad \bar{p}(\bar{x},\tau)=\frac{1}{\beta}\ln\left(\frac{\psi(\phi_f)}{\psi(\phi_0)}(\bar{x}-\bar{v}_{front}\tau)E_{C1}+1\right)+E_{C2},$$
$$\phi(\bar{x},\tau)=\phi_0 \qquad \text{(in the downstream region)}, \tag{7.61}$$

$$\bar{C}(\bar{x},\tau)=\left((\bar{x}-\bar{v}_{front}\tau)E_{C1}+1\right)^{-\frac{1}{\beta}},\quad \bar{p}(\bar{x},\tau)=\frac{1}{\beta}\ln\left((\bar{x}-\bar{v}_{front}\tau)E_{C1}+1\right)+E_{C2},$$
$$\phi(\bar{x},\tau)=\phi_f \qquad \text{(in the upstream region)}, \tag{7.62}$$

where E_{C1} and E_{C1} need to be determined using the related boundary conditions below.

Using the upstream boundary condition, namely $\partial\bar{p}/\partial\bar{x}=\bar{p}'_{fx}$ at $\bar{x}=0$, the following equation can be obtained:

$$\bar{p}'_{fx}=\frac{E_{C1}}{\beta(1-\bar{v}_{font}\tau E_{C1})}. \tag{7.63}$$

Similarly, using the downstream boundary condition (i.e. $\bar{p}=\bar{p}_0$ at $\bar{x}=\bar{L}_x$) yields the following equation:

$$\bar{p}_0=\frac{1}{\beta}\ln\left(\frac{\psi(\phi_f)}{\psi(\phi_0)}(\bar{L}_x-\bar{v}_{front}\tau)E_{C1}+1\right)+E_{C2}. \tag{7.64}$$

Solving Eqs. (7.63) and (7.64) yields the solutions for E_{C1} and E_{C2}.

$$E_{C1} = \frac{\bar{\beta}\bar{p}'_{fx}}{1 + \bar{\beta}\bar{p}'_{fx}\bar{v}_{front}\tau}, \tag{7.65}$$

$$E_{C2} = -\frac{1}{\bar{\beta}}\ln\left[\frac{\psi(\phi_f)\bar{\beta}\bar{p}'_{fx}(\bar{L}_x - \bar{v}_{front}\tau)}{\psi(\phi_0)\left(1 + \bar{\beta}\bar{p}'_{fx}\bar{v}_{front}\tau\right)} + 1\right] + \bar{p}_0, \tag{7.66}$$

where $\bar{p}_0$ is the prescribed pressure at the right boundary of the finite domain.

It is interesting to examine the behaviour of the solutions expressed by Eqs. (7.61) and (7.62) in a limit case when $\bar{\beta}$ approaches zero, namely when the pore-fluid becomes incompressible. Toward this end, the following mathematical equities need to be derived.

$$R = \frac{\psi(\phi_f)}{\psi(\phi_0)}, \tag{7.67}$$

$$\begin{aligned}
\lim_{\bar{\beta}\to 0}\frac{1}{\bar{\beta}}\ln(E_{C1}\xi + 1) &= \lim_{\bar{\beta}\to 0}\frac{1}{\bar{\beta}}\ln\frac{\bar{\beta}\bar{p}'_{fx}\xi + 1 + \bar{\beta}\bar{p}'_{fx}\bar{v}_{front}\tau}{1 + \bar{\beta}\bar{p}'_{fx}\bar{v}_{front}\tau} \\
&= \lim_{\bar{\beta}\to 0}\frac{\left(1 + \bar{\beta}\bar{p}'_{fx}\bar{v}_{front}\tau\right)\left[\left(\bar{p}'_{fx}\xi + \bar{p}'_{fx}\bar{v}_{front}\tau\right)\left(1 + \bar{\beta}\bar{p}'_{fx}\bar{v}_{front}\tau\right)\right]}{\left(1 + \bar{\beta}\bar{p}'_{fx}\xi + \bar{\beta}\bar{p}'_{fx}\bar{v}_{front}\tau\right)\left(1 + \bar{\beta}\bar{p}'_{fx}\bar{v}_{front}\tau\right)^2} \\
&\quad - \lim_{\bar{\beta}\to 0}\frac{\bar{p}'_{fx}\bar{v}_{front}\tau\left(1 + \bar{\beta}\bar{p}'_{fx}\xi + \bar{\beta}\bar{p}'_{fx}\bar{v}_{front}\tau\right)}{\left(1 + \bar{\beta}\bar{p}'_{fx}\xi + \bar{\beta}\bar{p}'_{fx}\bar{v}_{front}\tau\right)\left(1 + \bar{\beta}\bar{p}'_{fx}\bar{v}_{front}\tau\right)^2} \\
&= \bar{p}'_{fx}\xi + \bar{p}'_{fx}\bar{v}_{front}\tau - \bar{p}'_{fx}\bar{v}_{front}\tau = \bar{p}'_{fx}\xi,
\end{aligned} \tag{7.68}$$

$$\begin{aligned}
\lim_{\bar{\beta}\to 0}(\bar{p}_0 - E_{C2}) &= \lim_{\bar{\beta}\to 0}\frac{1}{\bar{\beta}}\ln\frac{1 + R\bar{\beta}\bar{p}'_{fx}(\bar{L}_x - \bar{v}_{front}\tau) + \bar{\beta}\bar{p}'_{fx}\bar{v}_{front}\tau}{1 + \bar{\beta}\bar{p}'_{fx}\bar{v}_{front}\tau} \\
&= \lim_{\bar{\beta}\to 0}\frac{\left(1 + \bar{\beta}\bar{p}'_{fx}\bar{v}_{front}\tau\right)\left\{\left[R\bar{p}'_{fx}(\bar{L}_x - \bar{v}_{front}\tau) + \bar{p}'_{fx}\bar{v}_{front}\tau\right]\left(1 + \bar{\beta}\bar{p}'_{fx}\bar{v}_{front}\tau\right)\right\}}{\left(1 + R\bar{\beta}\bar{p}'_{fx}(\bar{L}_x - \bar{v}_{front}\tau) + \bar{\beta}\bar{p}'_{fx}\bar{v}_{front}\tau\right)\left(1 + \bar{\beta}\bar{p}'_{fx}\bar{v}_{front}\tau\right)^2}, \\
&\quad - \lim_{\bar{\beta}\to 0}\frac{\left(1 + \bar{\beta}\bar{p}'_{fx}\bar{v}_{front}\tau\right)\left\{\bar{p}'_{fx}\bar{v}_{front}\tau\left[1 + R\bar{\beta}\bar{p}'_{fx}(\bar{L}_x - \bar{v}_{front}\tau) + \bar{\beta}\bar{p}'_{fx}\bar{v}_{front}\tau\right]\right\}}{\left(1 + R\bar{\beta}\bar{p}'_{fx}(\bar{L}_x - \bar{v}_{front}\tau) + \bar{\beta}\bar{p}'_{fx}\bar{v}_{front}\tau\right)\left(1 + \bar{\beta}\bar{p}'_{fx}\bar{v}_{front}\tau\right)^2} \\
&= R\bar{p}'_{fx}(\bar{L}_x - \bar{v}_{front}\tau) + \bar{p}'_{fx}\bar{v}_{front}\tau - \bar{p}'_{fx}\bar{v}_{front}\tau = R\bar{p}'_{fx}(\bar{L}_x - \bar{v}_{front}\tau)
\end{aligned} \tag{7.69}$$

Substituting Eqs. (7.68) and (7.69) into Eq. (7.61) yields the following equation:

$$\begin{aligned}\bar{p}(\bar{x},\tau) &= \frac{1}{\bar{\beta}}\ln\left(\frac{\psi(\phi_f)}{\psi(\phi_0)}(\bar{x}-\bar{v}_{front}\tau)E_{C1}+1\right)+E_{C2}\\ &= \frac{\psi(\phi_f)}{\psi(\phi_0)}\bar{p}'_{fx}(\bar{x}-\bar{v}_{front}\tau)-\frac{\psi(\phi_f)}{\psi(\phi_0)}\bar{p}'_{fx}(\bar{L}_x-\bar{v}_{front}\tau)+\bar{p}_0\\ &= -\frac{\psi(\phi_f)}{\psi(\phi_0)}\bar{p}'_{fx}(\bar{L}_x-\bar{x})+\bar{p}_0\\ &= -\bar{p}'_{f0}(\bar{L}_x-\bar{x})+\bar{p}_0,\end{aligned} \tag{7.70}$$

where $\bar{p}'_{f0}$ is the dimensionless pressure gradient of the pore-fluid in the downstream region of the chemical dissolution system.

Similarly, substituting Eqs. (7.68) and (7.69) into Eq. (7.62) yields the following equations:

$$\begin{aligned}\bar{p}(\bar{x},\tau) &= \frac{1}{\bar{\beta}}\ln\left((\bar{x}-\bar{v}_{front}\tau)E_{C1}+1\right)+E_{C2}\\ &= \bar{p}'_{fx}(\bar{x}-\bar{v}_{front}\tau)-\frac{\psi(\phi_f)}{\psi(\phi_0)}\bar{p}'_{fx}(\bar{L}_x-\bar{v}_{front}\tau)+\bar{p}_0\\ &= \bar{p}'_{fx}(\bar{x}-\bar{v}_{front}\tau)-\bar{p}'_{f0}(\bar{L}_x-\bar{v}_{front}\tau)+\bar{p}_0,\end{aligned} \tag{7.71}$$

$$\ln\bar{C}(\bar{x},\tau) = -\frac{1}{\bar{\beta}}\ln\left((\bar{x}-\bar{v}_{front}\tau)E_{C1}+1\right) = -\bar{p}'_{fx}(\bar{x}-\bar{v}_{front}\tau). \tag{7.72}$$

Equation (7.72) can be straightforwardly rewritten into the following form:

$$\bar{C}(\bar{x},\tau) = \exp\left[-\bar{p}'_{fx}(\bar{x}-\bar{v}_{front}\tau)\right]. \tag{7.73}$$

Thus, when $\bar{\beta}$ approaches zero, Eqs. (7.61) and (7.62) can degenerate to the following equations:

$$\bar{C}(\bar{x},\tau)=1,\quad \bar{p}(\bar{x},\tau)=-\bar{p}'_{0x}(\bar{L}_x-\bar{x})+\bar{p}_0,\quad \phi(\bar{x},\tau)=\phi_0\quad (\bar{x}>\bar{v}_{front}\tau), \tag{7.74}$$

$$\begin{aligned}&\bar{C}(\bar{x},\tau)=\exp[-\bar{p}'_{fx}(\bar{x}-\bar{v}_{front}\tau)],\quad \bar{p}(\bar{x},\tau)=\bar{p}'_{fx}(\bar{x}-\bar{v}_{front}\tau)-\bar{p}'_{0x}(\bar{L}_x-\bar{v}_{front}\tau)+\bar{p}_0,\\ &\phi(\bar{x},\tau)=\phi_f\qquad (\bar{x}\le\bar{v}_{front}\tau).\end{aligned} \tag{7.75}$$

The solutions expressed in Eqs. (7.74) and (7.75) are exactly the same as the previous solutions derived for incompressible pore-fluids (Zhao et al. 2008). This proves that the present solutions for compressible pore-fluids are correctly derived, so that they can be used to investigate the effect of pore-fluid compressibility on the base solution (i.e. trivial solution) for the fully-coupled nonlinear problem between porosity, pore-fluid pressure, pore-fluid density and reactive chemical-species transport within a fluid-saturated porous medium.

7.2.2 Evaluation of the Dimensionless Propagating Speed of the Chemical Dissolution Front Associated with the Derived Analytical Solutions for the Benchmark Problem

When the pore-fluid is compressible, the dimensionless propagating speed of the chemical dissolution front (i.e. $\bar{v}_{front}$) is unknown. To overcome this difficulty, it is necessary to divide the total period of the dimensionless time under consideration into many small time intervals. At each small dimensionless time interval, it is assumed that the pore-fluid density can be considered to be constant, so that the dimensionless propagating speed of the chemical dissolution front (i.e. $\bar{v}_{front}$) in the case of the pore-fluid being compressible can be determined on the basis of the analytical solution derived in the previous study (Zhao et al. 2008) or in Sect. 7.1.2. This is the reason why analytical solutions are derived for a special case where the pore-fluid density in the upstream region is considered to be constant. To determine the total traveling distance of a chemical dissolution front for a given dimensionless time, $\tau = \tau_k$, the following expressions are defined:

$$\tau_k = \sum_{i=1}^{k} \Delta\tau_i \quad (k = 1, 2, 3, \ldots, N), \tag{7.76}$$

$$\bar{v}_{front}(\tau_{k-1}) = -\frac{\bar{p}'_{fx}(\tau_{k-1})}{\phi_f - \phi_0} \quad (k = 1, 2, 3, \ldots, N), \tag{7.77}$$

where $\bar{v}_{front}(\tau_{k-1})$ is the dimensionless propagating speed of the chemical dissolution front at the beginning of the kth time interval (i.e. $\Delta\tau_k$), which is equal to the corresponding value at the end of the ($k-1$)th time interval (i.e. $\Delta\tau_{k-1}$).

For a given dimensionless time interval, $\Delta\tau_i$, the corresponding increase in the traveling distance of the chemical dissolution front can be expressed as follows:

$$\Delta\bar{s}_i = \bar{v}_{front}(\tau_{i-1})\Delta\tau_i \quad (i = k), \quad (k = 1, 2, 3, \ldots, N). \tag{7.78}$$

The total traveling distance of the chemical dissolution front at a given dimensionless time, $\tau = \tau_k$, can be calculated using the following equation:

$$\bar{s}_k = \sum_{i=1}^{k} \Delta\bar{s}_i = \sum_{i=1}^{k} \bar{v}_{front}(\tau_{i-1})\Delta\tau_i = -\frac{1}{\phi_f - \phi_0}\sum_{i=1}^{k} \bar{p}'_{fx}(\tau_{i-1})\Delta\tau_i \quad (k = 1, 2, 3, \ldots, N), \tag{7.79}$$

where $\bar{p}'_{fx}(\tau_{i-1})$ is the dimensionless pore-fluid pressure gradient at the beginning of the kth time interval (i.e. $\Delta\tau_k$), which is equal to the corresponding value at the end of the $(k-1)$th time interval (i.e. $\Delta\tau_{k-1}$).

To determine the term, $\bar{p}'_{fx}(\tau_{k-1})$, it is necessary to consider a vertical cross-section in Fig. 7.1. From the mass conservation point of view, the mass flux of pore-fluid should be identical at any vertical cross-sections in the conceptual model (Fig. 7.1). This means that the product of the Darcy velocity and pore-fluid density is equal to the same constant at any vertical cross-sections in the upstream direction of this model. Since the Darcy velocity is directly proportional to the pore-fluid pressure gradient, the product of the pore-fluid pressure gradient and pore-fluid density should be equal to the same constant at any vertical cross-sections of the same porosity. For this reason, the following equation holds.

$$\bar{p}'_{fx}(\tau_{k-1})\bar{\rho}_f(\tau_{k-1}) = (\bar{p}'_{fx})_{reference}(\bar{\rho}_f)_{reference}, \tag{7.80}$$

where $\bar{\rho}_f(\tau_{k-1})$ is the dimensionless density of the pore-fluid at the beginning of the kth time interval (i.e. $\Delta\tau_k$), which is equal to the corresponding value at the end of the (k-1)th time interval (i.e. $\Delta\tau_{k-1}$); $(\bar{p}'_{fx})_{reference}$ is the reference dimensionless pressure gradient of the pore-fluid; $(\bar{\rho}_f)_{reference}$ is the reference dimensionless density of the pore-fluid. Both $(\bar{p}'_{fx})_{reference}$ and $(\bar{\rho}_f)_{reference}$ can be determined using the previous analytical results when the pore-fluid density is considered to be constant (Zhao et al. 2008). Since $(\bar{\rho}_f)_{reference}$ is a constant, $(\bar{p}'_{fx})_{reference}$ should be equal to another constant in the upstream direction of the conceptual model, as discussed in Sect. 7.1.2 as well as in the previous study (Zhao et al. 2008). As shown in Eq. (7.23), $\bar{\rho}_f = 1$ if the pore-fluid density is considered to be constant. This leads to the following equation:

$$\bar{p}'_{fx}(\tau_{k-1}) = \frac{(\bar{p}'_{fx})_{reference}}{\bar{\rho}_f(\tau_{k-1})} \quad (k = 1, 2, 3, \ldots, N). \tag{7.81}$$

Substituting Eq. (7.81) into Eq. (7.79) yields the following equation for evaluating the total travel distance of the chemical dissolution front at the given time, namely $\tau = \tau_k$.

$$\bar{s}_k = \bar{v}_{front}(\tau_k)\tau_k = -\frac{(\bar{p}'_{fx})_{reference}}{\phi_f - \phi_0}\sum_{i=1}^{k}\frac{\Delta\tau_i}{\bar{\rho}_f(\tau_{k-1})} = (\bar{v}_{front})_{reference}\sum_{i=1}^{k}\frac{\Delta\tau_i}{\bar{\rho}_f(\tau_{k-1})}, \tag{7.82}$$

where $(\bar{v}_{front})_{reference}$ is the reference dimensionless propagating speed of the chemical dissolution front in the fluid-saturated porous medium, which has the pore-fluid of a constant density and the same porosity as that when the pore-fluid is considered to be compressible.

Note that $\bar{\rho}_f(\tau_{k-1})$ can be actually evaluated at the location of the chemical dissolution front (i.e. $\bar{x} = \bar{s}_{k-1}$) at the end of the $(k-1)$th time interval (i.e. $\Delta\tau_{k-1}$). As a result, the term (i.e. $\bar{v}_{front}\tau$) in Eqs. (7.61) and (7.62) can be replaced by the following expression:

$$\bar{v}_{front}\tau = \bar{v}_{front}(\tau_k)\tau_k = (\bar{v}_{front})_{reference} \sum_{i=1}^{k} \frac{\Delta\tau_i}{\bar{\rho}_f(\tau_{k-1})} \quad (k = 1, 2, 3, \ldots, N). \quad (7.83)$$

Equation (7.83) clearly indicates that the dimensionless propagating speed of the chemical dissolution front (i.e. $\bar{v}_{front}$) in the case of the pore-fluid being compressible is dependent on the dimensionless density of the pore-fluid. Since the pore-fluid density is a different constant for a different time interval, the overall effect of the pore-fluid compressibility on the dimensionless propagating speed of the chemical dissolution front is considered in the theoretical analysis. In fact, when the time interval approaches zero, the dimensionless propagating speed of the chemical dissolution front (i.e. $\bar{v}_{front}$) in the case of the pore-fluid being compressible can approach the exact solution. This means that Eq. (7.83) can be used to compute the analytical solutions for all the compressible pore-fluids in subcritical chemical dissolution systems. Nevertheless, Eq. (7.83) is not applicable to critical and supercritical chemical systems, for which analytical solutions are not available, so that computational methods need to be used for producing numerical solutions, as discussed later.

Since $\bar{\rho}_f(\tau_{k-1})$ is a function of the dimensionless pressure of the pore-fluid, it can be evaluated using Eq. (7.23) as follows:

$$\bar{\rho}_f(\tau_{k-1}) = \exp\{\bar{\beta}[\bar{p}(\tau_{k-1}) - \bar{p}_{ref}]\} \quad (k = 1, 2, 3, \ldots, N), \quad (7.84)$$

where $\bar{p}(\tau_{k-1})$ is the dimensionless pressure of the pore-fluid at the location of the chemical dissolution front (i.e. $\bar{x} = \bar{s}_{k-1}$) at the end of the (k-1)th time interval (i.e. $\Delta\tau_{k-1}$).

7.2.3 Effects of Pore-Fluid Compressibility on the Evolution of a Chemical Dissolution Front in Subcritical Chemical Dissolution Systems

The present analytical solutions can be used to investigate the effect of pore-fluid compressibility on the evolutions of chemical dissolution fronts in subcritical systems. Towards this goal, the following parameters are used in the corresponding computations. The dimensionless pore-fluid pressure gradient (i.e. $\bar{p}'_{fx} = -1$) is applied on the left boundary of the conceptual model, so that there is a horizontal throughflow from the left to the right of the conceptual model. The initial porosity of the porous medium is 0.1 (i.e. $\phi_0 = 0.1$), while the initial dimensionless chemical-species concentration is one within the computational domain that is defined as the space occupied by the conceptual model. The final porosity after the depletion of the soluble mineral is 0.2 (i.e. $\phi_f = 0.2$). The reference dimensionless density of the pore-fluid is unity, and the dimensionless pore-fluid pressure is equal to 100 (i.e. $\bar{p}_0 = 100$) at the right boundary of the model. To evaluate the value of the ratio, $\psi(\phi_f)/\psi(\phi_0)$, the permeability of the porous medium is calculated using the Carman-Kozeny formula, which has the power of 3 in the power law. The dimensionless time-step length (i.e. the dimensionless time-interval) is set to be a constant of 0.005 in the computation. Note that when pore-fluid is compressible, no criterion is available for assessing the instability of a chemical dissolution front in the fluid-saturated porous medium. In this case, the criterion that is derived when pore-fluid is incompressible can be used, as an approximation, to judge the instability of a chemical dissolution front in the fluid-saturated porous medium where pore-fluid is compressible. The critical Zhao number of the chemical dissolution system is equal to 1.77 when the pore-fluid is incompressible. Since the Zhao number of the system is equal to unity (i.e. $Zh = -\bar{p}'_{fx} = 1$), which is smaller than its corresponding critical value, the chemical dissolution system under consideration is subcritical so that a planar dissolution front remains planar during its propagation within the model. When the pore-fluid is incompressible, the dimensionless propagating speed of the chemical dissolution front (i.e. $(\bar{v}_{front})_{reference}$) is equal to 10, which can be used in Eq. (7.83).

Figure 7.2 shows the effects of pore-fluid compressibility on the evolution of the chemical dissolution front in the subcritical system at two dimensionless time instants, namely $\tau = 0.04$ and $\tau = 0.08$ respectively. To produce the analytical results shown in this figure, four different values of the dimensionless pore-fluid compressibility coefficient, namely $\bar{\beta} = 0$, 0.0001, 0.0005 and 0.001, have been used in the corresponding computations. In the case of $\bar{\beta} = 0$, the pore-fluid becomes incompressible, so that Eqs. (7.74) and (7.75) are used, instead of Eqs. (7.61) and (7.62), which are used for the other three values of the dimensionless pore-fluid compressibility coefficient. Since the mineral dissolution ratio is equal to zero (i.e. $\varepsilon = 0$), sharp chemical dissolution fronts have been obtained for all the four values of the dimensionless pore-fluid compressibility coefficient. It is noted that when the value of the dimensionless pore-fluid compressibility coefficient becomes smaller and smaller, the chemical dissolution front propagates slower and slower. This indicates that with an increase in pore-fluid compressibility, there is a decrease in the dimensionless propagating speed of the chemical dissolution front in a subcritical system.

Figure 7.3 shows the effect of pore-fluid compressibility on the distribution of the dimensionless pore-fluid pressure, while Fig. 7.4 shows the effect of pore-fluid compressibility on the distribution of the dimensionless pore-fluid density in the chemical dissolution system. In these two figures, the equal-value lines of either

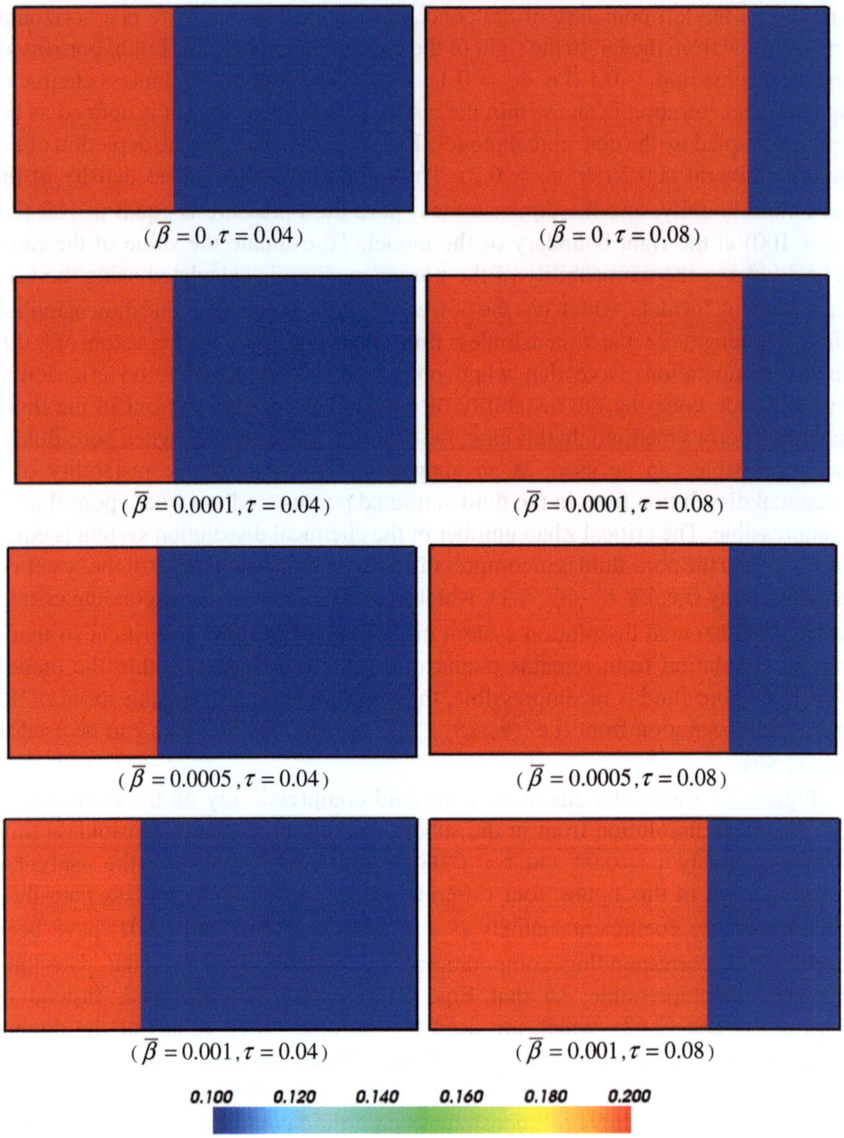

Fig. 7.2 Effects of pore-fluid compressibility on the evolution of the chemical dissolution front in a subcritical system (Porosity)

the dimensionless pore-fluid pressure or the dimensionless pore-fluid density have been displayed. It is observed that the space between two equal-value lines of the dimensionless pore-fluid pressure in the upstream direction of the chemical dissolution front is smaller than that in the downstream direction of the chemical dissolution front. This is because that the porosity of the porous medium in the upstream direction of the chemical dissolution front is greater than that in the

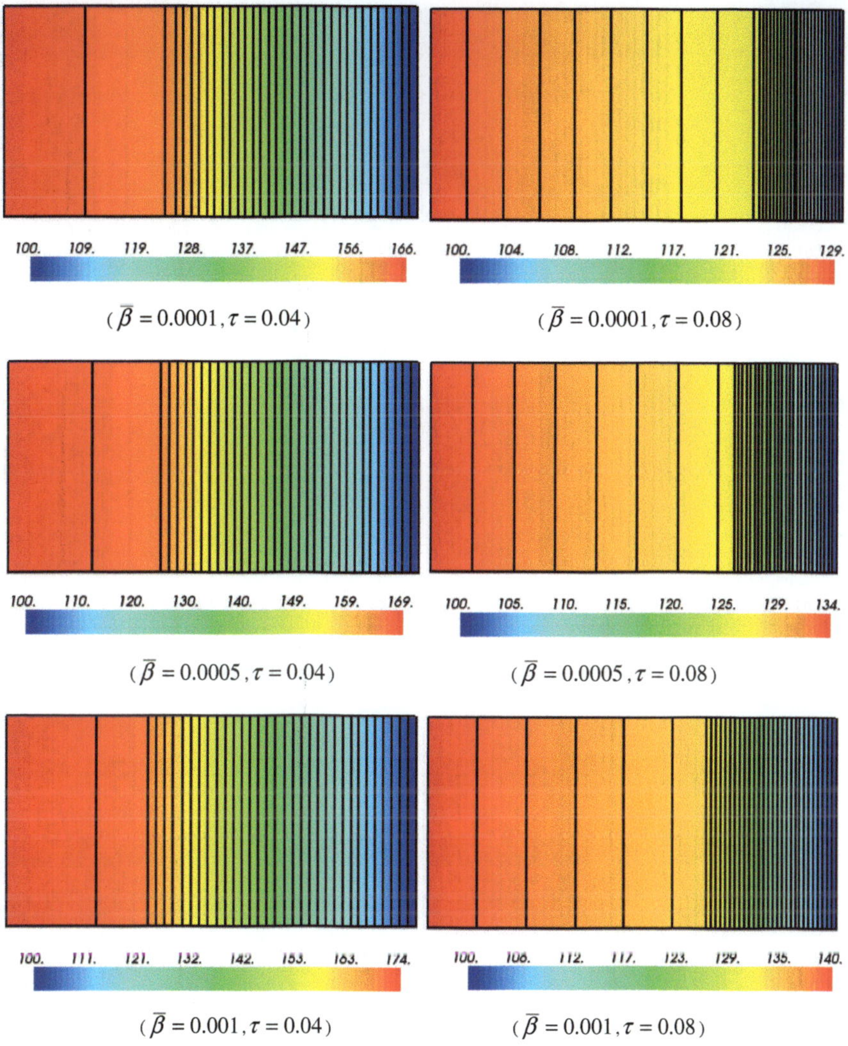

Fig. 7.3 Effects of pore-fluid compressibility on the distribution of the dimensionless pore-fluid pressure in a subcritical system

downstream direction of the chemical dissolution front. As a result, the absolute value of the dimensionless pressure gradient of the pore-fluid in the upstream direction of the chemical dissolution front is smaller than that in the downstream direction of the chemical dissolution front. Due to the strong dependence of the dimensionless pore-fluid pressure, the distribution pattern of the dimensionless pore-fluid density is exactly the same as that of the dimensionless pore-fluid pressure for all the four values of the dimensionless pore-fluid compressibility coefficients under consideration.

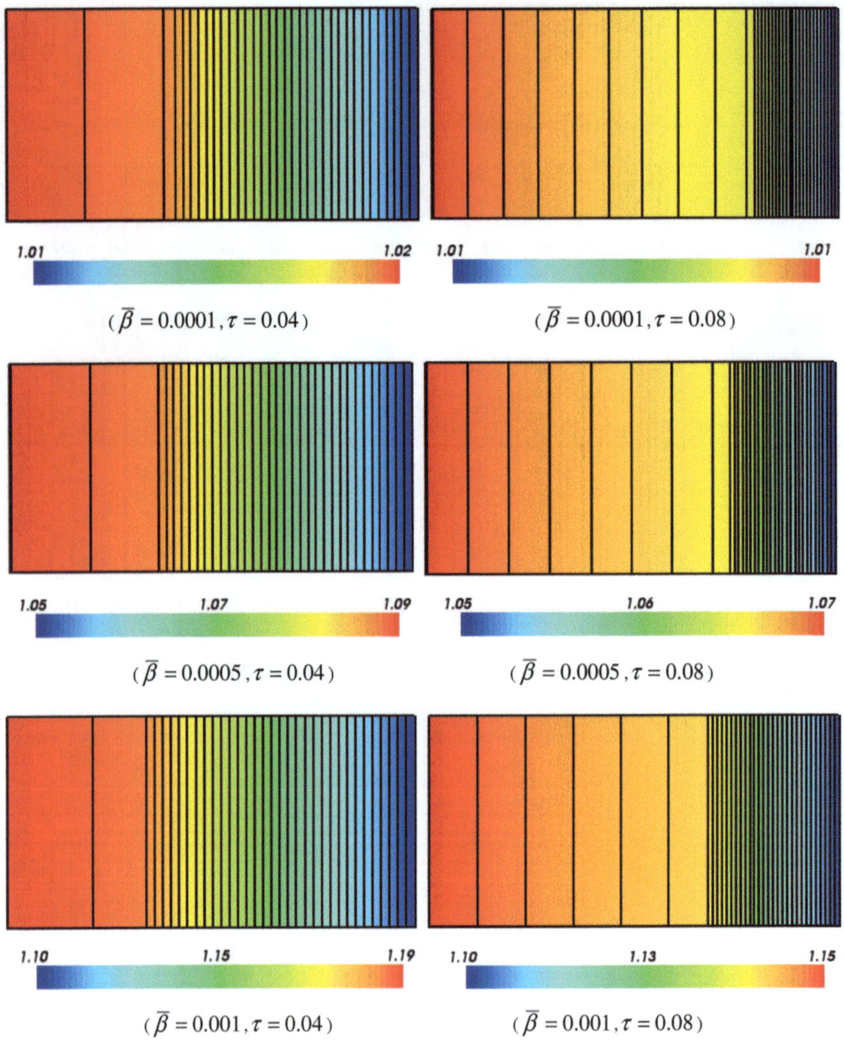

Fig. 7.4 Effects of pore-fluid compressibility on the distribution of the dimensionless pore-fluid density in a subcritical system

7.3 Computational Theory for Simulating the Morphological Evolution of a Chemical Dissolution Front with the Consideration of Pore-Fluid and Medium Compressibility

Although analytical solutions have been derived for subcritical chemical dissolution systems under some special circumstances, it is very difficult, if not impossible, to derived any analytical solutions for the morphological evolutions of chemical dissolution fronts in supercritical chemical dissolution systems. As a result, numerical solutions are alternatively produced in this situation. Since this study is an extension of the previous work where the pore-fluid is considered to be incompressible (Zhao et al. 2008), the previously-developed numerical methods need to be modified to include the consideration of pore-fluid and medium compressibility, so that they can be used to simulate morphological evolution of a chemical dissolution front in a supercritical chemical dissolution system where both pore-fluid and medium are compressible. For this purpose, the dimensionless governing equations of the problem to be solved, which have been derived in Sect. 7.1, can be summarized as follows:

$$\varepsilon \frac{\partial(\phi \bar{\rho}_f)}{\partial \tau} - \nabla \cdot [\bar{\rho}_f \psi^*(\phi) \nabla \bar{p}] = 0, \tag{7.85}$$

$$\varepsilon \frac{\partial}{\partial \tau}(\phi \bar{C}) - \nabla \cdot [D^*(\phi) \nabla \bar{C} + \bar{C} \psi^*(\phi) \nabla \bar{p}] \quad \frac{\partial(\phi - \bar{\alpha}\bar{p})}{\partial \tau} = 0, \tag{7.86}$$

$$\varepsilon \frac{\partial(\phi - \bar{\alpha}\bar{p})}{\partial \tau} + (\phi_f - \phi)(\bar{C} - 1) = 0, \tag{7.87}$$

$$\bar{\rho}_f = \exp\left[\bar{\beta}(\bar{p} - \bar{p}_{ref})\right]. \tag{7.88}$$

Since the pore-fluid density is only dependent on the pore-fluid pressure, Eq. (7.88) can be directly computed once the pore-fluid pressure is obtained from solving Eqs. (7.85)–(7.87). The procedure is comprised of two main methods, namely the finite difference method (Zhao et al. 1994) and finite element method (Zienkiewicz 1977). By using the finite difference method, Eqs. (7.85)–(7.87) are firstly turned into finite difference equations for a given time-step in the time domain. Then, by using the finite element method, the corresponding finite difference equations are further transformed into finite element equations in the spatial domain. If the pore fluid is incompressible, the resulting finite element equations can be solved using the conventional numerical algorithms (Chadam et al. 1986; Chen and Liu 2002; Zhao et al. 2008). However, if the pore-fluid is compressible, as considered in this investigation, the porosity gradient is involved in the solution process. For the chemical dissolution-front propagation problem under consideration, the porosity can have a jump at the location of the chemical dissolution front, from the theoretical point of view. This means that the porosity gradient is very large or even infinite in the case of the mineral dissolution ratio (ε) approaching zero. Consequently, the conventional numerical algorithms diverge when they are directly used to solve the resulting finite element equations of a chemical dissolution problem. To improve the solution convergence, a porosity-gradient replacement approach, in which the porosity gradient is replaced by the dimensionless pore-fluid pressure-gradient and dimensionless pore-fluid density-gradient in a strictly mathematical manner, is proposed (Zhao et al. 2012).

7.3.1 Derivation of the Finite Difference Equations of the Problem

For a given dimensionless time-step, $\tau + \Delta\tau$, the unknown porosity can be denoted by $\phi_{\tau+\Delta\tau} = \phi_\tau + \Delta\phi_{\tau+\Delta\tau}$, where ϕ_τ is the known porosity at the previous time-step and $\Delta\phi_{\tau+\Delta\tau}$ is the unknown porosity increment at the current time-step. Once $\Delta\phi_{\tau+\Delta\tau}$ is found, the unknown porosity (i.e. $\phi_{\tau+\Delta\tau}$) at the current time-step can be straightforwardly obtained. By using the backward difference scheme, Eq. (7.87) can be written as follows:

$$\left[\frac{\varepsilon}{\Delta\tau} + (1 - \bar{C}_{\tau+\Delta\tau})\right]\Delta\phi_{\tau+\Delta\tau} = (\phi_f - \phi_\tau)(1 - \bar{C}_{\tau+\Delta\tau}) + \frac{\varepsilon\bar{\alpha}}{\Delta\tau}(\bar{p}_{\tau+\Delta\tau} - \bar{p}_\tau), \quad (7.89)$$

where $\bar{C}_{\tau+\Delta\tau}$ is the dimensionless chemical-species concentration at the current time-step; $\bar{p}_{\tau+\Delta\tau}$ and $\bar{p}_\tau$ are the dimensionless pore-fluid pressures at the current and previous time-steps, respectively; $\Delta\tau$ is the dimensionless time increment at the current time-step.

Mathematically, there exist the following relationships in the finite difference sense.

$$\varepsilon\frac{\partial(\phi\bar{C})}{\partial\tau} = \varepsilon\frac{\Delta(\phi_{\tau+\Delta\tau}\bar{C}_{\tau+\Delta\tau})}{\Delta\tau} = \varepsilon\bar{C}_{\tau+\Delta\tau}\frac{\Delta\phi_{\tau+\Delta\tau}}{\Delta\tau} + \varepsilon\phi_{\tau+\Delta\tau}\frac{\Delta(\bar{C}_{\tau+\Delta\tau})}{\Delta\tau}, \quad (7.90)$$

$$\frac{\partial(\phi - \overline{\alpha p})}{\partial\tau} = \frac{\Delta(\phi_{\tau+\Delta\tau} - \overline{\alpha p}_{\tau+\Delta\tau})}{\Delta\tau} = \frac{1}{\varepsilon}(1 - \bar{C}_{\tau+\Delta\tau})(\phi_f - \phi_{\tau+\Delta\tau}), \quad (7.91)$$

$$\nabla\cdot[D^*(\phi)\nabla\bar{C}] = \nabla\cdot\left[D^*(\phi_{\tau+\Delta\tau})\nabla\bar{C}_{\tau+\Delta\tau}\right], \quad (7.92)$$

$$\begin{aligned}\nabla\cdot[\bar{C}\psi^*(\phi)\nabla\bar{p}] &= \bar{C}\nabla\cdot[\psi^*(\phi)\nabla\bar{p}] + \nabla\bar{p}\cdot[\psi^*(\phi)\nabla\bar{C}] \\ &= \bar{C}_{\tau+\Delta\tau}\nabla\cdot[\psi^*(\phi_{\tau+\Delta\tau})\nabla\bar{p}_{\tau+\Delta\tau}] + \nabla\bar{p}_{\tau+\Delta\tau}\cdot\left[\psi^*(\phi_{\tau+\Delta\tau})\nabla\bar{C}_{\tau+\Delta\tau}\right].\end{aligned} \quad (7.93)$$

Substituting Eqs. (7.90)–(7.93) into Eq. (7.86) yields the following finite difference equation:

$$\begin{aligned}&\left\{\frac{\varepsilon}{\Delta\tau}(2\phi_{\tau+\Delta\tau} - \phi_\tau) + \frac{1}{\varepsilon}(\phi_f - \phi_{\tau+\Delta\tau}) - \nabla\cdot[\psi^*(\phi_{\tau+\Delta\tau})\nabla\bar{p}_{\tau+\Delta\tau}]\right\}\bar{C}_{\tau+\Delta\tau} \\ &- \nabla\cdot\left[D^*(\phi_{\tau+\Delta\tau})\nabla\bar{C}_{\tau+\Delta\tau}\right] - \nabla\bar{p}_{\tau+\Delta\tau}\cdot\left[\psi^*(\phi_{\tau+\Delta\tau})\nabla\bar{C}_{\tau+\Delta\tau}\right] \\ &= \frac{\varepsilon}{\Delta\tau}\phi_{\tau+\Delta\tau}\bar{C}_\tau + \frac{1}{\varepsilon}(\phi_f - \phi_{\tau+\Delta\tau}).\end{aligned} \quad (7.94)$$

Equation (7.91) can be rewritten in the following discretized form:

$$\varepsilon\frac{\Delta\phi_{\tau+\Delta\tau}}{\Delta\tau} = (1 - \bar{C}_{\tau+\Delta\tau})(\phi_f - \phi_{\tau+\Delta\tau}) + \varepsilon\bar{\alpha}\frac{\Delta\bar{p}_{\tau+\Delta\tau}}{\Delta\tau}. \quad (7.95)$$

Therefore, the following equation exists mathematically:

$$\varepsilon\frac{\partial(\phi\bar{\rho}_f)}{\partial\tau} = \varepsilon\frac{\Delta[\phi_{\tau+\Delta\tau}(\bar{\rho}_f)_{\tau+\Delta\tau}]}{\Delta\tau} = \varepsilon(\bar{\rho}_f)_{\tau+\Delta\tau}\frac{\Delta\phi_{\tau+\Delta\tau}}{\Delta\tau} + \varepsilon\phi_{\tau+\Delta\tau}\frac{\Delta[(\bar{\rho}_f)_{\tau+\Delta\tau}]}{\Delta\tau}. \quad (7.96)$$

Substituting Eq. (7.95) into Eq. (7.96) yields the following equation:

$$\begin{aligned}\varepsilon\frac{\partial(\phi\bar{\rho}_f)}{\partial\tau} &= (\bar{\rho}_f)_{\tau+\Delta\tau}(1 - \bar{C}_{\tau+\Delta\tau})(\phi_f - \phi_{\tau+\Delta\tau}) \\ &\quad + \varepsilon\bar{\alpha}(\bar{\rho}_f)_{\tau+\Delta\tau}\frac{\Delta\bar{p}_{\tau+\Delta\tau}}{\Delta\tau} + \varepsilon\phi_{\tau+\Delta\tau}\frac{\Delta(\bar{\rho}_f)_{\tau+\Delta\tau}}{\Delta\tau}.\end{aligned} \quad (7.97)$$

Consequently, Eq. (7.85) can be rewritten in the following discretized form:

$$\begin{aligned}\nabla\cdot[\bar{\rho}_f\psi^*(\phi)\nabla\bar{p}] &= \nabla\cdot[(\bar{\rho}_f)_{\tau+\Delta\tau}\psi^*(\phi_{\tau+\Delta\tau})\nabla\bar{p}_{\tau+\Delta\tau}] \\ &= (\bar{\rho}_f)_{\tau+\Delta\tau}(1 - \bar{C}_{\tau+\Delta\tau})(\phi_f - \phi_{\tau+\Delta\tau}) + \varepsilon\bar{\alpha}(\bar{\rho}_f)_{\tau+\Delta\tau}\frac{\Delta\bar{p}_{\tau+\Delta\tau}}{\Delta\tau} + \varepsilon\phi_{\tau+\Delta\tau}\frac{\Delta(\bar{\rho}_f)_{\tau+\Delta\tau}}{\Delta\tau}.\end{aligned} \quad (7.98)$$

Equation (7.98) can be straightforwardly expressed as follows:

$$\begin{aligned}&\frac{\varepsilon\bar{\alpha}}{\Delta\tau}(\bar{\rho}_f)_{\tau+\Delta\tau}\bar{p}_{\tau+\Delta\tau} - \nabla\cdot[(\bar{\rho}_f)_{\tau+\Delta\tau}\psi^*(\phi_{\tau+\Delta\tau})\nabla\bar{p}_{\tau+\Delta\tau}] \\ &= \frac{\varepsilon\bar{\alpha}}{\Delta\tau}(\bar{\rho}_f)_{\tau+\Delta\tau}\bar{p}_{\tau} - (\bar{\rho}_f)_{\tau+\Delta\tau}(1 - \bar{C}_{\tau+\Delta\tau})(\phi_f - \phi_{\tau+\Delta\tau}) - \varepsilon\phi_{\tau+\Delta\tau}\frac{\Delta(\bar{\rho}_f)_{\tau+\Delta\tau}}{\Delta\tau}.\end{aligned} \quad (7.99)$$

To update the density of the pore-fluid, Eq. (7.88) can be also expressed in the following form:

$$(\bar{\rho}_f)_{\tau+\Delta\tau} = \exp\left[\bar{\beta}(\bar{p}_{\tau+\Delta\tau} - p_{ref})\right]. \quad (7.100)$$

7.3.2 Derivation of the Proposed Porosity-Gradient Replacement Approach

If the segregated scheme is used, the governing equations of the chemical dissolution problem can be solved one by one in a sequential manner. This can avoid to forming the global matrices of the system so that it can generally cause a significant reduction in the requirement for computer memory. To derive the proposed porosity-gradient replacement approach, the corresponding finite difference equations of the problem can be summarized as follows:

$$\left[\frac{\varepsilon}{\Delta\tau} + (1 - \bar{C}_{\tau+\Delta\tau})\right]\Delta\phi_{\tau+\Delta\tau} = (\phi_f - \phi_\tau)(1 - \bar{C}_{\tau+\Delta\tau}) + \frac{\varepsilon\bar{\alpha}}{\Delta\tau}(\bar{p}_{\tau+\Delta\tau} - \bar{p}_\tau), \quad (7.101)$$

$$\begin{aligned} &\left\{\frac{\varepsilon}{\Delta\tau}(2\phi_{\tau+\Delta\tau} - \phi_\tau) + \frac{1}{\varepsilon}(\phi_f - \phi_{\tau+\Delta\tau}) - \nabla\cdot[\psi^*(\phi_{\tau+\Delta\tau})\nabla\bar{p}_{\tau+\Delta\tau}]\right\}\bar{C}_{\tau+\Delta\tau} \\ &- \nabla\cdot\left[D^*(\phi_{\tau+\Delta\tau})\nabla\bar{C}_{\tau+\Delta\tau}\right] - \psi^*(\phi_{\tau+\Delta\tau})(\nabla\bar{p}_{\tau+\Delta\tau}\cdot\nabla\bar{C}_{\tau+\Delta\tau}) \\ &= \frac{\varepsilon}{\Delta\tau}\phi_{\tau+\Delta\tau}\bar{C}_\tau + \frac{1}{\varepsilon}(\phi_f - \phi_{\tau+\Delta\tau}), \end{aligned} \quad (7.102)$$

$$\begin{aligned} &\frac{\varepsilon\bar{\alpha}}{\Delta\tau}(\bar{\rho}_f)_{\tau+\Delta\tau}\bar{p}_{\tau+\Delta\tau} - \nabla\cdot[(\bar{\rho}_f)_{\tau+\Delta\tau}\psi^*(\phi_{\tau+\Delta\tau})\nabla\bar{p}_{\tau+\Delta\tau}] \\ &= \frac{\varepsilon\bar{\alpha}}{\Delta\tau}(\bar{\rho}_f)_{\tau+\Delta\tau}\bar{p}_\tau - (\bar{\rho}_f)_{\tau+\Delta\tau}(1 - \bar{C}_{\tau+\Delta\tau})(\phi_f - \phi_{\tau+\Delta\tau}) - \varepsilon\phi_{\tau+\Delta\tau}\frac{\Delta(\bar{\rho}_f)_{\tau+\Delta\tau}}{\Delta\tau}, \end{aligned} \quad (7.103)$$

$$(\bar{\rho}_f)_{\tau+\Delta\tau} = \exp\left[\bar{\beta}(\bar{p}_{\tau+\Delta\tau} - \bar{p}_{ref})\right]. \quad (7.104)$$

Note that since Eqs. (7.101)–(7.104) are a set of fully coupled nonlinear equations, they need to be solved sequentially and iteratively when the segregated scheme is used.

From Sect. 7.1, $\psi^*(\phi)$ and $D^*(\phi)$ can be expressed follows:

$$D^*(\phi) = \frac{\phi^{q+1}}{\phi_f^{q+1}} \quad \left(\frac{3}{2} \le q \le \frac{5}{2}\right), \quad (7.105)$$

$$\psi^*(\phi) = \frac{(1 - \phi_f)^2\phi^3}{\phi_f^3(1 - \phi)^2}. \quad (7.106)$$

When the finite element method is used to directly solve the difference equations (i.e. Eqs. (7.101)–(7.103)) of the problem, the porosity gradient must be involved, as can be demonstrated below.

Mathematically, a variable coefficient term, $\nabla\cdot[\psi^*(\phi_{\tau+\Delta\tau})\nabla\bar{p}_{\tau+\Delta\tau}]$, on the left hand side of Eq. (7.102) can be written in the following form:

$$\nabla\cdot[\psi^*(\phi_{\tau+\Delta\tau})\nabla\bar{p}_{\tau+\Delta\tau}] = \psi^*(\phi_{\tau+\Delta\tau})(\nabla\cdot\nabla\bar{p}_{\tau+\Delta\tau}) + \nabla\psi^*(\phi_{\tau+\Delta\tau})\cdot\nabla\bar{p}_{\tau+\Delta\tau}. \quad (7.107)$$

From Eq. (7.106), $\nabla\psi^*(\phi_{\tau+\Delta\tau})$ can be obtained as follows:

$$\nabla\psi^*(\phi_{\tau+\Delta\tau}) = \frac{2(1 - \phi_f)^2(\phi_{\tau+\Delta\tau})^2}{(\phi_f)^3(1 - \phi_{\tau+\Delta\tau})^3}\nabla\phi_{\tau+\Delta\tau}. \quad (7.108)$$

Equation (7.108) clearly demonstrates that in order to compute $\nabla\psi^*(\phi_{\tau+\Delta\tau})$, it is necessary to evaluate the porosity gradient, $\nabla\phi_{\tau+\Delta\tau}$. Although a different porosity-permeability relationship can be used to replace the current porosity-permeability model, the sharp porosity gradient cannot be avoided unless the permeability of the porous medium is independent of the porosity. In the case of dealing with the mineral dissolution problem in an ore forming system, the mineral dissolution ratio (ε) is usually a very small number (Chadam et al. 1986; Zhao et al. 2008, 2010). This will result in a very sharp chemical dissolution front, so that the porosity gradient at the location of the chemical dissolution front can be very large. The previous theoretical study has demonstrated that when the mineral dissolution ratio (ε) approaches zero in a chemical dissolution system of incompressible pore-fluid, there is a porosity discontinuity just at the location of the chemical dissolution front. This means that the porosity gradient can be very large at the location of the chemical dissolution front, even when the pore-fluid is compressible. For this reason, the conventional numerical algorithms may be hard to converge if they are directly used to solve the resulting finite element equations derived just from the corresponding finite difference equations (i.e. Eqs. (7.101)–(7.103)) of a chemical dissolution problem. To improve the convergence speed of solution, the left hand side of Eq. (7.107) is rewritten in the following form:

$$\nabla\cdot[\psi^*(\phi_{\tau+\Delta\tau})\nabla\bar{p}_{\tau+\Delta\tau}] = \frac{1}{(\bar{\rho}_f)_{\tau+\Delta\tau}}\Big\{\nabla\cdot[(\bar{\rho}_f)_{\tau+\Delta\tau}\psi^*(\phi_{\tau+\Delta\tau})\nabla\bar{p}_{\tau+\Delta\tau}] - \psi^*(\phi_{\tau+\Delta\tau})[\nabla\bar{p}_{\tau+\Delta\tau}\cdot\nabla(\bar{\rho}_f)_{\tau+\Delta\tau}]\Big\}. \quad (7.109)$$

Substituting Eq. (7.98) into Eq. (7.109) yields the following equation:

$$\nabla\cdot[\psi^*(\phi_{\tau+\Delta\tau})\nabla\bar{p}_{\tau+\Delta\tau}] = (1-\bar{C}_{\tau+\Delta\tau})(\phi_f-\phi_{\tau+\Delta\tau}) + \varepsilon\bar{\alpha}\frac{\Delta\bar{p}_{\tau+\Delta\tau}}{\Delta\tau} + \frac{\varepsilon\phi_{\tau+\Delta\tau}}{\Delta\tau}\frac{\Lambda(\bar{\rho}_f)_{\tau+\Delta\tau}}{(\bar{\rho}_f)_{\tau+\Delta\tau}} - \frac{\psi^*(\phi_{\tau+\Delta\tau})}{(\bar{\rho}_f)_{\tau+\Delta\tau}}[\nabla\bar{p}_{\tau+\Delta\tau}\cdot\nabla(\bar{\rho}_f)_{\tau+\Delta\tau}]. \quad (7.110)$$

Equation (7.110) clearly demonstrates that if the term of its right hand side is used to replace that of its left hand side, then the porosity gradient computation can be avoided in the finite difference equations of the chemical dissolution problem involving compressible pore-fluid. This is the reason why the proposed numerical treatment is called the porosity gradient replacement approach.

To facilitate the forthcoming derivation of the finite element equations of the chemical dissolution problem, Eq. (7.110) can be rewritten in the following form:

$$\nabla\cdot[\psi^*(\phi_{\tau+\Delta\tau})\nabla\bar{p}_{\tau+\Delta\tau}] = f\left[\phi_{\tau+\Delta\tau},\bar{C}_{\tau+\Delta\tau},(\bar{\rho}_f)_{\tau+\Delta\tau},\bar{p}_{\tau+\Delta\tau},(\bar{\rho}_f)_\tau,\bar{p}_\tau\right], \quad (7.111)$$

where

$$\begin{aligned} f &= f\left[\phi_{\tau+\Delta\tau}, \bar{C}_{\tau+\Delta\tau}, (\bar{\rho}_f)_{\tau+\Delta\tau}, \bar{p}_{\tau+\Delta\tau}, (\bar{\rho}_f)_{\tau}, \bar{p}_{\tau}\right] \\ &= (1-\bar{C}_{\tau+\Delta\tau})(\phi_f - \phi_{\tau+\Delta\tau}) + \frac{\varepsilon\bar{\alpha}}{\Delta\tau}(\bar{p}_{\tau+\Delta\tau} - \bar{p}_{\tau}) \\ &\quad + \frac{\varepsilon\phi_{\tau+\Delta\tau}}{\Delta\tau}\left(1 - \frac{(\bar{\rho}_f)_{\tau}}{(\bar{\rho}_f)_{\tau+\Delta\tau}}\right) - \frac{\psi^*(\phi_{\tau+\Delta\tau})}{(\bar{\rho}_f)_{\tau+\Delta\tau}}[\nabla\bar{p}_{\tau+\Delta\tau}\cdot\nabla(\bar{\rho}_f)_{\tau+\Delta\tau}]. \end{aligned} \quad (7.112)$$

7.3.3 Derivation of the Finite Element Equations of the Problem

Although the finite element method has been broadly used for solving many different types of scientific and engineering problems, the finite element equations related to the chemical dissolution-front propagation problem with compressible pore-fluid have not been available so far, owing to the high nonlinearity of the problem. The chemical dissolution-front propagation plays the following two critical roles: (1) changing flow channels through porosity creation and (2) generating mass source through chemical dissolution. Even though the finite element method itself is well known, the finite element equations of the chemical dissolution-front propagation problem with compressible pore-fluid are briefly derived below.

To derive the finite element equations of the problem, three unknown variables involved in the finite difference equations (i.e. Eqs. (7.101)–(7.103)) are defined as follows:

$$U_1 = \Delta\phi_{\tau+\Delta\tau}, \quad U_2 = \bar{C}_{\tau+\Delta\tau}, \quad U_3 = \bar{p}_{\tau+\Delta\tau}. \quad (7.113)$$

By using these new definitions, Eqs. (7.101)–(7.103) can be written in the following forms:

$$f_{U1}U_1 = f_{C1}, \quad (7.114)$$

$$f_{U2}U_2 - \psi^*(\phi_{\tau+\Delta\tau})(\nabla\bar{p}_{\tau+\Delta\tau}\cdot\nabla U_2) - \nabla\cdot[D^*(\phi_{\tau+\Delta\tau})\nabla U_2] = f_{C2}, \quad (7.115)$$

$$f_{U3a}U_3 - \nabla\cdot[f_{U3b}\nabla U_3] = f_{C3}, \quad (7.116)$$

where

$$f_{U1} = f_{U1}(\bar{C}_{\tau+\Delta\tau}) = \frac{\varepsilon}{\Delta\tau} + (1 - C_{\tau+\Delta\tau}), \quad (7.117)$$

$$f_{C1} = f_{C1}(\bar{C}_{\tau+\Delta\tau}, \bar{p}_{\tau+\Delta\tau}, \bar{p}_{\tau}) = (\phi_f - \phi_{\tau})(1 - \bar{C}_{\tau+\Delta\tau}) + \frac{\varepsilon\bar{\alpha}}{\Delta\tau}(\bar{p}_{\tau+\Delta\tau} - \bar{p}_{\tau}), \quad (7.118)$$

$$\begin{aligned} f_{U2} &= f_{U2}[\phi_{\tau+\Delta\tau}, \bar{C}_{\tau+\Delta\tau}, (\bar{\rho}_f)_{\tau+\Delta\tau}, \bar{p}_{\tau+\Delta\tau}, (\bar{\rho}_f)_{\tau}, \bar{p}_{\tau}] \\ &= \frac{\varepsilon}{\Delta\tau}(2\phi_{\tau+\Delta\tau} - \phi_{\tau}) + \frac{1}{\varepsilon}(\phi_f - \phi_{\tau+\Delta\tau}) - f], \end{aligned} \quad (7.119)$$

$$f_{C2} = f_{C2}(\phi_{\tau+\Delta\tau}, \bar{C}) = \frac{\varepsilon}{\Delta\tau}\phi_{\tau+\Delta\tau}\bar{C}_\tau + \frac{1}{\varepsilon}(\phi_f - \phi_{\tau+\Delta\tau}), \tag{7.120}$$

$$f_{U3a} = f_{U3}[(\bar{\rho}_f)_{\tau+\Delta\tau}] = \frac{\varepsilon\bar{\alpha}}{\Delta\tau}(\bar{\rho}_f)_{\tau+\Delta\tau}, \tag{7.121}$$

$$f_{U3b} = f_{U3b}[\phi_{\tau+\Delta\tau}, (\bar{\rho}_f)_{\tau+\Delta\tau}] = (\bar{\rho}_f)_{\tau+\Delta\tau}\psi^*(\phi_{\tau+\Delta\tau}), \tag{7.122}$$

$$\begin{aligned} f_{C3} &= f_{C3}\left[\phi_{\tau+\Delta\tau}, \bar{C}_{\tau+\Delta\tau}, (\bar{\rho}_f)_{\tau+\Delta\tau}, (\bar{\rho}_f)_\tau, \bar{p}_\tau\right] \\ &= \frac{\varepsilon\bar{\alpha}}{\Delta\tau}(\bar{\rho}_f)_{\tau+\Delta\tau}\bar{p}_\tau - (\bar{\rho}_f)_{\tau+\Delta\tau}(1 - \bar{C}_{\tau+\Delta\tau})(\phi_f - \phi_{\tau+\Delta\tau}) \\ &\quad - \frac{\varepsilon\phi_{\tau+\Delta\tau}}{\Delta\tau}[(\bar{\rho}_f)_{\tau+\Delta\tau} - (\bar{\rho}_f)_\tau]. \end{aligned} \tag{7.123}$$

Based on the finite element method, the distribution of the above-mentioned three unknown variables in a finite element can be described as follows:

$$U_1^e = [N]\{\Delta_1\}^e, \tag{7.124}$$

$$U_2^e = [N]\{\Delta_2\}^e, \tag{7.125}$$

$$U_3^e = [N]\{\Delta_3\}^e, \tag{7.126}$$

where U_1^e, U_2^e and U_3^e are the distribution fields of the three unknown variables within the finite element; $\{\Delta_1\}^e, \{\Delta_2\}^e$ and $\{\Delta_3\}^e$ are the corresponding nodal vectors of the element; $[N]$ is the shape function matrix of the element.

If the finite element under consideration has Q nodes, then the following expressions exist for this element:

$$(\{\Delta_1\}^e)^T = \{U_{11} \quad U_{12} \quad \ldots \quad U_{1Q}\}, \tag{7.127}$$

$$(\{\Delta_2\}^e)^T = \{U_{21} \quad U_{22} \quad \ldots \quad U_{2Q}\}, \tag{7.128}$$

$$(\{\Delta_3\}^e)^T = \{U_{31} \quad U_{32} \quad \ldots \quad U_{3Q}\}, \tag{7.129}$$

$$[N] = [N_1 \quad N_2 \quad \ldots \quad N_Q], \tag{7.130}$$

where N_i ($i = 1, 2, \ldots, Q$) is the shape function of node i; U_{1i}, U_{2i} and U_{3i} ($i = 1, 2, \ldots, Q$) are the nodal values of the unknown variables for the finite element under consideration.

Using the Galerkin weighted-residual method, Eqs. (7.114)–(7.116) can be rewritten, at the element level, as follows:

$$\left(\iint_A [N]^T f_{U1}[N]\,\mathrm{d}A\right)\{\Delta_1\}^e = \iint_A [N]^T f_{C1}\,\mathrm{d}A, \tag{7.131}$$

$$\left(\iint_A [N]^T f_{U2}[N]\,\mathrm{d}A\right)\{\Delta_2\}^e - \left(\iint_A \psi^*(\phi_\tau + \Delta_\tau)[N]^T(\nabla\bar{p}_{\tau+\Delta\tau}\cdot\nabla[N])\,\mathrm{d}A\right)\{\Delta_2\}^e$$
$$-\left(\iint_A [N]^T\nabla\cdot(D^*(\phi_{\tau+\Delta\tau})\nabla[N]\,\mathrm{d}A\right)\{\Delta_2\}^e = \iint_A [N]^T f_{C2}\,\mathrm{d}A, \tag{7.132}$$

$$\left(\iint_A [N]^T f_{U3a}[N]\,\mathrm{d}A\right)\{\Delta_3\}^e - \left(\iint_A [N]^T\nabla\cdot(f_{U3b}\nabla[N])\,\mathrm{d}A\right)\{\Delta_3\}^e$$
$$= \iint_A [N]^T f_{C3}\,\mathrm{d}A, \tag{7.133}$$

where A is the area of the element.

Note that the following expressions exist mathematically:

$$\left(\iint_A [N]^T\nabla\cdot(D^*(\phi_{\tau+\Delta\tau})\nabla[N])\,\mathrm{d}A\right)\{\Delta_2\}^e = \int_S [N]^T D^*(\phi_{\tau+\Delta\tau})\nabla([N]\{\Delta_2\}^e)\,\mathrm{d}S$$
$$-\left(\iint_A \nabla[N]^T\cdot(D^*(\phi_{\tau+\Delta\tau})\nabla[N])\,\mathrm{d}A\right)\{\Delta_2\}^e, \tag{7.134}$$

$$\left(\iint_A [N]^T\nabla\cdot(f_{U3b}\nabla[N])\,\mathrm{d}A\right)\{\Delta_3\}^e = \int_S [N]^T f_{U3b}\nabla([N]\{\Delta_3\}^e)\,\mathrm{d}S$$
$$-\left(\iint_A \nabla[N]^T\cdot(f_{U3b}\nabla[N])\,\mathrm{d}A\right)\{\Delta_3\}^e, \tag{7.135}$$

where S is the length of the element boundary.

Substituting Eqs. (7.134) and (7.135) into Eqs. (7.132) and (7.133) yields the following expressions:

$$\left(\iint_A [N]^T f_{U2}[N]\,\mathrm{d}A\right)\{\Delta_2\}^e - \left(\iint_A \psi^*(\phi_{\tau+\Delta\tau})[N]^T(\nabla\bar{p}_{\tau+\Delta\tau}\cdot\nabla[N])\,\mathrm{d}A\right)\{\Delta_2\}^e$$
$$+\left(\iint_A \nabla[N]^T\cdot(D^*(\phi_{\tau+\Delta\tau})\nabla[N])\,\mathrm{d}A\right)\{\Delta_2\}^e$$
$$= \iint_A [N]^T f_{C2}\,\mathrm{d}A + \int_S [N]^T D^*(\phi_{\tau+\Delta\tau})\nabla([N]\{\Delta_2\}^e)\,\mathrm{d}S, \tag{7.136}$$

$$\left(\iint_A [N]^T f_{U3a}[N]\,\mathrm{d}A\right)\{\Delta_3\}^e + \left(\iint_A \nabla[N]^T \cdot (f_{U3b}\nabla[N])\,\mathrm{d}A\right)\{\Delta_3\}^e = \iint_A [N]^T f_{C3}\,\mathrm{d}A + \int_S [N]^T f_{U3b}\nabla([N]\{\Delta_3\}^e)\,\mathrm{d}S. \quad (7.137)$$

Consideration of Eqs. (7.131), (7.136) and (7.137) leads to the discretized equations of the single finite element as follows:

$$[M_1]^e\{\Delta_1\}^e = \{P_1\}^e, \quad (7.138)$$

$$([M_2]^e - [H]^e + [K_1]^e)\{\Delta_2\}^e = \{P_2\}^e, \quad (7.139)$$

$$([M_3]^e + [K_2]^e)\{\Delta_3\}^e = \{P_3\}^e, \quad (7.140)$$

where $[M_1]^e$, $[M_2]^e$, $[M_3]^e$, $[H]^e$, $[K_1]^e$ and $[K_2]^e$ are the property matrices of the single finite element; $\{P_1\}^e$, $\{P_2\}^e$ and $\{P_3\}^e$ are the "load" vectors of the single finite element. These matrices and vectors can be expressed as follows:

$$[M_1]^e = \iint_A [N]^T f_{U1}[N]\,\mathrm{d}A, \quad (7.141)$$

$$[M_2]^e = \iint_A [N]^T f_{U2}[N]\,\mathrm{d}A, \quad (7.142)$$

$$[M_3] = \iint_A [N]^T f_{U3a}[N]\,\mathrm{d}A, \quad (7.143)$$

$$[H]^e = \iint_A \psi^*(\phi_{\tau+\Delta\tau})[N]^T(\nabla\bar{p}_{\tau+\Delta\tau} \cdot \nabla[N])\,\mathrm{d}A, \quad (7.144)$$

$$[K_1]^e = \iint_A \nabla[N]^T \cdot (D^*(\phi_{\tau+\Delta\tau})\nabla[N])\,\mathrm{d}A, \quad (7.145)$$

$$[K_2]^e = \iint_A \nabla[N]^T \cdot (f_{U3b}\nabla[N])\,\mathrm{d}A, \quad (7.146)$$

$$\{P_1\}^e = \iint_A [N]^T f_{C1}\,\mathrm{d}A, \quad (7.147)$$

$$\{P_2\}^e = \iint_A [N]^T f_{C2}\,\mathrm{d}A + \int_S [N]^T D^*(\phi_{\tau+\Delta\tau})\nabla([N]\{\Delta_2\}^e)\,\mathrm{d}S = \iint_A [N]^T f_{C2}\,\mathrm{d}A + \int_S [N]^T D^*(\phi_{\tau+\Delta\tau})\nabla(\bar{C}_{\tau+\Delta\tau})^e\,\mathrm{d}S, \quad (7.148)$$

$$\begin{aligned}\{P_3\}^e &= \iint_A [N]^T f_{C3}\,\mathrm{d}A + \int_S [N]^T f_{U3b} \nabla([N]\{\Delta_3\}^e)\mathrm{d}S \\ &= \iint_A [N]^T f_{C3}\mathrm{d}A + \int_S [N]^T f_{U3b} \nabla(\bar{p}_{\tau+\Delta\tau})^e \mathrm{d}S,\end{aligned} \tag{7.149}$$

where $\nabla(\bar{C}_{\tau+\Delta\tau})^e$ and $\nabla(\bar{p}_{\tau+\Delta\tau})^e$ are the dimensionless chemical-species concentration-gradient and pore-fluid pressure-gradient at the boundary, *S*, of the finite element.

Assembling the property matrices and vectors of all the elements in a system yields the following discretized governing equations of the system:

$$[M_1]\{\Delta_1\} = \{P_1\}, \tag{7.150}$$

$$([M_2] - [H] + [K_1])\{\Delta_2\} = \{P_2\}, \tag{7.151}$$

$$([M_3] + [K_2])\{\Delta_3\} = \{P_3\}, \tag{7.152}$$

where $[M_1], [M_2], [M_3], [H], [K_1]$ and $[K_2]$ are the global property matrices of the system; $\{P_1\}, \{P_2\}$ and $\{P_3\}$ are the global "load" vectors of the system; $\{\Delta_1\}, \{\Delta_2\}$ and $\{\Delta_3\}$ are the corresponding global vectors of the system.

As mentioned previously, Eqs. (7.127)–(7.129) are a set of fully coupled nonlinear equations, so that they need to be solved, both sequentially and iteratively, for the porosity, dimensionless chemical-species concentration and dimensionless pore-fluid pressure at the current time-step. Once the dimensionless pore-fluid pressure is obtained, Eq. (7.104) can be used to compute the dimensionless pore-fluid density at the current time-step. However, when Eq. (7.127) is solved for the porosity increment (i.e. $\Delta\phi_{\tau+\Delta\tau}$) at the current time-step, both the dimensionless chemical-species concentration and the dimensionless pore-fluid pressure at the current time-step are not known. Similarly, when Eq. (7.128) is solved for the chemical-species concentration at the current time-step, both the dimensionless pore-fluid pressure and the dimensionless pore-fluid density at the current time-step remain unknown. Even when Eq. (7.129) is solved for the dimensionless pore-fluid pressure at the current time-step, the dimensionless pore-fluid density at the current time-step is still not known. This indicates that Eqs. (7.127)–(7.129) are fully coupled, so that an iteration scheme needs to be used to solve them sequentially and iteratively. At the first iteration step, both the dimensionless chemical-species concentration and the dimensionless pore-fluid pressure at the previous time-step are used as a reasonable guess for them at the current time-step when Eq. (7.127) is solved for the porosity increment (i.e. $\Delta\phi_{\tau+\Delta\tau}$). When $\Delta\phi_{\tau+\Delta\tau}$ is found, the porosity (i.e. $\phi_{\tau+\Delta\tau}$) at the current time-step can be straightforwardly obtained. In the similar way, both the dimensionless pore-fluid pressure and the dimensionless pore-fluid density at the previous time-step are used as a reasonable guess at the current time-step when Eq. (7.128) is solved for the dimensionless chemical-species concentration. Except for the dimensionless pore-fluid density at the previous time-step being used as a reasonable guess at the current time-step, the resulting approximate porosity and dimensionless chemical-species concentration can be used when Eq. (7.129) is solved for the dimensionless pore-fluid pressure at the current time-step. When the dimensionless pore-fluid pressure is obtained, Eq. (7.104) can be used to compute the dimensionless pore-fluid density at the current time-step. At the second iteration step, the same procedure as used in the first iteration step is followed, so that the following convergence criterion can be established after the second iteration step.

$$E = \mathrm{Max}\left(\sqrt{\sum_{i=1}^{N_\phi}\left(\phi_{i,\tau+\Delta\tau}^{k} - \phi_{i,\tau+\Delta\tau}^{k-1}\right)^2},\ \sqrt{\sum_{i=1}^{N_{\bar{C}}}\left(\bar{C}_{i,\tau+\Delta\tau}^{k} - \bar{C}_{i,\tau+\Delta\tau}^{k-1}\right)^2},\ \sqrt{\sum_{i=1}^{N_{\bar{p}}}\left(\bar{p}_{i,\tau+\Delta\tau}^{k} - \bar{p}_{i,\tau+\Delta\tau}^{k-1}\right)^2}\right) < \bar{E}, \tag{7.153}$$

where E and $\bar{E}$ are the maximum error at the kth iteration step and the allowable error limit; $N_\phi, N_{\bar{C}}$ and $N_{\bar{p}}$ are the total numbers of the degree-of-freedom for the porosity, dimensionless chemical-species concentration and dimensionless pore-fluid pressure respectively; k is the index number at the current iteration step and $k-1$ is the index number at the previous iteration step; $\phi_{i,\tau+\Delta\tau}^{k}$, $\bar{C}_{i,\tau+\Delta\tau}^{k}$ and $\bar{p}_{i,\tau+\Delta\tau}^{k}$ are the porosity, dimensionless chemical-species concentration and dimensionless pore-fluid pressure of node i at both the current time-step and the current iteration step; $\phi_{i,\tau+\Delta\tau}^{k}$, $\bar{C}_{i,\tau+\Delta\tau}^{k-1}$ and $\bar{p}_{i,\tau+\Delta\tau}^{k-1}$ are the porosity, dimensionless chemical-species concentration and dimensionless pore-fluid pressure of node i at the current time-step but at the previous iteration step. It is noted that $k \geq 2$ in Eq. (7.153).

The convergence criterion is checked after the second iteration step. If the convergence criterion is not met, then the iteration is repeated at the current time-step. Otherwise, the convergent solution is obtained at the current time-step and the solution procedure goes to the next time-step until the final time-step is reached.

7.3.4 Verification of the Proposed Numerical Procedure

For the purpose of verifying the proposed porosity-gradient replacement approach and the related numerical procedure, it is necessary to produce the corresponding numerical solutions. Towards this goal, the following parameters are used in the related computations. The dimensionless pore-fluid pressure-gradient applied on the left boundary is equal to -1 (i.e. $\bar{p}'_{fx} = -1$), implying a horizontal throughflow from the left to the right of the problem domain. The initial porosity of the porous medium is 0.1 (i.e. $\phi_0 = 0.1$), while the initial dimensionless chemical-species concentration is unity within the problem domain. To reflect the appropriate boundary condition in the benchmark problem, a time-dependent boundary condition of the dimensionless chemical-species concentration (i.e. $\bar{C}(\tau) = \left((-\bar{v}_{front}\tau)E_{C1} + 1\right)^{-\frac{1}{\beta}}$), which can be determined using the analytical solution expressed in Eq. (7.62), is applied on the left boundary of the problem domain. The final porosity after the depletion of the soluble mineral is 0.2 (i.e. $\phi_f = 0.2$). The dimensionless reference pore-fluid density is unity, and the dimensionless pore-fluid pressure is equal to 100 (i.e. $\bar{p}_0 = 100$) at the right boundary of the problem domain. The dimensionless compressibility coefficient of the porous medium is zero (i.e. $\bar{\alpha} = 0$). To evaluate the ratio, $\psi(\phi_f)/\psi(\phi_0)$, the permeability of the porous medium is calculated using the Carman-Kozeny formula, which has the power of 3 in the power law. The mineral dissolution ratio is assumed to be 0.001. The dimensionless time-step length is set to be a constant of 0.005 in the computation. If the pore-fluid is incompressible, the critical Zhao number of the chemical dissolution system is equal to 1.77 (Zhao et al. 2008). Since the Zhao number of the system is equal to unity (i.e. $Zh = -\bar{p}'_{fx} = 1$), which is smaller than its corresponding critical value, the chemical dissolution system under consideration is subcritical when the compressibility of the pore-fluid is neglected. To evaluate the dimensionless propagating speed of the chemical dissolution front (i.e. $\bar{v}_{front}$), the reference dimensionless propagating speed is assumed to be 10 in the computation. Note that this reference dimensionless propagating speed is determined theoretically when the pore-fluid is assumed to be incompressible (Zhao et al. 2008, 2009). The problem domain is first simulated using 19,701 four-node square elements of 20,000 nodal points in total. This generates a fine mesh that is called the basic mesh in this investigation, unless it is otherwise clearly stated.

Figures 7.5, 7.6 and 7.7 shows the comparison of the numerical results and analytical solutions for the porosity, dimensionless chemical-species concentration and dimensionless pore-fluid pressure evolution of the benchmark problem in the case of $\bar{\beta} = 0.0001$ respectively. In these three figures, four dimensionless time instants, namely $\tau = 0.2, \tau = 0.4, \tau = 0.6$ and $\tau = 0.8$, are used to display the numerical results (that are represented by thick lines) and the corresponding analytical solutions (that are represented by thin lines). It is observed that the numerical results agree well with the corresponding analytical solutions, especially for the dimensionless chemical-species concentration and dimensionless pore-fluid pressure (see Figs. 7.6 and 7.7). In the case of obtaining the analytical solutions, the mineral dissolution ratio is equal to zero (i.e. $\varepsilon = 0$), so that the porosity gradient at the chemical dissolution front approaches infinity. However, in the case of obtaining the numerical results, the mineral dissolution ratio is equal to 0.001 (i.e. $\varepsilon = 0.001$), so that the porosity gradients at the chemical dissolution fronts can reach very large values for the chemical dissolution-front propagation problem under consideration. This is the reason why there is a small difference between the theoretically-predicted chemical dissolution front (represented by porosity) and the numerically-simulated one. Nevertheless, compared with the related analytical solutions, the maximum relative error of the numerical results for both the dimensionless chemical-species concentration and the dimensionless pore-fluid

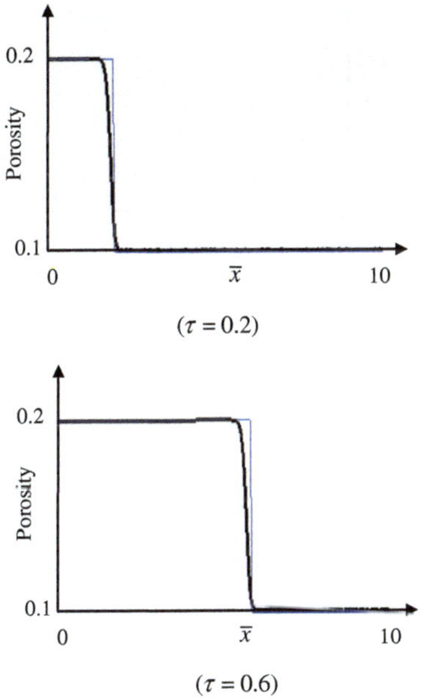

Fig. 7.5 Comparisons of numerical and analytical solutions for porosity evolution in the benchmark problem ($\bar{\beta} = 0.0001$)

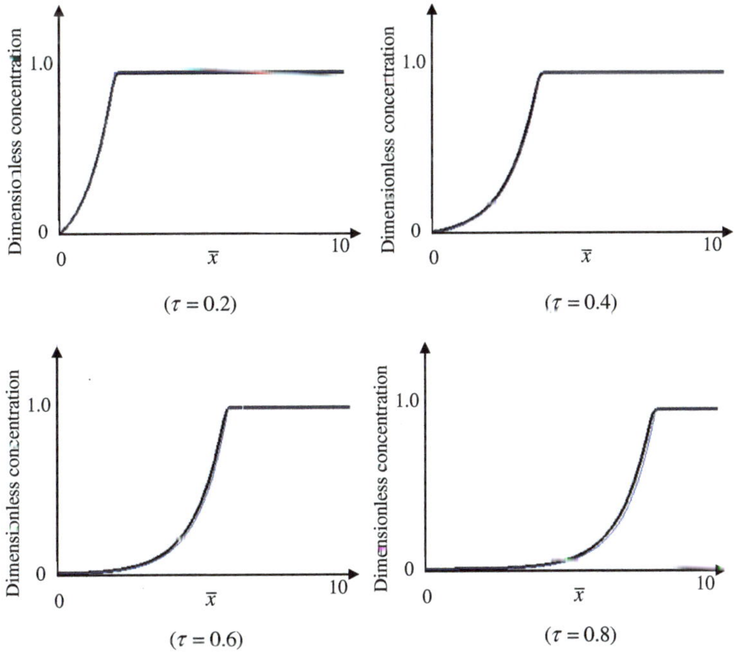

Fig. 7.6 Comparisons of numerical and analytical solutions for dimensionless chemical-species concentration evolution in the benchmark problem ($\bar{\beta} = 0.0001$)

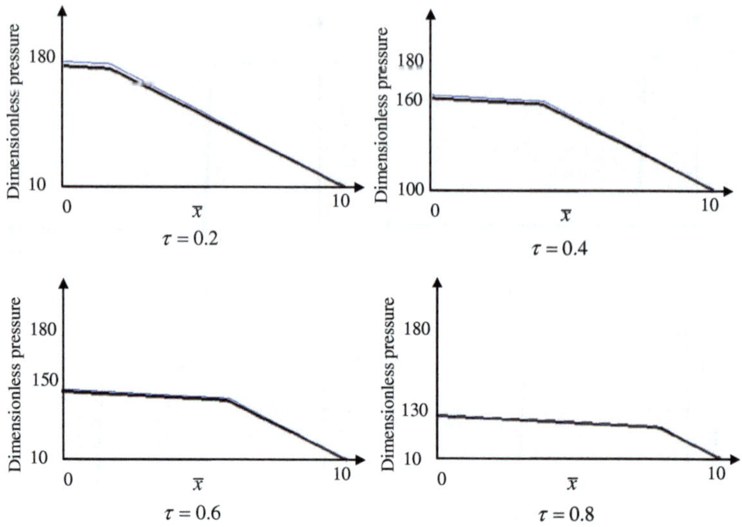

Fig. 7.7 Comparisons of numerical and analytical solutions for dimensionless pore-fluid pressure evolution in the benchmark problem ($\bar{\beta} = 0.0001$)

pressure is smaller than 3 %. This indicates that even though the porosity gradient at the chemical dissolution front is very large, the proposed porosity-gradient replacement approach and the related numerical procedure can be used to produce both correct and accurate numerical results for solving the coupled nonlinear problem between porosity, pore-fluid density, pore-fluid flow and reactive chemical-species transport in the fluid-saturated porous medium with the consideration of pore-fluid compressibility.

Figures 7.8 and 7.9 shows the comparison of the numerical results and analytical solutions for the porosity and dimensionless chemical-species evolution of the benchmark problem when the dimensionless pore-fluid compressibility coefficient (i.e. $\bar{\beta}$) is increased from 0.0001 to 0.0005. In these two figures, thick lines are used to represent numerical simulation results, while thin lines are used to represent the corresponding analytical solutions at four different time instants. Again, the numerical results agree well with the corresponding analytical solutions. This further indicates that both correct and accurate numerical results can be obtained from using the proposed porosity-gradient replacement approach and the related numerical procedure to simulate chemical dissolution-front propagation problems in fluid-saturated porous media including pore-fluid compressibility.

To investigate the solution convergence with mesh sizes, other three meshes of 80000, 5000 and 1250 nodes have been respectively used to simulate the benchmark problem. This means that four different meshes are used in this investigation. If mesh 1, 2, 3 and 4 is used to represent the mesh of 1250, 5000, 20000 and 80000 nodes respectively, then one element of mesh 1 is divided into four smaller

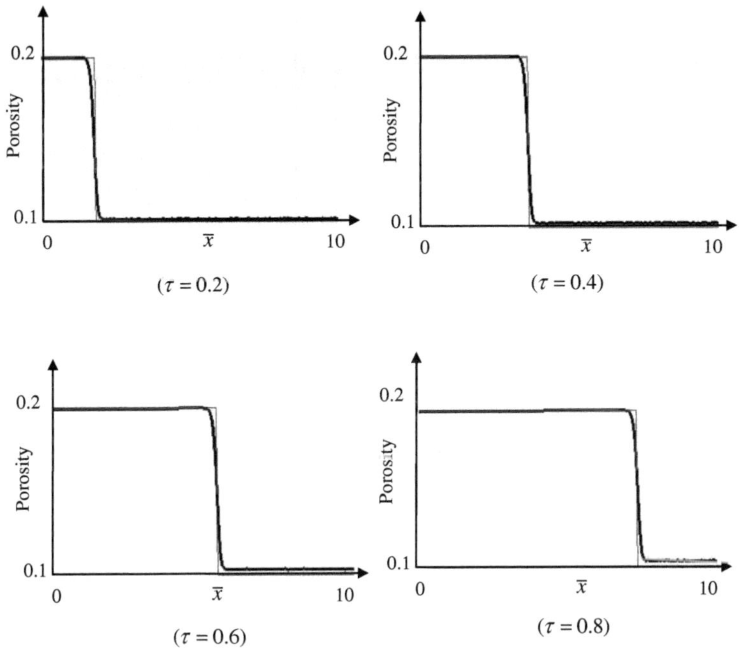

Fig. 7.8 Comparisons of numerical and analytical solutions for porosity evolution in the benchmark problem ($\bar{\beta} = 0.0005$)

elements to form mesh 2. The same division rule has been applied not only between meshes 2 and 3 but between meshes 3 and 4 as well. This means that from mesh 1 to mesh 4, the mesh becomes finer and finer so that the element sizes become smaller and smaller. Figure 7.10 shows the effect of mesh sizes on the solution convergence for the porosity evolution in the benchmark problem due to two different values of the dimensionless pore-fluid compressibility, namely $\bar{\beta} = 0.0001$ and $\bar{\beta} = 0.001$. In this figure, the blue thin line represents the analytical solution, while the black, red, green and blue thick lines represent the numerical solutions obtained from mesh 4, 3, 2 and 1 respectively. It is observed that with the decrease of the mesh size, the numerical solution gradually converge to the analytical solution. Since the black thick line is almost overlapped with the red thick line, it is indicated that the mesh of 20,000 nodes (i.e. the basic mesh) is fine enough if the proposed porosity-gradient replacement approach is used to simulate the benchmark problem. For the purpose of comparing the solution convergence of the proposed porosity-gradient replacement approach with that of the previous method, the conventional finite element method (Zhao et al. 2008) and mesh 2 are used to simulate the same benchmark problem in the case of $\bar{\beta} = 0.001$. The yellow thick line (in Fig. 7.10) shows the related numerical results obtained from the conventional finite element method. Since the green thick line is considerably

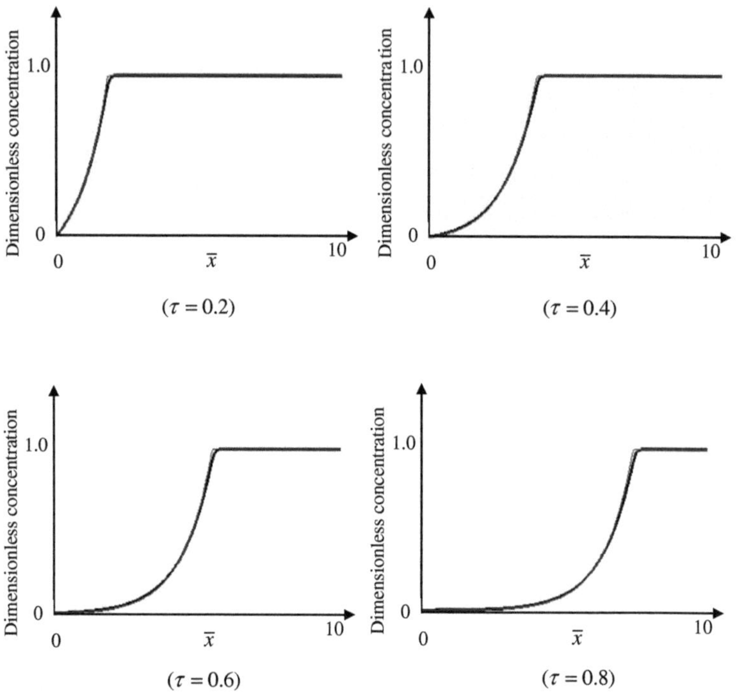

Fig. 7.9 Comparisons of numerical and analytical solutions for dimensionless chemical-species concentration evolution in the benchmark problem ($\bar{\beta} = 0.0005$)

close to the analytical solution than the yellow thick line, it is indicated that the solution convergence of the proposed porosity-gradient replacement approach is faster than that of the conventional finite element method if the same mesh is used.

7.4 Effect of Pore-Fluid Compressibility on the Morphological Evolution of a Chemical Dissolution Front in the Supercritical Chemical Dissolution System

To investigate the effect of pore-fluid compressibility on the evolution of chemical dissolution fronts in a supercritical system, the conceptual model (shown in Fig. 7.1) without considering medium compressibility is simulated using 19,701 four-node rectangular elements of 20,000 nodal points in total. The following parameters are used in the resulting computational model. The dimensionless pore-fluid pressure gradient (i.e. $\bar{p}'_{fx} = -10$) is applied at the left boundary, implying

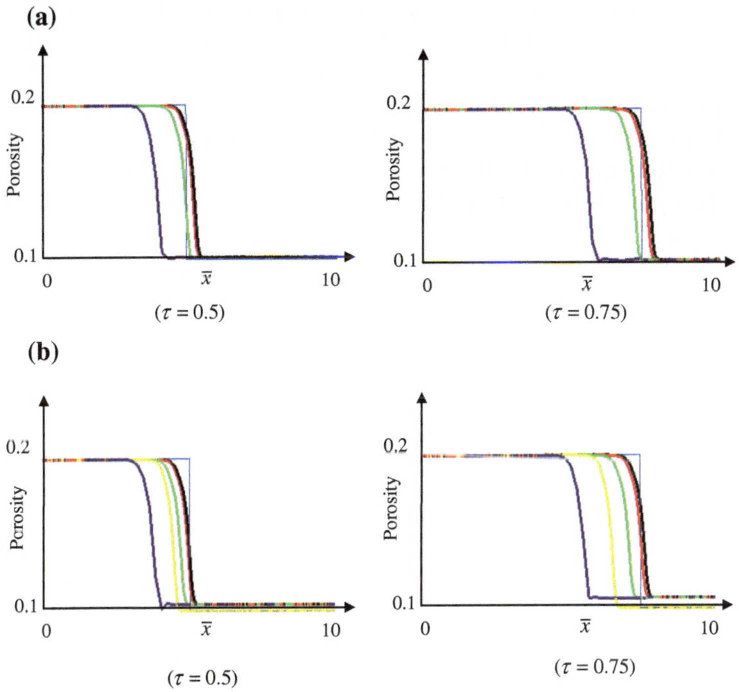

Fig. 7.10 Effect of mesh sizes on the solution convergence for porosity evolution in the benchmark problem: **a** $\bar{\beta} = 0.0001$; **b** $\bar{\beta} = 0.001$

that the Zhao number of the chemical dissolution system is equal to 10. The reference dimensionless density of the pore-fluid is unity, and the dimensionless pore-fluid pressure is equal to 100 (i.e. $\bar{p}_0 = 100$) at the right boundary of the model. Both the top and the bottom boundaries are assumed to be impermeable for the pore-fluid and chemical species. The mineral dissolution ratio is assumed to be 0.001, while the dimensionless time-step length is set to be 0.001 in the computation. In addition, the dimensionless medium compressibility coefficient is assumed to be zero. Other parameters are exactly the same as those used in the verification example of the subcritical Zhao number. Since no criterion is available for assessing the instability of a chemical dissolution front in the fluid-saturated porous medium with a compressible pore-fluid, the existing criterion that is derived when pore-fluid is incompressible can be approximately used to judge the instability of a chemical dissolution front in the fluid-saturated porous medium where the pore-fluid is compressible. Using the above-mentioned parameters, the critical Zhao number of the chemical dissolution system is approximately equal to 1.77 when the pore-fluid is incompressible. Since the dimensionless Zhao number of the system is greater than its critical value, the chemical dissolution system under consideration is supercritical so that a planar dissolution front can evolve into different morphologies during its propagation within the system. In order to simulate the instability of the chemical dissolution front, a small perturbation of 1 % initial porosity is randomly added to the initial porosity field in the computational domain.

Figures 7.11 and 7.12 shows the effect of pore-fluid compressibility on the evolution of the chemical dissolution front in the supercritical system. To examine the effect of pore-fluid compressibility on the response of the chemical dissolution system, four values of the pore-fluid compressibility coefficient, namely $\bar{\beta} = 0$, 0.0001, 0.0005 and 0.0009, have been used in the corresponding computations. From these simulation results, it is observed that with a gradual increase in the value of the pore-fluid compressibility coefficient, the propagating speed of the chemical dissolution front decreases accordingly. The same phenomenon has been also observed from the theoretical solutions for a subcritical chemical dissolution system, which is represented in the previous part of this section. It can be concluded that the propagating speed of a chemical dissolution front in a fluid-saturated porous medium with compressible pore-fluid is smaller than that with incompressible pore-fluid.

The value of the pore-fluid compressibility coefficient can have a significant effect on both the occurring time and the amplitude of the unstable irregular chemical dissolution-front. For instance, the first occurring time of the unstable irregular chemical dissolution-front is about 0.035, 0.035, 0.06 and 0.08 for $\bar{\beta} = 0$, 0.0001, 0.0005 and 0.0009 respectively. Note that both the occurring time and the amplitude of the unstable irregular chemical dissolution-front when $\bar{\beta} = 0$ is almost identical to those when $\bar{\beta} = 0.0001$. This indicates that it is only necessary to consider the compressibility of pore-fluid when it has a dimensionless compressibility coefficient that is greater than 0.0001. Generally, the smaller the pore-fluid compressibility, the faster the unstable irregular chemical dissolution-front grows. This means that the consideration of pore-fluid compressibility stabilizes the growth of an unstable irregular chemical dissolution-front. In other words, the consideration of pore-fluid compressibility makes it difficult for an unstable irregular chemical dissolution-front to grow in a fluid-saturated porous medium.

7.5 Effect of Medium Compressibility on the Morphological Evolution of a Chemical Dissolution Front in the Supercritical Chemical Dissolution System

The same numerical algorithms and procedures as those employed in the previous section can be used to investigate the effect of medium compressibility on the morphological evolution of a chemical dissolution front in the fluid-saturated porous medium of supercritical Zhao numbers. To enable the investigation to be focused on the effect of medium compressibility, the pore-fluid is considered to be

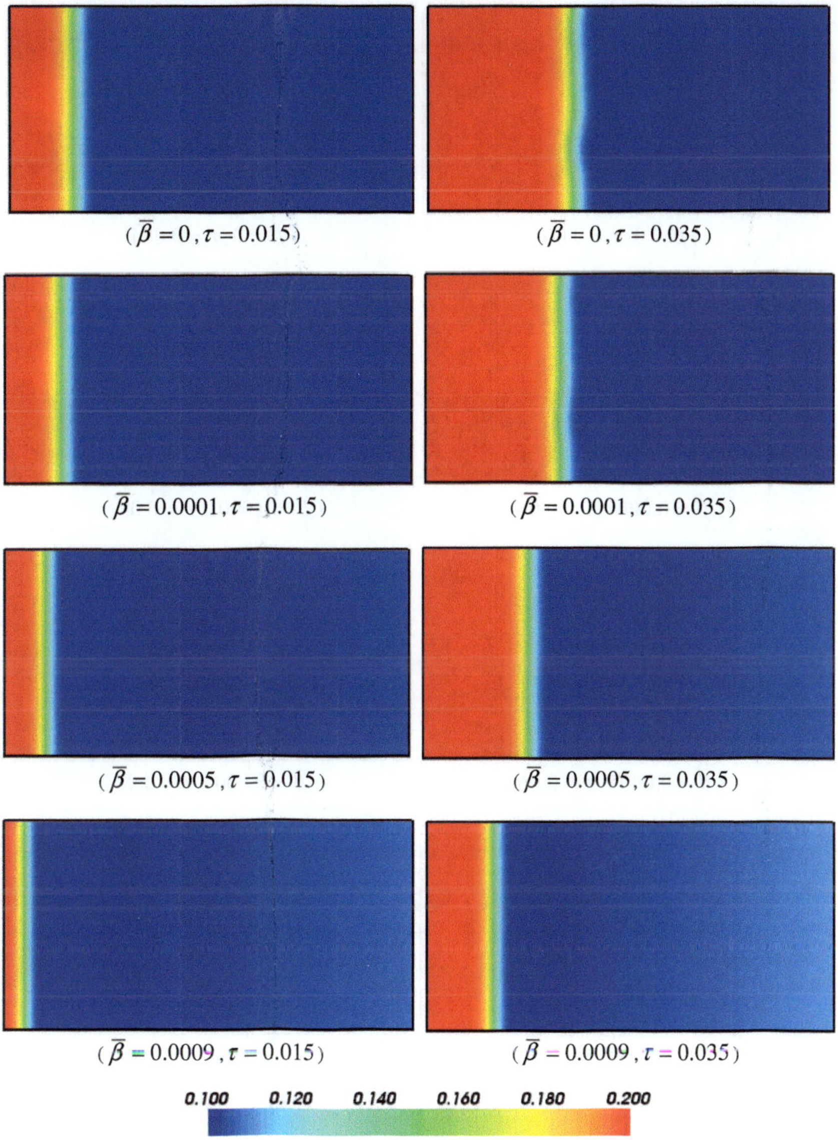

Fig. 7.11 Effects of pore-fluid compressibility on the evolution of the chemical dissolution front in a supercritical system (Porosity)

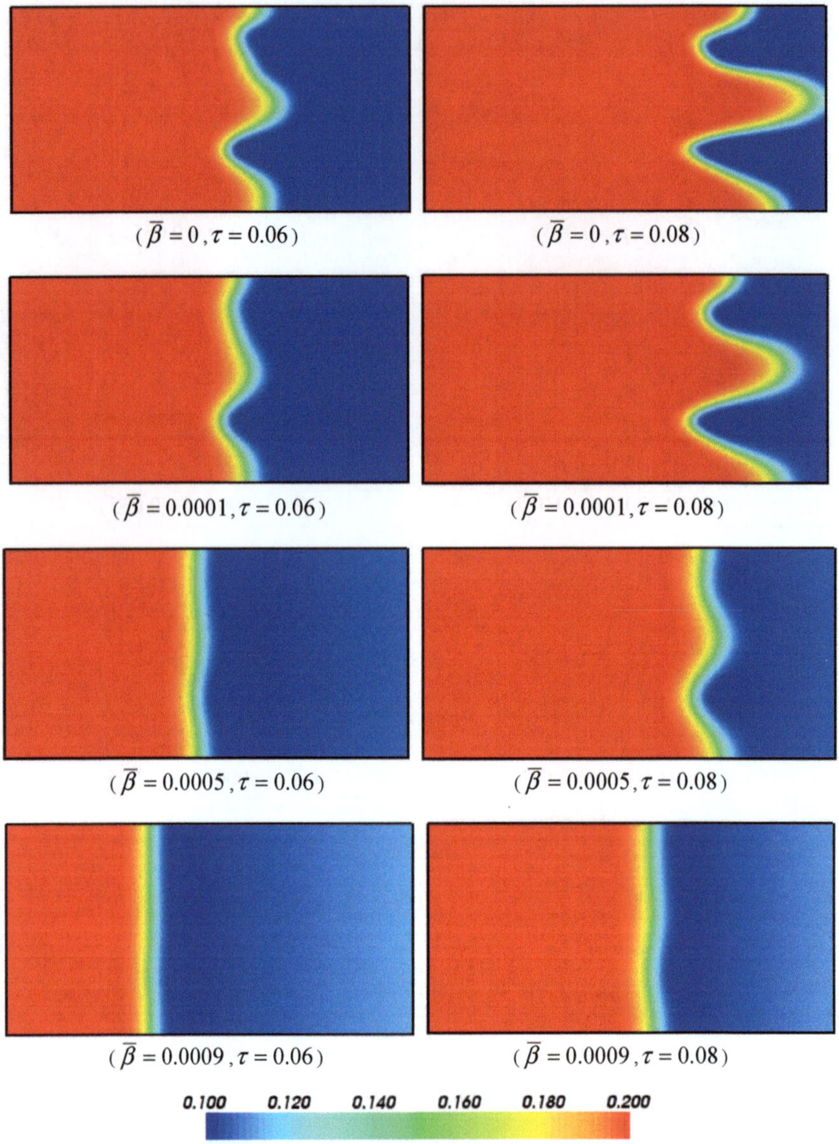

Fig. 7.12 Effects of pore-fluid compressibility on the evolution of the chemical dissolution front in a supercritical system (Porosity)

incompressible in this section. Except for the following changes, the same parameters as those employed in the previous section have been used here. The dimensionless pore-fluid pressure is zero at the right boundary of the computational model, while the dimensionless chemical-species concentration is zero at the left boundary of the computational model. Four values of the dimensionless medium compressibility coefficient, namely $\bar{\alpha} = 0, 1 \times 10^{-5}, 5 \times 10^{-5}$ and 1×10^{-4}, have been used in the corresponding computational models.

Figures 7.13 and 7.14 shows the effect of medium compressibility on the evolution of the chemical dissolution front, which is represented by the porosity of the porous medium, in the supercritical system. From these simulation results, it is observed that medium compressibility has a little influence on the dimensionless propagating speed of the chemical dissolution front, although it may remarkably affect the amplitude of the unstable irregular chemical dissolution-front in the supercritical chemical dissolution system. With the increase of the dimensionless time, the amplitude of the resulting irregular chemical dissolution-front increases significantly, especially for the cases where the medium compressibility coefficient has smaller values. This demonstrates that the chemical dissolution front is morphologically unstable during its propagation within the computational model. Generally, the amplitude of the unstable irregular chemical dissolution-front gradually increases with a gradual decrease in the value of the medium compressible coefficient, indicating that the consideration of medium compressibility stabilizes the growth of an unstable irregular chemical dissolution-front in a supercritical chemical dissolution system. In other words, the consideration of medium compressibility enables an unstable irregular chemical dissolution-front to grow difficultly in the fluid-saturated porous medium of a supercritical Zhao number. Since both the occurring time and the amplitude of the unstable irregular chemical dissolution-front when $\bar{\alpha} = 0$ is almost identical to those when $\bar{\alpha} = 1 \times 10^{-5}$, it has been recognized that it is only necessary to consider medium compressibility for soft materials, rather than for hard rocks, because the latter may have the dimensionless medium compressibility coefficients that are usually smaller than the value of 1×10^{-5}. However, when soft materials are considered, the first approach (as mentioned in the beginning of this chapter), in which the detailed deformation process including a stress-strain constitutive relationship in a porous medium can be considered, may be more appropriate for simulating the deformation of the porous medium associated with the computational simulation of chemical dissolution-front instability problems. This issue needs to be further investigated in a future study.

In summary, the related theoretical and numerical results have demonstrated that: (1) not only can pore-fluid compressibility affect the propagating speeds of chemical dissolution fronts in both subcritical and supercritical chemical dissolution systems, but also it can affect the growth and amplitudes of irregular chemical dissolution-fronts in supercritical chemical dissolution systems; (2) medium compressibility may have a little influence on the propagating speeds of chemical dissolution fronts, but it can have significant effects on the growth and

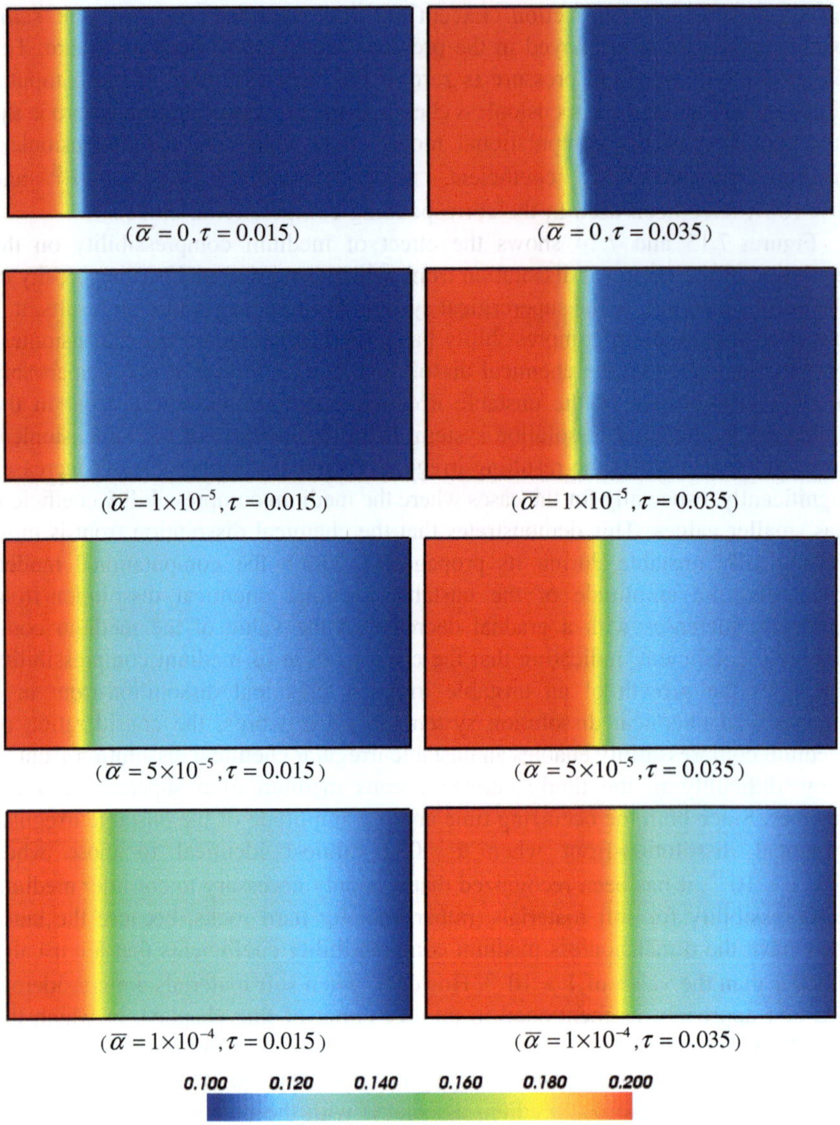

Fig. 7.13 Effects of medium compressibility on the evolution of the chemical dissolution front in a supercritical system (Porosity)

amplitudes of irregular chemical dissolution fronts in supercritical chemical dissolution systems; and (3) both medium and pore-fluid compressibility may stabilize irregular chemical dissolution-fronts in supercritical chemical dissolution systems.

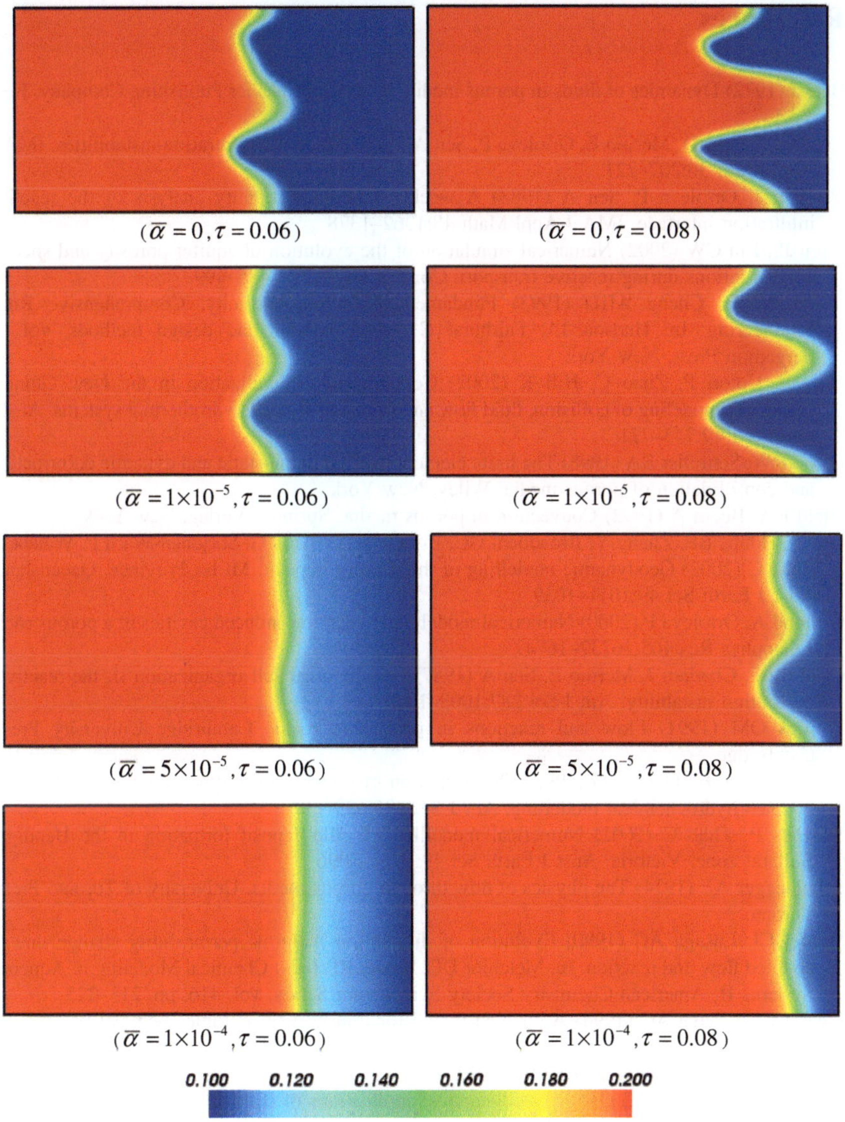

Fig. 7.14 Effects of medium compressibility on the evolution of the chemical dissolution front in a supercritical system (Porosity)

References

Bear J (1972) Dynamics of fluids in porous media. American Elsevier Publishing Company, New York

Chadam J, Hoff D, Merino E, Ortoleva P, Sen A (1986) Reactive infiltration instabilities. IMA J Appl Math 36:207–221

Chadam J, Ortoleva P, Sen A (1988) A weekly nonlinear stability analysis of the reactive infiltration interface. IMA J Appl Math 48:1362–1378

Chen JS, Liu CW (2002) Numerical simulation of the evolution of aquifer porosity and species concentrations during reactive transport. Comput Geosci 28:485–499

Detournay E, Cheng AHD (1993) Fundamentals of poroelasticity, Comprehensive Rock Engineering. In: Hudson JA, Fairhurst C (eds) Analysis and design methods, vol. 2. Pergamon Press, New York

Gow P, Upton P, Zhao C, Hill K (2002) Copper-gold mineralization in the New Guinea: numerical modeling of collision, fluid flow and intrusion-related hydrothermal systems. Aust J Earth Sci 49:753–771

Lewis RW, Schrefler BA (1998) The finite element method in the static and dynamic deformation and consolidation of porous media. Wiley, New York

Nield DA, Bejan A (1992) Convection in porous media. Springer-Verlag, New York

Ord A, Hobbs BE, Zhang Y, Broadbent GC, Brown M, Willetts G, Sorjonen-Ward P, Walshe J, Zhao C (2002) Geodynamic modelling of the century deposit, Mt Isa Province, Queensland. Aust J Earth Sci 49:1011–1039

Ormond A, Ortoleva P (2000) Numerical modeling of reaction-induced cavities in a porous rock. J Geophys Res 105:16737–16747

Ortoleva P, Chadam J, Merino E, Sen A (1987) Geochemical self-organization II: the reactive-infiltration instability. Am J Sci 287:1008–1040

Phillips OM (1991) Flow and reactions in permeable rocks. Cambridge University Press, Cambridge

Raffensperger JP, Garven G (1995) The formation of unconformity-type uranium ore deposits: coupled hydrochemical modelling. Am J Sci 295:639–696

Schaubs P, Zhao C (2002) Numerical modelling of gold-deposit formation in the Bendigo-Ballarat zone, Victoria. Aust J Earth Sci 49:1077–1096

Scheidegger AE (1974) The physics of flow through porous media. University of Toronto Press, Toronto

Steefel CI, Lasaga AC (1990) Evolution of dissolution patterns: permeability change due to coupled flow and reaction. In: Melchior DC, Basset RL (eds) Chemical Modeling in Aqueous Systems II, American Chemistry Society Symposium Series, vol. 416, pp. 213–225

Steefel CI, Lasaga AC (1994) A coupled model for transport of multiple chemical species and kinetic precipitation/dissolution reactions with application to reactive flow in single phase hydrothermal systems. Am J Sci 294:529–592

Turcotte DL, Schubert G (1982) Geodynamics: applications of continuum physics to geological problems. Wiley, New York

Yeh GT, Tripathi VS (1991) A model for simulating transport of reactive multispecies components: model development and demonstration. Water Resour Res 27:3075–3094

Zhao C, Xu TP, Valliappan S (1994) Numerical modelling of mass transport problems in porous media: a review. Comput Struct 53:849–860

Zhao C, Mühlhaus HB, Hobbs BE (1997) Finite element analysis of steady-state natural convection problems in fluid-saturated porous media heated from below. Int J Numer Anal Meth Geomech 21:863–881

Zhao C, Hobbs BE, Mühlhaus HB (1998) Finite element modelling of temperature gradient driven rock alteration and mineralization in porous rock masses. Comput Methods Appl Mech Eng 165:175–187

Zhao C, Hobbs BE, Mühlhaus HB, Ord A, Lin G (2000) Numerical modelling of double diffusion driven reactive flow transport in deformable fluid-saturated porous media with particular consideration of temperature-dependent chemical reaction rates. Eng Comput 17:367–385

Zhao C, Hobbs BE, Mühlhaus HB, Ord A, Lin G (2002) Analysis of steady-state heat transfer through mid-crustal vertical cracks with upward throughflow in hydrothermal systems. Int J Numer Anal Meth Geomech 26:1477–1491

Zhao C, Hobbs BE, Ord A, Hornby P, Peng S, Liu L (2007) Mineral precipitation associated with vertical fault zones: the interaction of solute advection, diffusion and chemical kinetics. Geofluids 7:3–18

Zhao C, Hobbs BE, Hornby P, Ord A, Peng S, Liu L (2008) Theoretical and numerical analyses of chemical-dissolution front instability in fluid-saturated porous rocks. Int J Numer Anal Meth Geomech 32:1107–1130

Zhao C, Hobbs BE, Ord A (2009) Fundamentals of computational geoscience: numerical methods and algorithms. Springer, Berlin

Zhao C, Hobbs BE, Ord A, Peng S (2010) Effects of mineral dissolution ratios on chemical-dissolution front instability in fluid-saturated porous media. Transp Porous Media 82:317–335

Zhao C, Reid LB, Regenauer-Lieb K, Poulet T (2012) A porosity-gradient replacement approach for computational simulation of chemical-dissolution front propagation in fluid-saturated porous media including pore-fluid compressibility. Comput Geosci 16:735–755

Zienkiewicz OC (1977) The finite element method. McGraw-Hill, London

Chapter 8
Computational Simulation of Three-Dimensional Behaviour of Chemical Dissolution-Front Instability in Fluid-Saturated Porous Media

Numerical simulation of chemical dissolution-front instability in fluid-saturated porous media is an important topic in the field of computational geosciences, which is a beautiful marriage between the contemporary computational mechanics and traditional geosciences. If fresh water is injected into a fluid-saturated porous medium with the solute (i.e. chemical-species) being in an equilibrium state, it can break the equilibrium state of the solute so that the whole system becomes chemically far from equilibrium. To drive the system towards a new chemical equilibrium state, the solid part of the solute is dissolved within the porous medium. This process causes an increase in the porosity of the porous medium. The resulting porosity increase can cause a corresponding increase in the permeability of the medium so that the pore-fluid flow can be enhanced within the porous medium. This means that both permeability and diffusivity are dependent on the porosity of a porous medium (Bear 1972). When the injected fresh water flow is relatively weak, the chemical dissolution front is stable so that a planar chemical dissolution-front remains the planar shape during its propagation within the porous medium. However, when the injected fresh water flow is strong enough, the chemical dissolution front becomes unstable. In this case, a planar chemical dissolution-front can be changed into a complicated and complex morphology during its propagation within the porous medium. This is the scientific problem, known as the chemical dissolution-front instability problem in the fluid-saturated porous medium, to be considered in this investigation. It is noted that there is another mechanism that can also cause pore-fluid channeling instability due to the rheological asymmetry associated with compaction and decompaction in ductile rocks (Connolly and Podladchikov 2007). This instability is closely associated with porosity waves and can be referred to as the mechanical instability. Owing to a significant mechanism difference between a chemical dissolution instability problem and a mechanical instability one, this chapter focuses on the consideration of chemical dissolution instability problems. However, a possible interaction between these two different mechanisms may need to be considered in a future investigation.

C. Zhao, *Physical and Chemical Dissolution Front Instability in Porous Media*, Lecture Notes in Earth System Sciences,
DOI: 10.1007/978-3-319-08461-9_8

In essence, a chemical dissolution-front instability problem in the fluid-saturated porous medium can be considered as a complex-system science problem because it has the following two fundamental characteristics: an emergence and an attractor. The emergence describes the sudden change in the behaviour of the system and can be called the singularity of the system, from the mathematical point of view. The attractor describes the corresponding new pattern when the system reaches or passes the emergent point. From the thermodynamics point of view, an attractor is often called the dissipative structure of the system. There are many complex-system science problems in nature. For example, convective instability of pore-fluid flow within the crust of the Earth is a typical problem in the field of complex system science. For this particular problem, the emergent point of a system is expressed by the critical Rayleigh number, which is a dimensionless number to represent both the geometry and the thermodynamic characteristics of the system, while the attractor of the system is expressed by the convective flow pattern, which is a direct consequence when the system becomes critical or supercritical (Phillips 1991; Nield and Bejan 1992; Zhao et al. 1997, 1998, 1999, 2000). A common characteristic of this kind of complex-system science problem is that although the emergent point of the system may be found analytically, the attractor of the system needs to be determined using the numerical method (Zhao et al. 1997, 1998, 1999; Simms and Garven 2004; Yang et al. 2004). From the complex-system science point of view, the emergent point of a chemical dissolution-front instability problem in fluid-saturated porous media can be expressed by the critical value of a dimensionless number. This dimensionless number, known as the Zhao number, should be capable of representing the geometrical, hydrodynamic, thermodynamic and chemical kinetic characteristics of the system in a comprehensive manner (Zhao et al. 2009). If the value of the Zhao number of a reactive transport system with the effect of positive porosity feedback is less than its corresponding critical value, then the shape of a planar chemical dissolution-front remains unchanged during the propagation of the chemical dissolution-front. However, if the Zhao number of a reactive transport system is greater than its corresponding critical value, then the shape of a planar chemical dissolution-front can be changed into a complicated morphology during its propagation in the reactive transport system, where the effect of positive porosity feedback plays an important role in the system.

Since the chemical dissolution-front instability problem in fluid-saturated porous media belongs to a coupled nonlinear problem with some kind of singularity, the solution of the problem losses its uniqueness at the corresponding singularity point. This poses a challenging issue when a conventional numerical method is used to solve a chemical dissolution-front instability problem in fluid-saturated porous media. The reason for this is that when the conventional numerical method is used to solve the chemical dissolution-front instability problem of a homogenous domain, the resulting numerical solution cannot switch automatically from a trivial solution state (e.g. a planar chemical dissolution-front) into a nontrivial solution state (e.g. a complicated morphological chemical dissolution-front). This challenging issue was discussed in detail for the finite element simulation of convective instability problems within fluid-saturated porous media (Zhao et al. 1997). For the numerical simulation of the chemical dissolution-front instability problem of a homogenous domain, it is common practice to use a planar chemical dissolution-front with some kind of small perturbation for generating a chemical dissolution front of complicated morphological shape during the propagation of the perturbed chemical dissolution-front (Chadam et al. 1986, 1988; Ortoleva et al. 1987; Chen and Liu 2002). In this case, the resulting chemical dissolution-front may be strongly dependent on the perturbed chemical dissolution-front. To remove this dependency, an alternative procedure, in which a small randomly-generated perturbation of porosity is added into the initial porosity of the whole homogenous domain, is proposed (Zhao et al. 2008a). The maximum value of the perturbed porosity is assumed to be equal to 1 % the initial porosity of the whole homogenous domain. In the proposed alternative procedure, no perturbation is applied to the initial planar chemical dissolution-front, so that the initial chemical dissolution-front can freely evolve into a chemical dissolution front of complicated morphological shape during its propagation within the system.

Because the coupling effect of pore-fluid flow and chemical reactions plays an important role in the process of ore body formation and mineralization (Steefel and Lasage 1990, 1994; Phillips 1991; Yeh and Tripathi 1991; Raffensperger and Garven 1995; Xu et al. 1999, 2004; Ormond and Ortoleva 2000; Chen and Liu 2002; Gow et al. 2002; Ord et al. 2002; Schaubs and Zhao 2002; Zhao et al. 1997, 2001), the chemical dissolution-front instability in fluid-saturated porous media can become one of the potential mechanisms to control ore body formation and mineralization within the upper crust of the Earth. For example, in some geological situations associated with decarbonation fronts in contact aureoles (Heinrich et al. 1995; Ferry 1996), BIF-hosted iron ores (Morris 1980; Klein and Beukes 1992), roll-front uranium (Wilde and Wall 1987), Irish-style and/or MVT base metal scenarios (Hitzman 1995; Everett et al. 1999; Garven et al. 1999; Appold and Garven 2000), permeability-enhancement along dissolution fronts, potentially with finger-like behaviour, is likely to be important for understanding the related ore forming mechanisms. Although a considerable amount of research has been conducted on the instability of chemical dissolution-front propagation within two-dimensional pore-fluid saturated porous media (Chadam et al. 1986, 1988; Ortoleva et al. 1987; Chen and Liu 2002), limited work has been conducted to simulate the three-dimensional morphological evolution of a chemical dissolution front during its propagation within three-dimensional fluid-saturated porous media (Zhao et al. 2008b). Since most problems in the real world are three-dimensional in geometry, understanding of the behaviour of a problem in three-dimensional cases can usually produce superior insight over two-dimensional models. Thus, the development of a numerical method for simulating the three-dimensional morphological evolution of a chemical dissolution front during its propagation within three-dimensional fluid-saturated porous media becomes one of the main purposes of this chapter.

8.1 Governing Equations of the Problem and the Proposed Numerical Procedure

8.1.1 Governing Equations of the Problem

For a fluid-saturated porous medium, Darcy's law is often used to describe pore-fluid flow, while Fick's law and the related chemical kinetics (Steefel and Lasaga 1994) are commonly used to describe reactive mass transport phenomena. If the pore-fluid (i.e. water in this investigation) is incompressible, the governing equations of the coupled nonlinear problem between porosity, pore-fluid flow and reactive single-chemical-species transport in the fluid-saturated porous medium can be expressed as follows.

$$\frac{\partial \phi}{\partial t} - \nabla \cdot [\psi(\phi)\nabla p] = 0, \tag{8.1}$$

$$\frac{\partial}{\partial t}(\phi C) - \nabla \cdot [\phi D(\phi)\nabla C + C\psi(\phi)\nabla p] + \rho_s k_{Echemical} \frac{\bar{A}_p}{\bar{V}_p}(\phi_f - \phi)(C - C_{eq}) = 0, \tag{8.2}$$

$$\frac{\partial \phi}{\partial t} + k_{Echemical} \frac{\bar{A}_p}{\bar{V}_p}(\phi_f - \phi)(C - C_{eq}) = 0, \tag{8.3}$$

$$\psi(\phi) = \frac{k(\phi)}{\mu}, \tag{8.4}$$

where p and C are the pore-fluid pressure and the concentration of the chemical species; C_{eq} is the equilibrium concentration of the chemical species; μ is the dynamic viscosity of the pore-fluid; ϕ is the porosity of the porous medium; ϕ_f is the final (i.e. maximum) porosity of the porous medium after the completion of soluble mineral dissolution; $D(\phi)$ is the diffusivity of chemical species; $k(\phi)$ is the permeability of the porous medium. $\bar{V}_p$ is the average volume of a soluble grain; $\bar{A}_p$ is the average surface area of the soluble grain; $k_{Echemical}$ is the comprehensive rate constant of the chemical reaction (as defined in Eq. (2.15) previously); ρ_s is the molar density (i.e. moles per volume) of the soluble grains.

The physical meanings of Eqs. (8.1)–(8.3) are as follows: Eq. (8.1) expresses the mass conservation of the pore-fluid in the system, while Eq. (8.2) expresses the reactive transport of a chemical species within the system. Equation (8.3) is used to express the porosity evolution within the system. For the coupled nonlinear system described by Eqs. (8.1)–(8.3), the chemical species concentration, pore-fluid pressure and porosity are three basic independent variables to be solved. Since three independent variables are used to describe three independent equations, the problem related to this system is well posed so that it can be solved either analytically or numerically, from the mathematical point of view.

It is noted that dimensionless variables and governing equations are always preferred in the study of the problem involved in multiple scales and multiple processes. The main advantages in using dimensionless variables and governing equations are as follows: (1) Solutions for dimensionless variables describe the behaviour of a family of problem rather than that of a particular problem. This makes the solutions of more general meanings. (2) Dimensionless variables can be used to measure the relative importance of various terms in governing equations, so that the dominant physical phenomenon can be identified for the problem. This provides a clear focus for the effective and efficient modelling of the problem. (3) Dimensionless equations can result in a significant reduction in the large differences between orders of magnitude for some terms in the corresponding dimensional equations, just like the partial differential equations that are used to describe the three-dimensional chemical dissolution-front instability problem in this study. This generally makes the numerical solution more accurate and stable. For these reasons, dimensionless variables and governing equations are used in the following theoretical and numerical analysis.

In the case of the density of a soluble mineral greatly exceeding the equilibrium concentration of the soluble mineral, a small parameter, known as the mineral dissolution ratio (Zhao et al. 2008c), can be defined as follows:

$$\varepsilon = \frac{C_{eq}}{\rho_s} \ll 1. \tag{8.5}$$

In this particular situation, Eqs. (8.1)–(8.3) can be rewritten into the following dimensionless equations.

$$\varepsilon \frac{\partial \phi}{\partial \tau} - \nabla \cdot \left[\bar{\psi}(\phi) \nabla \bar{p} \right] = 0, \tag{8.6}$$

$$\varepsilon \frac{\partial}{\partial \tau}(\phi \bar{C}) - \nabla \cdot \left[\bar{D}(\phi) \nabla \bar{C} + \bar{C} \bar{\psi}(\phi) \nabla \bar{p} \right] - \frac{\partial \phi}{\partial \tau} = 0, \tag{8.7}$$

$$\varepsilon \frac{\partial \phi}{\partial \tau} + (\phi_f - \phi)(\bar{C} - 1) = 0. \tag{8.8}$$

Note that in the process of deriving the above three dimensionless equations, the following dimensionless parameters and variables have been used.

$$\begin{aligned} &\bar{x} = \frac{x}{L^*}, \quad \bar{y} = \frac{y}{L^*}, \quad \bar{z} = \frac{z}{L^*}, \quad \bar{C} = \frac{C}{C_{eq}}, \quad \bar{p} = \frac{p}{p^*}, \quad \bar{L} = \frac{L}{L^*}, \\ &\tau = \frac{t}{t^*}\varepsilon, \quad \bar{D}(\phi) = \frac{\phi D(\phi)}{\phi_f D(\phi_f)}, \quad \bar{\psi}(\phi) = \frac{\psi(\phi)}{\psi(\psi_f)} \end{aligned} \tag{8.9}$$

where τ is a slow dimensionless time to describe the slowness of the chemical dissolution that takes place in the system. Other characteristic parameters used in Eq. (8.9) can be expressed as follows:

$$t^* = \frac{\bar{V}_p}{k_{Echemical}\bar{A}_p C_{eq}}, \quad L^* = \sqrt{\phi_f D(\phi_f) t^*}, \quad p^* = \frac{\phi_f D(\phi_f)}{\psi(\phi_f)} \tag{8.10}$$

where t^* is the characteristic time of the system; L^* is the characteristic length of the system; p^* is the characteristic pore-fluid pressure of the system. These characteristic parameters are used to scale the dimensional partial differential equations (i.e. Eqs. (8.1)–(8.3)) into the corresponding dimensionless ones (i.e. Eqs. (8.6)–(8.8)). For this reason, these characteristic parameters are the key to translate the dimensionless results obtained from using Eqs. (8.6)–(8.8) into the dimensional results suitable for dealing with the real problem described by Eqs. (8.1)–(8.3).

8.1.2 The Proposed Numerical Procedure

Generally, there are two kinds of solution methodologies that are commonly used to solve a set of coupled nonlinear equations, such as those described by Eqs. (8.6)–(8.8). One is called the fully coupled solver, while another is called the segregated solver. Since the fully coupled algorithm solves the whole set of coupled nonlinear equations in a simultaneous manner, it requires the formation of the global system matrix that includes all the unknown degrees of freedom associated with the discretized problem of the system. A large amount of numerical experience has demonstrated that the fully coupled solver is cost-effective for most two-dimensional problems. However, for extremely large two-dimensional and most three-dimensional problems in the real world, the peripheral storage required for the global matrix of the whole system may exceed the available computer resources, namely the combination of CPU time and peripheral disk storage. On the contrary, since the segregated algorithm solves each of the coupled nonlinear equations of the system separately in a sequential manner, it avoids the direct formation of the global system matrix that includes all the unknown degrees of freedom associated with the discretized problem of the system. In the segregated algorithm, the global system matrix is decomposed into smaller sub-matrices, each of which is related to the nodal unknowns associated with only one governing equation of the system. These smaller sub-matrices are then solved in a sequential manner using either the direct Gaussian elimination scheme or other conjugate type schemes. Because the storage requirement for the individual sub-matrices is significantly less than that needed to store the global matrix of the whole system, the segregated algorithm requires much less computer storage than the fully coupled algorithm. In particular, the savings in computer storage becomes much more significant as the size of the system and the number of the coupled nonlinear equations involved in the system increases. The previous numerical simulations have demonstrated that for the reactive transport system of a single chemical-species, both the fully coupled solver and the segregated solver can produce the numerical results of the similar accuracy (Fluid Dynamics International 1997). For the above-mentioned reasons, the segregated algorithm is used to solve the coupled nonlinear problem between porosity, pore-fluid flow and reactive single-chemical-species transport in the three-dimensional fluid-saturated porous medium in this chapter.

To ensure the accuracy of the numerical solution, Eqs. (8.6)–(8.8) need to be solved in an iterative manner unless a much smaller dimensionless time-step length is used in the computation. For a given dimensionless time-step, $\tau + \Delta\tau$, the porosity at the current time-step can be denoted by $\phi_{\tau+\Delta\tau} = \phi_\tau + \Delta\phi_{\tau+\Delta\tau}$, where ϕ_τ is the porosity at the previous time-step and $\Delta\phi_{\tau+\Delta\tau}$ is the porosity increment at the current time-step. Using the backward difference scheme, Eq. (8.8) can be written as follows:

$$\left[\frac{\varepsilon}{\Delta\tau} + (1 - \bar{C}_{\tau+\Delta\tau})\right]\Delta\phi_{\tau+\Delta\tau} = (\phi_f - \phi_\tau)(1 - \bar{C}_{\tau+\Delta\tau}), \tag{8.11}$$

where $\bar{C}_{\tau+\Delta\tau}$ is the dimensionless chemical-species concentration at the current time-step; $\Delta\tau$ is the dimensionless time increment at the current time-step.

Mathematically, Eq. (8.7) can be expressed, in the finite difference sense, as follows (Zhao et al. 2008a):

$$\left[\frac{\varepsilon}{\Delta\tau}\phi_{\tau+\Delta\tau} + \frac{1}{\varepsilon}(\phi_f - \phi_{\tau+\Delta\tau})\right]\bar{C}_{\tau+\Delta\tau} - \nabla\cdot\left[\bar{D}(\phi_{\tau+\Delta\tau})\nabla\bar{C}_{\tau+\Delta\tau}\right] - \nabla\bar{p}_{\tau+\Delta\tau}\cdot\left[\bar{\psi}(\phi_{\tau+\Delta\tau})\nabla\bar{C}_{\tau+\Delta\tau}\right]$$
$$= \frac{\varepsilon}{\Delta\tau}\phi_{\tau+\Delta\tau}\bar{C}_\tau + \frac{1}{\varepsilon}(\phi_f - \phi_{\tau+\Delta\tau}). \tag{8.12}$$

Similarly, Eq. (8.6) can be rewritten in the following discretized form:

$$\nabla\cdot\left[\bar{\psi}(\phi)\nabla\bar{p}\right] = \nabla\cdot\left[\bar{\psi}(\phi_{\tau+\Delta\tau})\nabla\bar{p}_{\tau+\Delta\tau}\right] - (1 - \bar{C}_{\tau+\Delta\tau})(\phi_f - \phi_{\tau+\Delta\tau}) \tag{8.13}$$

In order to avoid the nonlinear term of the porosity at the current time step, the porosity increment, instead of the porosity itself, is used for the finite difference discretization of the porosity evolution equation, namely Eq. (8.11). On the other hand, the dimensionless chemical-species concentration and pore-fluid pressure are used for the finite difference discretization of both the mass transport equation and the continuity equation, namely Eqs. (8.12) and (8.13). This means that a combination of the incremental variable method and the full variable method is used, in this chapter, to solve the fully coupled nonlinear problem between porosity, pore-fluid flow and reactive chemical-species transport within fluid-saturated porous media. After the above-mentioned approximation is used,

Eqs. (8.11)–(8.13) can be viewed as linear partial differential equations, so that the finite element method (Zienkiewicz 1977) can be straightforwardly used to discretize Eqs. (8.11)–(8.13) in space (Zhao et al. 2012).

8.2 Verification of the Proposed Numerical Procedure

Since the proposed numerical procedure in the previous section belongs to a kind of approximate solution method, its correctness and accuracy need to be assessed before it can be used to simulate chemical dissolution-front instability problems in three-dimensional fluid-saturated porous media. For this purpose, a three-dimensional benchmark problem, whose governing equations are exactly the same as those expressed in Eqs. (8.6)–(8.8), is established on the basis of the rigorous analytical solution, in which the problem domain is geometrically infinite. To ensure comparability between a numerical solution and the corresponding analytical solution, the three-dimensional benchmark problem of a finite computational domain should be constructed in such a way that the geometry and boundary conditions of the three-dimensional benchmark problem can be equivalently considered in both the theoretical analysis, which results in a rigorous analytical solution, and the numerical simulation, which produces the approximate numerical solution. For this reason, appropriate boundary conditions have to be applied on the boundaries of the three-dimensional benchmark problem of a finite computational domain.

Figure 8.1 shows the geometry of the three-dimensional benchmark problem, namely a fully coupled nonlinear problem between porosity, pore-fluid flow and reactive chemical-species transport within a three-dimensional fluid-saturated porous medium. For this three-dimensional benchmark problem, the dimensionless pore-fluid pressure-gradient of a negative unit value (i.e. $\bar{p}'_{fx} = -1$) is applied on the left surface boundary, implying that there is an injected horizontal flow from the left surface to the right surface of the three-dimensional computational model. In this case, the Zhao number of the chemical dissolution system is unity. All other four surface boundaries, namely the front, the back, the top and the bottom surface boundaries, are assumed to be impermeable for the pore-fluid and chemical species, implying that there is no escape of both the pore-fluid and the chemical-species on these four surface boundaries. Since the final porosity (i.e. ϕ_f) after the depletion of the soluble mineral is assumed to be 0.2, it can be applied on the left surface boundary as a boundary condition of the three-dimensional computational domain. In addition, the dimensionless chemical-species concentration is assumed to be zero on the left surface boundary, to reflect the depletion of the soluble mineral on this boundary.

To get the distribution of the dimensionless pore-fluid pressure determined in the computational model, the dimensionless pore-fluid pressure is assumed to be a constant of zero on the right surface boundary of the computational model. It must

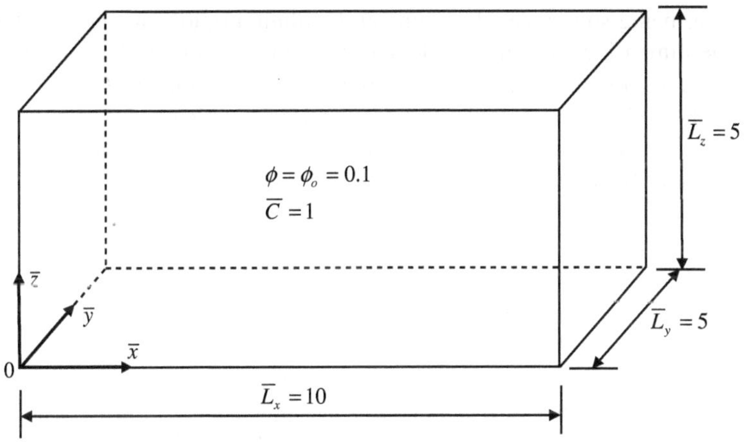

Fig. 8.1 Geometry of the three-dimensional chemical dissolution-front instability problem

be pointed out that since the governing equation for the dimensionless pore-fluid pressure is a second-order equation (Zhao et al. 2008b), any constant value of the dimensionless pore-fluid pressure can be prescribed on the right surface boundary. From the mathematical point of view, the total value of the dimensionless pore-fluid pressure within the computational model is equal to this constant plus the computed value of the dimensionless pore-fluid pressure obtained from applying zero pore-fluid pressure on the right surface boundary of the computational model. This indicates that the computed dimensionless pore-fluid pressure obtained in this investigation plus any constant remains the solution for the dimensionless pore-fluid pressure of the considered problem. Although the prescribed value of the dimensionless pore-fluid pressure on the right surface boundary affects the computed value of the dimensionless pore-fluid pressure, it dose not affect the values of the dimensionless pore-fluid pressure-gradient, the porosity and the dimensionless chemical-species concentration within the computational model.

The dimensionless height and width of the three-dimensional computational model are assumed to be 5, while the dimensionless length of the three-dimensional computational model is assumed to be 10. Except for the left surface boundary, the initial porosity (i.e. ϕ_0) of the porous medium is 0.1, while the initial dimensionless chemical-species concentration is unity within the three-dimensional computational domain. In order to test the robustness of the proposed numerical procedure, a small perturbation of 1 % the initial porosity is randomly added to the initial porosity field in the three-dimensional computational domain, resulting in a new initial porosity field prior to running the computational model. This means that the new initial porosity field of the computational model becomes three-dimensional at the beginning of the numerical simulation. From the geometrical shape and material distribution points of view, the benchmark problem considered here is three-dimensional, rather than one-dimensional, even though the input planar chemical dissolution-front remains its planar shape in the case of the system being subcritical. The mineral dissolution ratio is assumed to be 0.01, while the dimensionless time-step length is set to be 0.005 in the computation.

The permeability of the porous medium is calculated using the Carman–Kozeny formula (Scheidegger 1974; Nield and Bejan 1992), which has the power of 3 in the power law. The diffusivity of chemical species is calculated using the power law (Bear 1972), which has the power of 2.

$$k(\phi) = \frac{k_0(1-\phi_0)^2\phi^3}{\phi_0^3(1-\phi)^2}, \quad D(\phi) = D_0\phi^2, \tag{8.14}$$

where ϕ_0 and k_0 are the initial reference porosity and permeability of the porous medium respectively; D_0 is the diffusivity of the chemical species in pure water.

Both the dimensional permeability and the dimensional diffusivity are further translated, in the numerical analysis, into the following dimensionless quantities using the related expressions in Eq. (8.9).

$$\bar{\psi}(\phi) = \frac{\psi(\phi)}{\psi(\phi_f)} = \frac{k(\phi)}{k(\phi_f)} = \frac{(1-\phi_f)^2\phi^3}{(1-\phi)^2\phi_f^3}, \quad \bar{D}(\phi) = \frac{\phi D(\phi)}{\phi_f D(\phi_f)} = \left(\frac{\phi}{\phi_f}\right)^3. \tag{8.15}$$

Because the computational domain of the three-dimensional benchmark problem is of finite size, the following time-dependent boundary condition of the dimensionless chemical-species concentration needs to be applied on the left surface boundary.

$$\bar{C}(\tau) = \exp\left(\bar{p}'_{fx}\bar{v}_{front}\tau\right), \tag{8.16}$$

where τ is the dimensionless time; $\bar{p}'_{fx}$ is the dimensionless pore-fluid pressure-gradient that is applied on the left surface boundary; $\bar{v}_{front} = -\bar{p}'_{fx}\big/(\phi_f - \phi_0)$ is the dimensionless speed of the chemical dissolution-front propagation within the computational domain of the three-dimensional benchmark problem.

If the dimensionless pore-fluid pressure is prescribed as zero at the right boundary surface of the three-dimensional benchmark problem, then the analytical solutions for this benchmark problem can be derived and expressed as follows (Zhao et al. 2008b):

$$\bar{C}(\bar{x},\tau) = 1, \quad \bar{p}(\bar{x},\tau) = -\frac{k(\phi_f)}{k(\phi_0)}\bar{p}'_{fx}(\bar{L}_x - \bar{x}), \quad \phi(\bar{x},\tau) = \phi_0, \quad (\bar{x} > \bar{v}_{front}\tau), \tag{8.17}$$

$$\bar{C}(\bar{x},\tau)=\exp\left[-\bar{p}'_{fx}(\bar{x}-\bar{v}_{front}\tau)\right],\quad \bar{p}(\bar{x},\tau)=\bar{p}'_{fx}(\bar{x}-\bar{v}_{front}\tau)-\frac{k(\phi_f)}{k(\phi_0)}\bar{p}'_{fx}(\bar{L}_x-\bar{v}_{front}\tau),$$
$$\phi(\bar{x},\tau)=\phi_f,\qquad (\bar{x}\le\bar{v}_{front}\tau),\tag{8.18}$$

where ϕ_0 and ϕ_f are the initial and final porosity of the three-dimensional benchmark problem; $k(\phi_0)$ and $k(\phi_f)$ are the corresponding initial and final permeability of the three-dimensional computational domain respectively; $\bar{L}_x$ is the dimensionless length of the three-dimensional benchmark problem; Other quantities are of the same meaning as mentioned previously.

For this three-dimensional benchmark problem, it turns out that the critical Zhao number of the chemical dissolution system can be expressed as follows (Zhao et al. 2008b):

$$Zh_{critical}=-\bar{p}'_{fx}\Big|_{critical}=\frac{\sqrt{2}(3-\beta)(1+\beta)}{2(1-\beta)},\quad \beta=\frac{\psi(\phi_0)}{\psi(\phi_f)}=\frac{k(\phi_0)}{k(\phi_f)}\tag{8.19}$$

where $Zh_{critical}$ is the critical Zhao number of the three-dimensional benchmark problem; ψ and k have the same meanings as those defined in Eq. (8.4).

Using the dimensionless quantities defined in Eqs. (8.9) and (8.10), the Zhao number of this three-dimensional benchmark problem can be defined as follows:

$$Zh=-\bar{p}'_{fx}=-\frac{p'_{fx}L^*}{p^*}=-\frac{k(\phi_f)L^*p'_{fx}}{\phi_f\mu D(\phi_f)}=-\frac{k(\phi_f)p'_{fx}}{\mu\sqrt{\phi_f D(\phi_f)}}\sqrt{\frac{\bar{V}_p}{k_{Echemical}\bar{A}_p C_{eq}}}$$
$$=F_{Advective}F_{Diffusive}F_{Chemical}F_{Shape}\tag{8.20}$$

where

$$F_{Advection}=-\frac{k(\phi_f)p'_{fx}}{\mu}\ \text{(representing the solute advection)},\tag{8.21}$$

$$F_{Diffusion}=\frac{1}{\sqrt{\phi_f D(\phi_f)}}\ \text{(representing the solute diffusion/dispersion)},\tag{8.22}$$

$$F_{Chemical}=\sqrt{\frac{1}{k_{Echemical}C_{eq}}}\ \text{(representing the chemical dissolution kinetics)},\tag{8.23}$$

$$F_{Shape}=\sqrt{\frac{\bar{V}_p}{\bar{A}_p}}\ \text{(representing the shape factor of the dissolvable mineral)}.\tag{8.24}$$

When the characteristic quantities (i.e. t^*, L^* and p^* in Eq. (8.10)) are used, the dimensional governing equations (i.e. Eqs. (8.1)–(8.3)) in the real time–space coordinate system have been translated into the dimensionless ones (i.e. Eqs. (8.6)–(8.8)) in a generalized time–space coordinate system. Since the generalized time–space coordinate system is an abstract time–space coordinate system, the physical meaning of the pore-fluid pressure-gradient in the real time–space coordinate system is totally different from that of the (generalized) dimensionless pore-fluid pressure-gradient in the generalized time–space coordinate system. As indicated in Eq. (8.20), the (generalized) dimensionless pore-fluid pressure-gradient is equal to the negative value of the Zhao number, which a comprehensive dimensionless number to represent the interaction between the solute advection (represented by the term, $F_{Advection}$), the solute diffusion (represented by the term, $F_{Diffusion}$), the chemical dissolution kinetics (represented by the term, $F_{Chemical}$) and the shape factor of the soluble mineral (represented by the term, F_{Shape}) in the reactive transport system.

Using the concept of the Zhao number, a criterion can be established to judge the instability likelihood of a chemical dissolution front associated with this three-dimensional benchmark problem. If the Zhao number of a chemical dissolution system is greater than its corresponding critical value, then the chemical dissolution front of the reactive transport system becomes unstable, while if the Zhao number of a chemical dissolution system is smaller than its corresponding critical value, then the chemical dissolution front of the reactive transport system is stable. The case of $Zh = Zh_{critical}$ represents a situation where the chemical dissolution front of the reactive transport system is neutrally unstable, implying that the applied small perturbation can be maintained but it does not grow in the corresponding reactive transport system.

It is noted from Eq. (8.20) that for the reactive transport system considered, an increase in permeability causes an increase in the Zhao number, implying that the dissolution-enhanced permeability can destabilize the instability of the chemical dissolution front. On the contrary, an increase in diffusivity causes a decrease in the Zhao number, implying that the dissolution-enhanced diffusivity can stabilize the instability of the chemical dissolution front. Similarly, an increase in either the equilibrium concentration of the chemical species or the chemical reaction constant of the dissolution reaction can cause the stabilization of the chemical dissolution front, for the reactive transport system considered in this chapter.

The use of Eq. (8.16) as a boundary condition on the left surface boundary of the three-dimensional benchmark problem ensures that the obtained numerical solutions can be compared with the analytical solutions expressed in Eqs. (8.17) and (8.18). According to the above-mentioned parameters, the theoretical value of the critical Zhao number of the three-dimensional system is equal to 2.501, while the theoretical value of the corresponding dimensionless speed of the chemical dissolution-front propagation is equal to 10. Since the Zhao number of the three-dimensional system is less than its corresponding critical value, the coupled nonlinear system considered in this investigation is subcritical so that a planar chemical dissolution-front remains the planar shape during its propagation within the three-dimensional system, even though a small perturbation of porosity is added into the whole computational domain of the coupled nonlinear system. To appropriately simulate the propagation of the chemical dissolution front, the whole three-dimensional computational domain is simulated by 57,600 eight-node brick elements of 63,125 nodal points in total.

Figures 8.2, 8.3 and 8.4 show the comparison of numerical solutions with analytical ones for the porosity, dimensionless chemical-species concentration and dimensionless pore-fluid pressure distributions within the three-dimensional computational domain at three different time instants. In these figures, the left column shows the numerical results obtained from using the proposed numerical procedure, while the right column shows the corresponding analytical solutions obtained from using Eqs. (8.17) and (8.18). It is noted that although a small perturbation of 1 % the initial porosity is randomly added to the initial porosity field in the case of the coupled nonlinear system being subcritical, it does not affect the initial planar shape of the chemical dissolution front during its propagation within the computational domain of the three-dimensional benchmark problem, indicating that the proposed numerical procedure is numerically stable and robust. Because the numerical solutions agree very well with the analytical ones, it is demonstrated that the proposed numerical procedure is capable of simulating the planar chemical dissolution-front propagation within the three-dimensional fluid-saturated porous medium. Although there are some smooth effects on the numerically-simulated propagation fronts due to numerical dispersion, the propagation speed of the numerically-simulated propagation front is in good coincidence with that of analytically-predicted propagation front. For this three-dimensional benchmark problem, the overall accuracy of the numerical results is indicated by the dimensionless pore-fluid pressure. The maximum relative error of the numerically-simulated dimensionless pore-fluid pressure is 3.8, 5.4 and 8.1 % in the case of the dimensionless time being 0.25, 0.5 and 0.75 respectively. It is worth noting that if necessary, this relative error can be further reduced by taking one of the following measures: (1) using smaller elements and time-step length; (2) using upwind scheme to reduce the numerical diffusion near the chemical dissolution front; and (3) developing a stepwise element to simulate the discontinuity of the chemical dissolution front. Because the maximum relative error of the numerical solution can be controlled within a reasonable range, it has quantitatively demonstrated that the proposed numerical procedure can produce accurate numerical solutions for the planar dissolution-front propagation problem within a three-dimensional fluid-saturated porous medium.

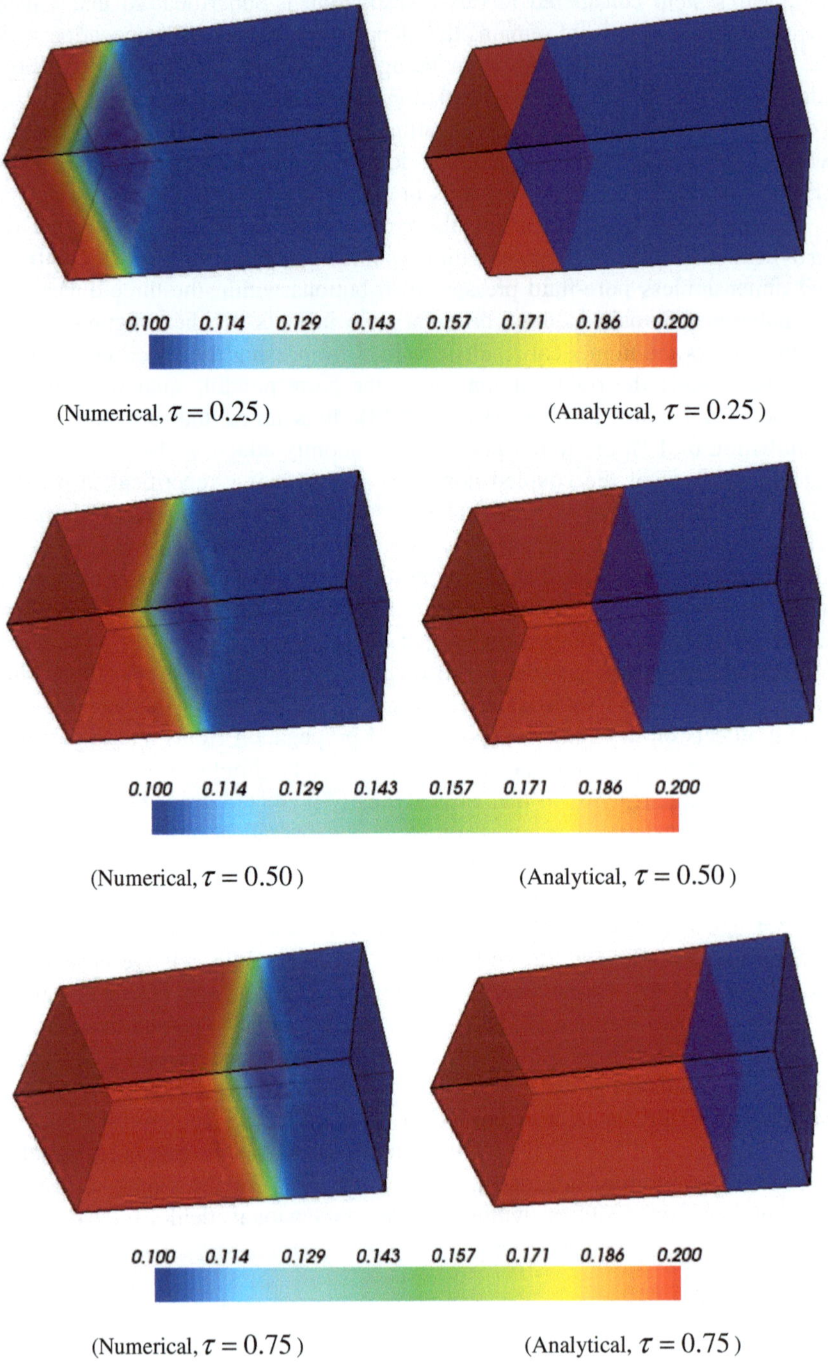

Fig. 8.2 Comparison of numerical solutions with analytical ones at different time instants (porosity)

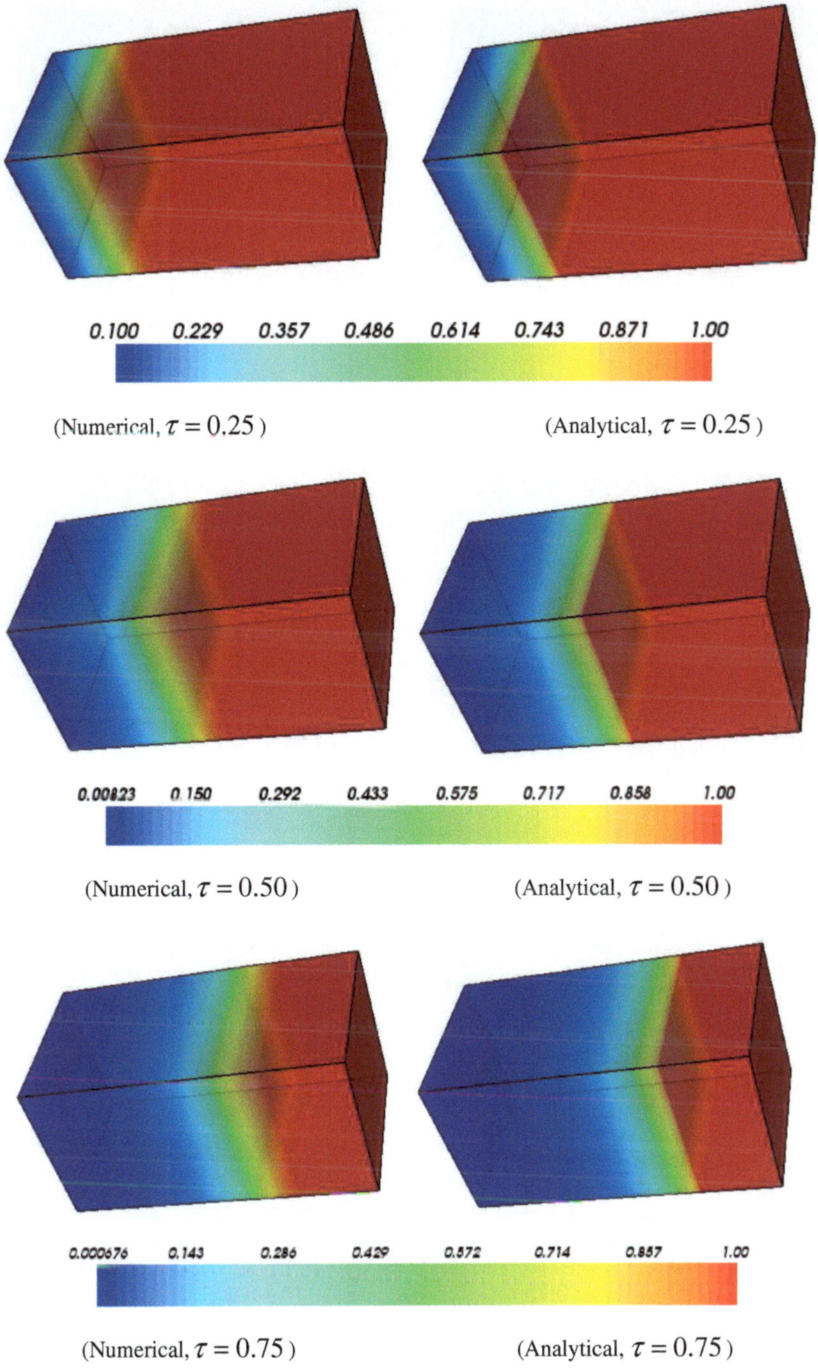

Fig. 8.3 Comparison of numerical solutions with analytical ones at different time instants (dimensionless chemical-species concentration)

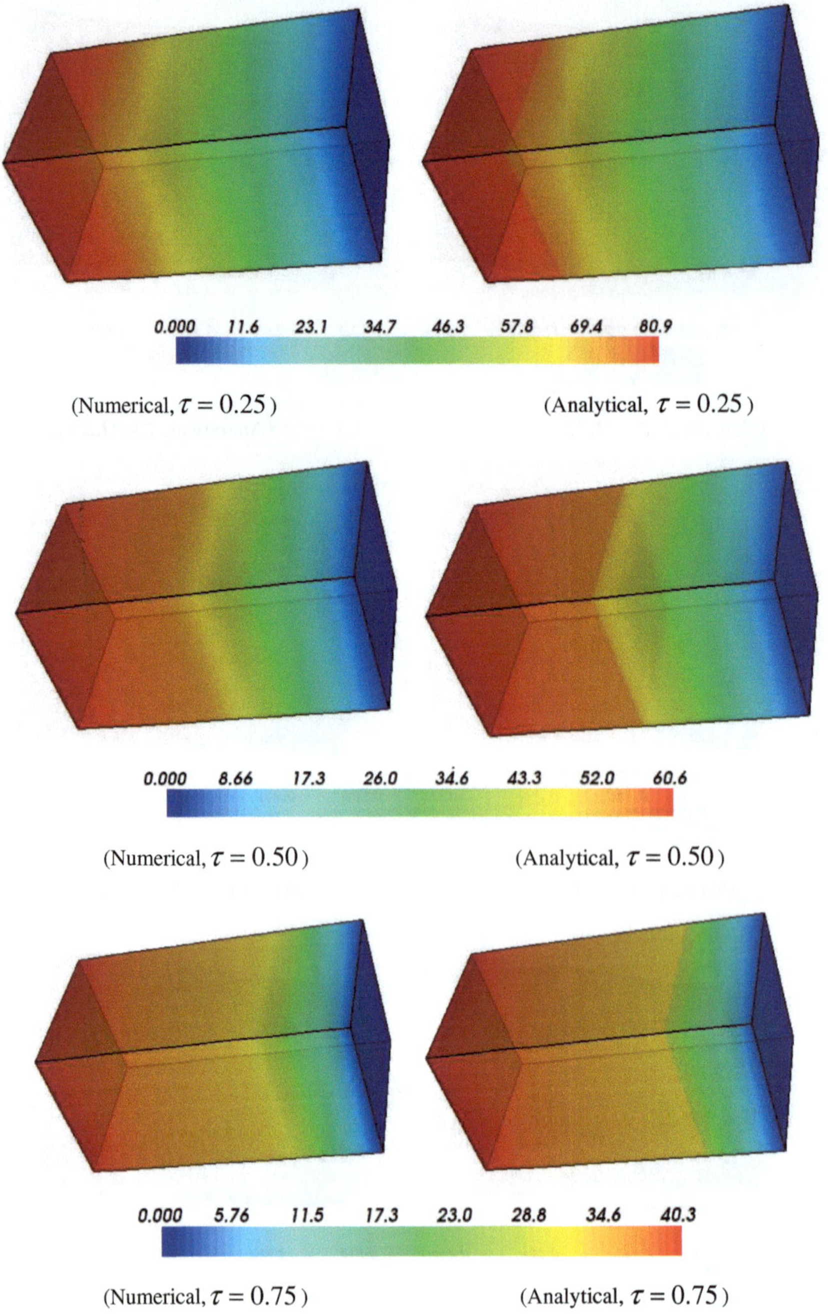

Fig. 8.4 Comparison of numerical solutions with analytical ones at different time instants (dimensionless pore-fluid pressure)

8.3 Morphological Evolution of Three-Dimensional Chemical-Dissolution Fronts in Fluid-Saturated Porous Media

After the proposed numerical procedure is verified using the three-dimensional benchmark problem, it is applied to simulate the morphological evolution of a three-dimensional chemical dissolution front within a supercritical system, which has a (generalized) dimensionless pore-fluid pressure-gradient (i.e. $\bar{p}'_{fx} = -10$) on the left boundary of the three-dimensional computational domain, implying that the Zhao number of the chemical dissolution system under consideration is equal to 10. In this case, the dimensionless speed of the chemical dissolution-front propagation is equal to 100. The mineral dissolution ratio is 0.001, while the dimensionless time-step length is 0.001 in the computation. The values of other parameters are exactly the same as those used in the verification example of the previous section. Since the Zhao number of the chemical dissolution system under consideration is greater than its corresponding critical value, the initial planar chemical dissolution-front can evolve into a complicated morphology during its propagation within the three-dimensional computational domain. In order to simulate the instability of the chemical dissolution front, a small perturbation of 1 % the initial porosity is randomly added to the initial porosity field in the whole three-dimensional computational domain.

Figures 8.5 and 8.6 show the porosity and dimensionless chemical-species concentration distributions associated with the evolution of the chemical dissolution front in the three-dimensional fluid-saturated porous medium. It needs to be pointed out that in order to display the porosity propagation front clearly, both the constant porosity of 0.2 on the left-hand side of the three-dimensional chemical dissolution front and the constant porosity of 0.1 on the right-hand side of the three-dimensional chemical dissolution front are not shown in Fig. 8.5. Similarly, in order to display the dimensionless chemical-species-concentration propagation front clearly, both the constant dimensionless chemical-species concentration of zero on the left-hand side of the three-dimensional chemical dissolution front and the constant dimensionless chemical-species concentration of 1.0 on the right-hand side of the three-dimensional chemical dissolution front are not shown in Fig. 8.6. It is observed that after the dimensionless time is greater than 0.02, the planar chemical dissolution-front is gradually changed into an irregular three-dimensional one. With the increase of the dimensionless time, the irregularity of the resulting three-dimensional chemical dissolution front grows significantly. For example, the finger-like chemical dissolution front can be clearly generated after the dimensionless time is greater than 0.06. Although both the porosity and the dimensionless chemical-species concentration have the similar three-dimensional propagation front, the distribution of their maximum values along the chemical dissolution front is in a complementary manner. This can be observed from the fact that the peak value of the porosity is in good correspondence with the trough value of the dimensionless chemical-species concentration in the three-dimensional

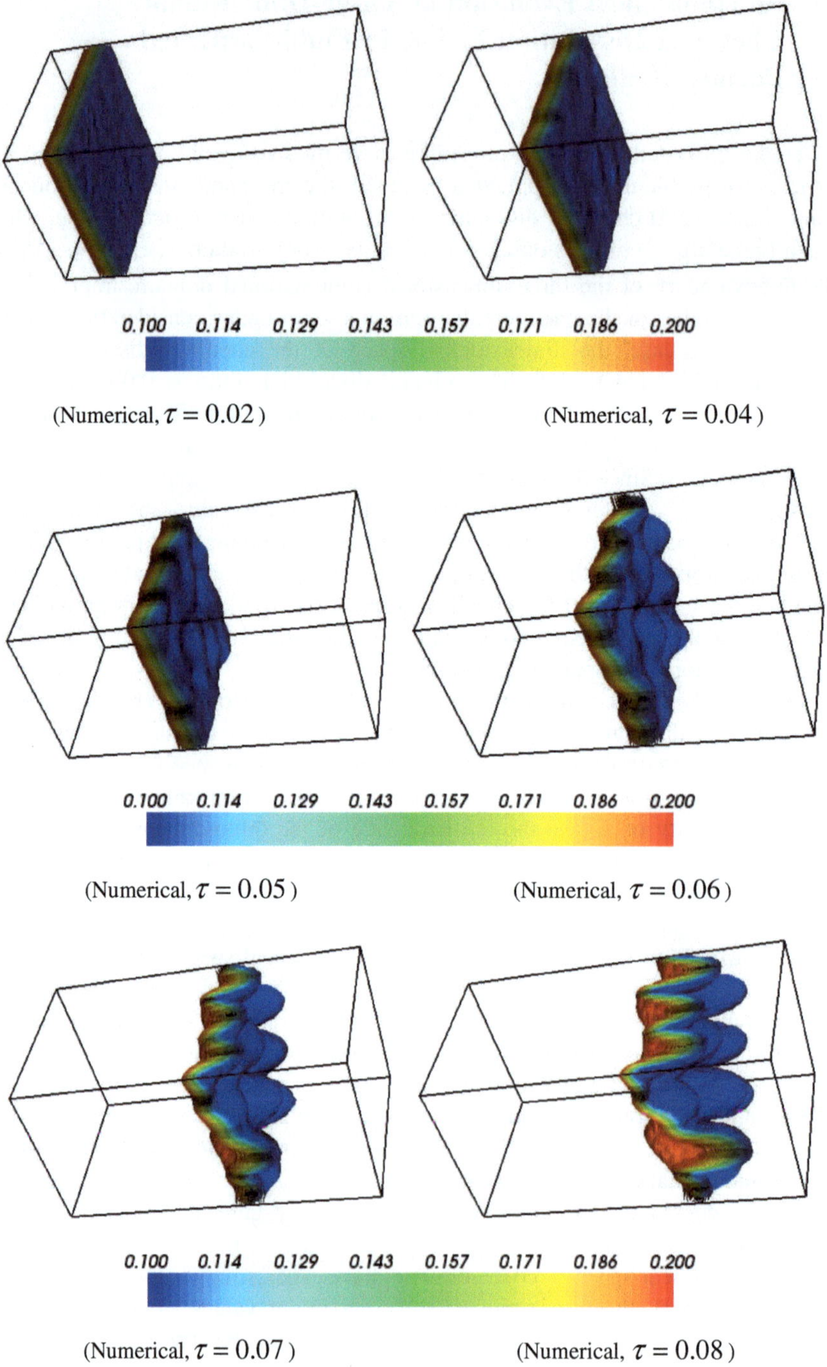

Fig. 8.5 Porosity distributions due to morphological evolution of the three-dimensional chemical dissolution front in the fluid-saturated porous medium

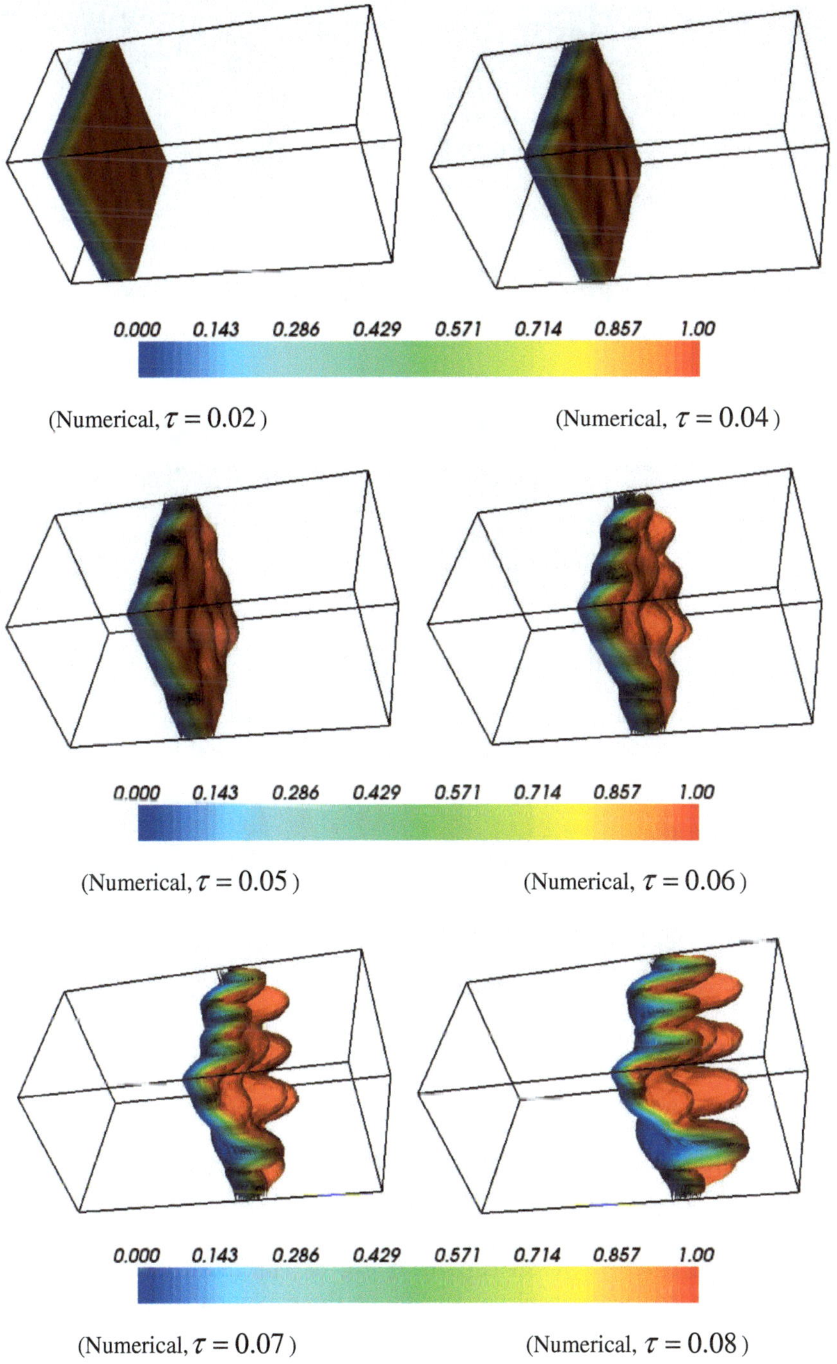

Fig. 8.6 Dimensionless concentration distributions due to morphological evolution of the three-dimensional chemical dissolution front in the fluid-saturated porous medium

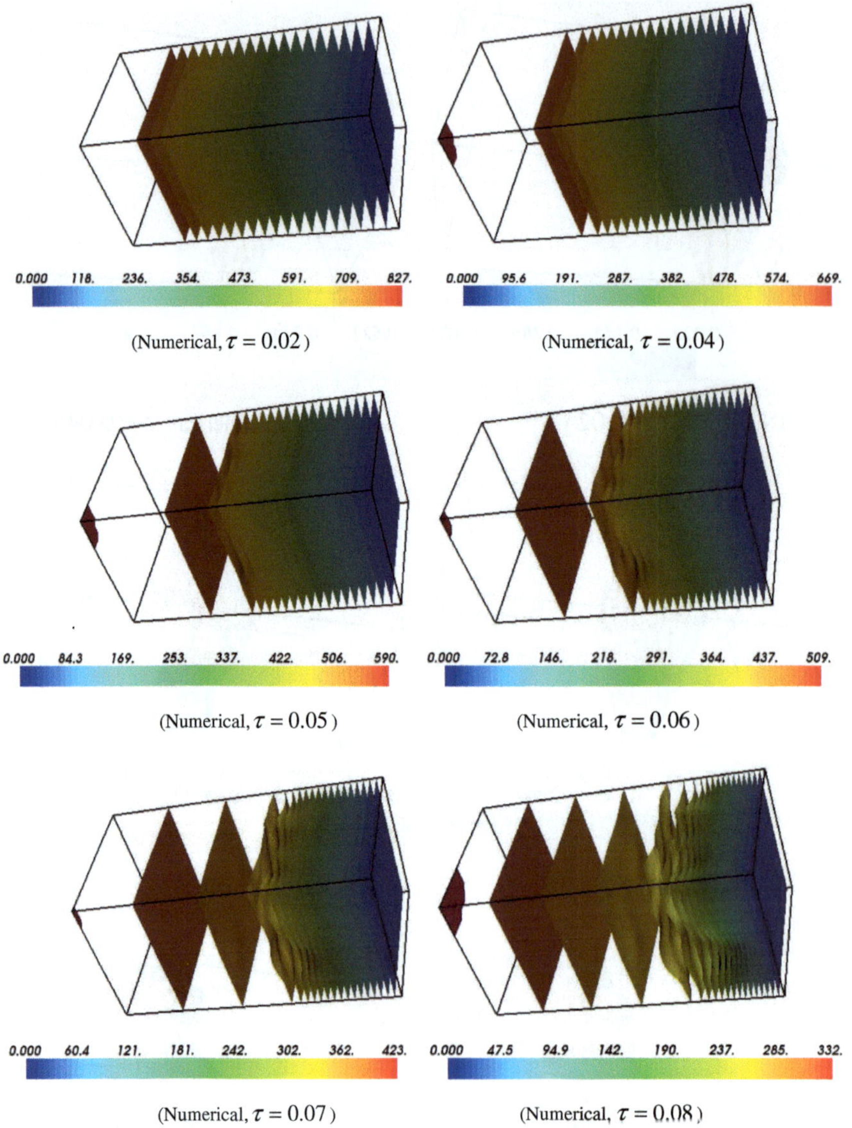

Fig. 8.7 Dimensionless pore-fluid pressure distributions due to morphological evolution of the three-dimensional chemical dissolution front in the fluid-saturated porous medium

computational domain. In this regard, it can be concluded that the proposed numerical procedure is capable of simulating the evolution of the three-dimensional chemical dissolution front in a fluid-saturated porous medium in the case of the coupled nonlinear system becoming supercritical.

Except for simulating the evolution patterns of both the porosity and the dimensionless chemical-species concentration, the dimensionless pore-fluid-pressure evolution pattern during the propagation of the unstable chemical dissolution-front can be also simulated in the three-dimensional computational model. Figure 8.7 shows the dimensionless pore-fluid-pressure evolution pattern during the propagation of the chemical dissolution front in the three-dimensional computational model. It is observed that even though the dimensionless pore-fluid pressure is continuous, there exists a clear transition for the (generalized) dimensionless pore-fluid pressure-gradient distribution in the three-dimensional computational model. This transition takes place just on the three-dimensional chemical dissolution front (i.e. the three-dimensional porosity propagation front) in a fluid-saturated porous medium. Because the porosity creation and pore-fluid flow focusing play an important role in ore body formation and mineralization, the proposed numerical procedure can be used for simulating the potential physical and chemical processes, which are closely associated with the generation of giant ore deposits within the upper crust of the Earth.

References

Appold MS, Garven G (2000) Reactive flow models of ore formation in the Southeast Missouri district. Econ Geol Bull Soc Econ Geol 95:1605–1626

Bear J (1972) Dynamics of fluids in porous media. American Elsevier Publishing Company, New York

Chadam J, Hoff D, Merino E, Ortoleva P, Sen A (1986) Reactive infiltration instabilities. IMA J Appl Math 36:207–221

Chadam J, Ortoleva P, Sen A (1988) A weekly nonlinear stability analysis of the reactive infiltration interface. IMA J Appl Math 48:1362–1378

Chen JS, Liu CW (2002) Numerical simulation of the evolution of aquifer porosity and species concentrations during reactive transport. Comput Geosci 28:485–499

Connolly JAD, Podladchikov YY (2007) Decompaction weakening and channeling instability in ductile porous media: implications for asthenospheric melt segregation. J Geophys Res 112(B10205):1–15

Everett CE, Wilkinson JJ, Rye DM (1999) Fracture-controlled fluid flow in the Lower Palaeozoic basement rocks of Ireland: implications for the genesis of Irish-type Zn–Pb deposits. In: McCaffrey KJW, Lonergan L, Wilkinson JJ (eds) Fractures, fluid flow and mineralization, vol 155. Geological Society, London, pp 247–276 (Special Publications)

Ferry JM (1996) Prograde and retrograde fluid flow during contact metamorphism of siliceous carbonate rocks from the Ballachulish aureole Scotland. Contrib Miner Petrol 124:235–254

Fluid Dynamics International (1997) Fluid dynamics analysis package: FIDAP. Fluid Dynamics International Inc, Illinois

Garven G, Appold MS, Topygina VI, Hazlett TJ (1999) Hydrologic modeling of the genesis of carbonate-hosted lead–zinc ores. Hydrogeol J 7:108–126

Gow P, Upton P, Zhao C, Hill K (2002) Copper–gold mineralization in the New Guinea: numerical modeling of collision, fluid flow and intrusion-related hydrothermal systems. Aust J Earth Sci 49:753–771

Heinrich W, Hoffbauer R, Hubberten HW (1995) Contrasting fluid flow patterns at the Bufa del Diente contact metamorphic aureole, north-east Mexico: evidence from stable isotopes. Contrib Miner Petrol 119:362–376

Hitzmann MW (1995) Mineralisation in the Irish Zn–Pb–(Ba–Ag) Orefield. In: Anderson K, Ashton J, Earls G, Hitzman M, Tear S (eds) Irish Carbonate-hosted deposits. Society of Economic Geologists, Guidebook Series, vol 21, pp 25–61

Klein C, Beukes NJV (1992) Proterozoic iron-formation. In: Kondie KC (ed) Proterozoic crustal evolution. Elsevier, Amsterdam

Morris RCV (1980) A textural and mineralogical study of the relationship of iron ore to banded iron-formation in the Hamersley Iron Province of Western Australia. Econ Geol 75:184–209

Nield DA, Bejan A (1992) Convection in porous media. Springer-Verlag, New York

Ord A, Hobbs BE, Zhang Y, Broadbent GC, Brown M, Willetts G, Sorjonen-Ward P, Walshe J, Zhao C (2002) Geodynamic modelling of the century deposit, Mt Isa Province, Queensland. Aust J Earth Sci 49:1011–1039

Ormond A, Ortoleva P (2000) Numerical modeling of reaction-induced cavities in a porous rock. J Geophys Res 105:16737–16747

Ortoleva P, Chadam J, Merino E, Sen A (1987) Geochemical self-organization. II: The reactive-infiltration instability. Am J Sci 287:1008–1040

Phillips OM (1991) Flow and reactions in permeable rocks. Cambridge University Press, Cambridge

Raffensperger JP, Garven G (1995) The formation of unconformity-type uranium ore deposits: coupled hydrochemical modelling. Am J Sci 295:639–696

Schaubs P, Zhao C (2002) Numerical modelling of gold-deposit formation in the Bendigo-Ballarat zone, Victoria. Aust J Earth Sci 49:1077–1096

Scheidegger AE (1974) The physics of flow through porous media. University of Toronto Press, Toronto

Simms MA, Garven G (2004) Thermal convection in faulted extensional sedimentary basins: theoretical results from finite-element modeling. Geofluids 4:109–130

Steefel CI, Lasaga AC (1990) Evolution of dissolution patterns: permeability change due to coupled flow and reaction. In: Melchior DC, Basset RL Chemical modeling in aqueous systems II. American Chemistry Society Symposium Series, vol 416. pp 213–225

Steefel CI, Lasaga AC (1994) A coupled model for transport of multiple chemical species and kinetic precipitation/dissolution reactions with application to reactive flow in single phase hydrothermal systems. Am J Sci 294:529–592

Wilde AR, Wall VJ (1987) Geology of the Nabarlek Uranium deposit, Northern-territory, Australia. Econ Geol 82:1152–1168

Xu TF, Samper J, Ayora C, Manzano M, Custodio E (1999) Modelling of non-isothermal multi-component reactive transport in field scale porous media flow systems. J Hydrol 214:144–164

Xu TF, Apps JA, Pruess K (2004) Numerical simulation of CO_2 disposal by mineral trapping in deep aquifers. Appl Geochem 19:917–936

Yang J, Large RR, Bull SW (2004) Factors controlling free thermal convection in faults in sedimentary basins: implications for the formation of zinc–lead mineral deposits. Geofluids 4:237–247

Yeh GT, Tripathi VS (1991) A model for simulating transport of reactive multispecies components: model development and demonstration. Water Resour Res 27:3075–3094

Zhao C, Mühlhaus HB, Hobbs BE (1997) Finite element analysis of steady-state natural convection problems in fluid-saturated porous media heated from below. Int J Numer Anal Meth Geomech 21:863–881

Zhao C, Hobbs BE, Mühlhaus HB (1998) Finite element modelling of temperature gradient driven rock alteration and mineralization in porous rock masses. Comput Methods Appl Mech Eng 165:175–187

Zhao C, Hobbs BE, Mühlhaus HB, Ord A (1999) Finite element analysis of flow patterns near geological lenses in hydrodynamic and hydrothermal systems. Geophys J Int 138:146–158

Zhao C, Hobbs BE, Mühlhaus HB, Ord A (2000) Finite element modelling of dissipative structures for nonequilibrium chemical reactions in fluid-saturated porous media. Comput Methods Appl Mech Eng 184:1–14

Zhao C, Lin G, Hobbs BE, Mühlhaus HB, Ord A, Wang, Y (2001) Finite element modelling of heat transfer through permeable cracks in hydrothermal systems with upward throughflow. In: Engineering computations, vol 18, pp 996–1011

Zhao C, Hobbs BE, Hornby P, Ord A, Peng S, Liu L (2008a) Theoretical and numerical analyses of chemical dissolution front instability in fluid-saturated porous rocks. Int J Numer Anal Meth Geomech 32:1107–1130

Zhao C, Hobbs BE, Ord A, Hornby P, Peng S (2008b) Morphological evolution of three-dimensional chemical-dissolution front in fluid-saturated porous media: a numerical simulation approach. Geofluids 8:113–127

Zhao C, Hobbs BE, Ord A, Hornby P, Peng S (2008c) Effect of reactive surface areas associated with different particle shapes on chemical dissolution front instability in fluid-saturated porous rocks. Transp Porous Media 73:75–94

Zhao C, Hobbs BE, Ord A (2009) Fundamentals of computational geoscience: numerical methods and algorithms. Springer, Berlin

Zhao C, Reid LB, Regenauer-Lieb K, Poulet T (2012) A porosity-gradient replacement approach for computational simulation of chemical-dissolution front propagation in fluid-saturated porous media including pore-fluid compressibility. Comput Geosci 16:735–755

Zienkiewicz OC (1977) The finite element method. McGraw-Hill, London

Chapter 9
Fundamental Theory for Nonaqueous-Phase-Liquid Dissolution-Front Instability Problems in Fluid-Saturated Porous Media

The transport of nonaqueous phase liquids (NAPLs) in contaminated subsurface is an important problem in geoenvironmental engineering (Miller et al. 1990, 1998; Geller and Hunt 1993; Powers et al. 1994; Imhoff et al. 1994, 1996, 2002; Soerens et al. 1998; Willson et al. 1999; Seyedabbasi et al. 2008). The nonaqueous phase liquid that is usually trapped at residual saturation in the pore space of a fluid-saturated porous medium can become long-term sources of groundwater contamination as a result of its low solubility. Compared with the density of a NAPL such as trichloroethylene, the equilibrium concentration of the NAPL in the aqueous phase fluid is much smaller. For instance, the ratio of the equilibrium concentration to the density of the NAPL can be as small as 8.7×10^{-4} (Imhoff and Miller 1996). This means that it may take a long period of time to dissolve a NAPL in the aqueous phase fluid, even if the dissolution rate of the NAPL is high. One of the common techniques used to remove the trapped NAPL from contaminated soils is to flush the contaminated soils by fresh water flow.

If fresh water passes through a fluid-saturated porous medium that is contaminated by a NAPL, it dilutes the equilibrium concentration of the NAPL so that the solution of the NAPL is changed from an equilibrium state into a nonequilibrium state. To maintain the equilibrium state of the solution, the trapped NAPL within the pore space of the fluid-saturated porous medium is dissolved into the aqueous phase fluid (i.e. water in this chapter). This indicates that as the first role played in a NAPL dissolution system, fresh water flow is the trigger to break the initial equilibrium state of the NAPL dissolution system. Since water flow can transport the dissolved NAPL from one place to another, the second role of fresh water flow is to maintain solute advection within the NAPL dissolution system. Thus, solute advection is the first important mechanism involved in a NAPL dissolution system. Another important mechanism involved in a NAPL dissolution system is the dissolution reaction of the NAPL, which can also play two roles in the NAPL dissolution system. Firstly, the dissolution reaction can cause a reduction in the trapped NAPL so that the NAPL can be removed from the contaminated porous medium by solute advection. Secondly,

C. Zhao, *Physical and Chemical Dissolution Front Instability in Porous Media*, Lecture Notes in Earth System Sciences,
DOI: 10.1007/978-3-319-08461-9_9

the dissolution reaction can cause an increase in fluid flow pathways so that the permeability of the porous medium can be increased as a result of the NAPL dissolution. In addition, solute can be transported from a high concentration region to a low concentration one through solute diffusion/dispersion, which is the third important mechanism involved in a NAPL dissolution system. Due to the involvement of the above-mentioned mechanisms, a NAPL dissolution problem can be attributed to a typical reactive mass transport problem with a permeability enhancement effect. Although considerable progress has been made in the study of reactive mass transport problems (Steefel and Lasage 1990, 1994; Yeh and Tripathi 1991; Raffensperger and Garven 1995; Schafer et al. 1998a, b; Xu et al. 1999; Ormond and Ortoleva 2000; Alt-Epping and Smith 2001; Zhao et al. 1998, 2003, 2005, 2006), the theoretical consideration of NAPL dissolution-front instability is still limited (Zhao et al. 2010b).

For the theoretical analysis of instability problems, it is desirable to establish a criterion that can be used to assess whether or not the considered system is stable. Toward this end, a comprehensive dimensionless number is derived to represent the driving mechanisms and controlling characteristics of a nonlinear dynamic system. For example, in the case of convective flow instability problems (Nield and Bejan 1992; Zhao et al. 2008a), the Rayleigh number is used to represent the buoyancy that is the driving force and other controlling characteristics of the system. In the case of mineral dissolution-front instability problems (Chadam et al. 1986, 1988; Ortoleva et al. 1987; Imhoff and Miller 1996; Renard et al. 1998; Imhoff et al. 2003; Chen and Liu 2002; Chen et al. 2009; Zhao et al. 2008b, c, 2009), the Zhao number is used to represent the pore-fluid flow that is the driving force, and other controlling characteristics such as solute diffusion/dispersion, chemical kinetics and particle shapes of minerals within the mineral dissolution system. Although Imhoff and Miller (1996) conducted some theoretical analyses for NAPL-dissolution instability problems, as indicated by Eqs. (9.32)–(9.34) in their paper (Imhoff and Miller 1996), their analytical results derived in a limit case as the ratio of the equilibrium concentration to the density of the NAPL approaches zero are questionable because they are independent of the main driving force of the NAPL dissolution system. Thus, it is necessary to redo theoretical analysis, in the limit case and when solute dispersion is considered as the second-order tensor, for the instability of NAPL dissolution fronts in fluid-saturated porous media.

9.1 Mathematical Modeling of NAPL Dissolution Problems in Two-Dimensional Fluid-Saturated Porous Media

To describe NAPL dissolution problems in a two-dimensional fluid-saturated porous medium, Imhoff and Miller (1996) made the following assumptions: (1) although NAPLs are mixtures of several species, only a single-species NAPL is considered; (2) water is negligible in the NAPL; (3) the NAPL is at residual saturation, existing as discrete ganglia or blobs trapped by capillary forces within the porous medium; (4) in the absence of the NAPL, the porous medium is homogeneous and isotropic; (5) the NAPL is immobile and incompressible; (6) water is incompressible and the solid matrix of the porous medium is rigid; (7) two-dimensional flow is in a horizontal plane so that gravity is neglected; and (8) the aqueous phase moves through the region of the residual NAPL and slowly dissolves NAPL ganglia. Under these assumptions, considering the mass conservation of the aqueous phase, nonaqueous phase and NAPL species, respectively, yields the following governing equations:

$$\frac{\partial}{\partial t}[\phi(1-S_n)\rho_a] + \nabla\cdot[\phi(1-S_n)\rho_a\mathbf{v_a}] = R, \tag{9.1}$$

$$\frac{\partial}{\partial t}(\phi S_n\rho_n) = -R, \tag{9.2}$$

$$\frac{\partial}{\partial t}[\phi(1-S_n)\rho_a\chi] + \nabla\cdot[\phi(1-S_n)\rho_a\chi\mathbf{v_a}] = \nabla\cdot[\phi(1-S_n)\rho_a\mathbf{D_h}\cdot\nabla\chi] + R, \tag{9.3}$$

where ϕ is the porosity; S_n is the NAPL saturation (i.e. the fraction of the void space occupied by the NAPL); ρ_a and ρ_n are the aqueous phase and nonaqueous phase densities; R is the mass exchange of the NAPL species from the nonaqueous phase to the aqueous phase; χ is the mass fraction of the NAPL species in the aqueous phase; $\mathbf{D_h}$ is the general dispersion tensor; and $\mathbf{v_a}$ is the averaged linear (i.e. interstitial) velocity vector of the aqueous phase.

Based on Darcy's law, the averaged linear velocity vector of the aqueous phase can be expressed as follows (Bear 1972):

$$\mathbf{v_a} = \frac{-k(S_n)}{\phi(1-S_n)\mu_a}\nabla p_a, \tag{9.4}$$

where $k(S_n)$ is the saturation-dependent permeability of the porous medium to aqueous phase flow; μ_a is the dynamic viscosity of the aqueous phase; p_a is the aqueous phase pressure.

Imhoff and Miller (1996) proposed to use the following expressions for the saturation-dependent permeability, the dispersion tensor and the mass exchange of the NAPL species from the nonaqueous phase to the aqueous phase, respectively:

$$k(S_n) = k_f\left(\frac{1-S_n-S_{ai}}{1-S_{ai}}\right)^3, \tag{9.5}$$

$$\mathbf{D_h} = \begin{bmatrix} \tau D_m + \alpha_L v_a & 0 \\ 0 & \tau D_m + \alpha_T v_a \end{bmatrix}, \tag{9.6}$$

$$R = K(C_{eq} - C), \tag{9.7}$$

where k_f is the intrinsic permeability of the porous medium after the NAPL is completely dissolved; S_{ai} is the irreducible saturation of the aqueous phase; τ is the tortuosity of the porous medium; D_m is the molecular diffusivity of the NAPL species in the aqueous phase; α_T and α_L are the transversal and longitudinal dispersivities of the NAPL species in the aqueous phase; v_a is the amplitude of the averaged linear velocity vector of the aqueous phase; C is the concentration of the NAPL species in the bulk aqueous phase ($C = \rho_a \chi$); C_{eq} is the equilibrium concentration of the NAPL species in the aqueous phase; and K is the mass transfer rate coefficient as follows:

$$K = \beta_0 S_n^{\beta_1}, \tag{9.8}$$

where β_0 is a function of the porous medium, the NAPL, and the velocity, viscosity and density of the aqueous phase fluid; β_1 is a constant (Imhoff and Miller 1996).

Substituting Eqs. (9.4)–(9.7) into (9.1)–(9.3) yields the following governing equations for the NAPL dissolution system:

$$\phi \frac{\partial S_n}{\partial t} = -\nabla \cdot \left(\frac{k(S_n)}{\mu_a} \nabla p_a \right) - \frac{K}{\rho_a} (C_{eq} - C), \tag{9.9}$$

$$\phi \frac{\partial S_n}{\partial t} = -\frac{K}{\rho_n} (C_{eq} - C), \tag{9.10}$$

$$\phi \frac{\partial}{\partial t} [(1 - S_n)C] = \nabla \cdot [\phi(1 - S_n)\mathbf{D_h} \cdot \nabla C] + \nabla \cdot \left(\frac{k(S_n)}{\mu_a} C \nabla p_a \right) + K(C_{eq} - C). \tag{9.11}$$

Note that S_n, C and p_a are three independent variables in these three equations so that the NAPL dissolution problem is well posed mathematically.

As the boundary conditions expressed in the work of Imhoff and Miller (1996) are incorrect for the NAPL dissolution problem of an infinite domain, they need to be corrected as follows:

$$\lim_{x \to -\infty} C = 0, \quad \lim_{x \to -\infty} \frac{\partial p_a}{\partial x} = p'_{axf} \quad \text{(upstream boundary)}, \tag{9.12}$$

$$\lim_{x \to \infty} C = C_{eq}, \quad \lim_{x \to \infty} \frac{\partial p_a}{\partial x} = p'_{ax0} \quad \text{(downstream boundary)}, \tag{9.13}$$

where p'_{axf} is the pressure gradient of the aqueous phase as x approaches negative infinity in the upstream of the aqueous phase fluid flow; p'_{ax0} is the unknown pressure gradient of the aqueous phase fluid as x approaches positive infinity in the downstream of the aqueous phase fluid flow. Since p'_{axf} drives the aqueous phase fluid flow continuously along the positive x direction, it has a negative algebraic value (i.e. $p'_{axf} < 0$) for the problem under consideration.

The initial condition for this problem is: $S_n(x, 0) = S_{n0}$ expect at the negative infinity, where $\lim_{x \to -\infty} S_n(x, 0) = 0$. Note that S_{n0} is the initial saturation of the NAPL.

9.2 Theoretical Analysis of NAPL Dissolution Induced Instability Problems in Two-Dimensional Fluid-Saturated Porous Media

9.2.1 Previous Approach

In the previous approach, Imhoff and Miller (1996) considered a typical theoretical problem involving the propagation of a planar NAPL dissolution-front in a full plane (i.e. an infinite domain), which is comprised of a fluid-saturated homogeneous porous medium. For this problem, the aqueous phase fluid flow is only in the x direction when the planar NAPL dissolution-front is in a stable state, while it remains dominant in the x direction even when the planar NAPL dissolution-front is at the early stages of an unstable state. In the former case, the aqueous phase fluid velocity is exactly equal to zero in the y direction, whereas in the latter case, the aqueous phase fluid velocity in the y direction is relatively small to that in the x direction so that it can be neglected, from the theoretical analysis point of view.

In the work presented by Imhoff and Miller (1996), the following dimensionless quantities are used:

$$\bar{x} = \frac{x}{l^*}, \quad \bar{y} = \frac{\bar{y}}{l^*}, \quad \bar{t} = \frac{t}{t^*}, \quad \bar{C} = \frac{C}{C_{eq}}, \quad \bar{p}_a = \frac{p_a}{p_a^*}, \tag{9.14}$$

$$\bar{\mathbf{D}}_\mathbf{h} = \frac{\mathbf{D}_\mathbf{h}}{D_{Lf}}, \quad \bar{k}(S_n) = \frac{k(S_n)}{k_f}, \quad \bar{K} = \frac{K}{K_0}, \quad \bar{\rho}_a = \frac{\rho_a}{\rho_n}, \tag{9.15}$$

where $\bar{x}$ and $\bar{y}$ are the dimensionless coordinates; $\bar{t}$ is the dimensionless time; $\bar{C}$ is the dimensionless concentration of the NAPL species in the aqueous phase; $\bar{p}_a$ is the dimensionless aqueous phase pressure; $\bar{\mathbf{D}}_\mathbf{h}$ is the dimensionless (general) dispersion tensor; $\bar{k}(S_n)$ is the dimensionless saturation-dependent permeability of the porous medium to aqueous phase flow; K_0 is the value of K at $S_n = S_{n0}$; $\bar{K}$ is the dimensionless mass transfer rate coefficient of the NAPL species from the nonaqueous phase to the aqueous phase; $\bar{\rho}_a$ is the dimensionless aqueous phase density; D_{Lf}, l^*, t^* and p_a^* are defined as follows:

$$D_{Lf} = \tau D_m + \alpha_L v_{af}, \tag{9.16}$$

$$l^* = \frac{D_{Lf}}{v_{af}}, \tag{9.17}$$

$$t^* = \frac{D_{Lf} S_{n0}}{\varepsilon v_{af}^2}, \tag{9.18}$$

$$p_a^* = \frac{\phi D_{Lf} \mu_a}{k_f}, \tag{9.19}$$

where $\varepsilon = C_{eq}/\rho_n$; v_{af} is the final value for the amplitude of the linear averaged velocity vector of the aqueous phase after the NAPL is completely dissolved from the porous medium; and other quantities have the same meanings as defined previously.

Based on Eqs. (9.14), (9.17) and (9.19), the dimensionless pressure gradient of the aqueous phase in the $\bar{x}$ direction can be expressed as follows:

$$\frac{\partial \bar{p}_a}{\partial \bar{x}} = \left(\frac{l^*}{p_a^*}\right) \frac{\partial p_a}{\partial x} = -\frac{1}{p'_{axf}} \frac{\partial p_a}{\partial x}, \tag{9.20}$$

where p'_{axf} is the pressure gradient of the aqueous phase after the NAPL is completely dissolved from the porous medium.

Considering Eq. (9.20) and Darcy's law yields the following equation:

$$\bar{v}_a = \frac{v_a}{v_{af}} = -\bar{k}(S_n) \frac{1}{1 - S_n} \frac{\partial \bar{p}_a}{\partial \bar{x}}. \tag{9.21}$$

Substituting above dimensionless quantities into Eqs. (9.9)–(9.11) yields the following dimensionless equations:

$$\frac{\varepsilon}{S_{n0}} \frac{\partial S_n}{\partial \bar{t}} = -\nabla \cdot \bar{k}(S_n) \nabla \bar{p}_a - \Gamma \frac{\varepsilon \bar{K}}{\phi \bar{\rho}_a} (1 - \bar{C}), \tag{9.22}$$

$$\frac{1}{S_{n0}} \frac{\partial S_n}{\partial \bar{t}} = -\Gamma \frac{\bar{K}}{\phi} (1 - \bar{C}), \tag{9.23}$$

$$\frac{\varepsilon}{S_{n0}} \frac{\partial}{\partial \bar{t}} [(1 - S_n)\bar{C}] = \nabla \cdot [(1 - S_n)\bar{\mathbf{D}}_{\mathbf{h}} \cdot \nabla \bar{C}] + \nabla \cdot [(1 - S_n)\bar{C}\bar{k}(S_n)\nabla \bar{p}_a] - \frac{1}{S_{n0}} \frac{\partial S_n}{\partial \bar{t}}, \tag{9.24}$$

where

$$\Gamma = \frac{K_0 D_{Lf}}{v_{af}^2}. \tag{9.25}$$

The boundary conditions associated with the above dimensionless quantities can be expressed as follows:

$$\lim_{\bar{x}\to-\infty} \bar{C} = 0, \quad \lim_{\bar{x}\to-\infty} \frac{\partial \bar{p}_a}{\partial \bar{x}} = \bar{p}'_{axf} = -1 \quad \text{(upstream boundary)}, \tag{9.26}$$

$$\lim_{\bar{x}\to\infty} \bar{C} = 1, \quad \lim_{\bar{x}\to\infty} \frac{\partial \bar{p}_a}{\partial \bar{x}} = \bar{p}'_{ax0} = -\frac{1}{\bar{k}(S_{n0})} \quad \text{(downstream boundary)}. \tag{9.27}$$

The use of the above dimensionless quantities implies that in the previous approach (Imhoff and Miller 1996), the dimensionless pressure gradients of the aqueous phase are equal to two *fixed* constants as $\bar{x} \to -\infty$ and $\bar{x} \to +\infty$. This may be the first weakness of the previous approach since the main driving force, namely the dimensionless pressure gradient of the aqueous phase, was not used as an independent variable in the corresponding theoretical analysis.

Similarly, the corresponding initial condition is: $S_n(\bar{x}, 0) = S_{n0}$ expect at the negative infinity, where $\lim_{\bar{x}\to-\infty} S_n(\bar{x}, 0) = 0$. Note that S_{n0} is the initial saturation of the NAPL.

In the limit case as ε approaches zero, a planar NAPL dissolution-front divides the problem domain into two regions, an upstream region and a downstream region, relative to the propagation location of the planar NAPL dissolution-front. Across the planar NAPL dissolution-front, the NAPL saturation undergoes a step jump from its initial value into its final value. As a result, the planar NAPL dissolution-front propagation problem can be treated as a Stefan moving boundary problem (Chadam et al. 1986; Imhoff and Miller 1996). In this limit case, the corresponding dimensionless governing equations of the NAPL dissolution problem in both the downstream region and the upstream region can be expressed as follows:

$$\bar{C} = 1, \quad \nabla^2 \bar{p}_a = 0, \quad S_n = S_{n0} \quad \text{(in the downstream region)}, \tag{9.28}$$

$$\nabla \cdot \left\{ \left[\left(\frac{m}{D_{Lf}} \mathbf{I} - \frac{\partial \bar{p}_a}{\partial \bar{x}} \bar{\mathbf{D}}^*_{dispersion} \right) \cdot \nabla \bar{C} \right] + \bar{C} \nabla \bar{p}_a \right\} = 0, \quad \nabla^2 \bar{p}_a = 0, \quad S_n = 0$$
$$\text{(in the upstream region)}, \tag{9.29}$$

where

$$\bar{\mathbf{D}}^*_{dispersion} = \begin{bmatrix} \bar{\alpha}_L & 0 \\ 0 & \bar{\alpha}_T \end{bmatrix}, \tag{9.30}$$

where $\mathbf{I}$ is a unit second-order tensor (i.e. a 2 by 2 unit matrix); $\bar{\alpha}_L = \alpha_L/l^*$ and $\bar{\alpha}_T = \alpha_T/l^*$.

If the planar NAPL dissolution-front is denoted by $S(\bar{x}, \bar{t}) = 0$, then the dimensionless pressure, dimensionless NAPL concentration and the mass fluxes of both the NAPL and the aqueous phase fluid should be continuous on $S(\bar{x}, \bar{t}) = 0$. This leads to the following interface conditions for the planar NAPL dissolution-front propagation problem:

$$\lim_{S\to 0^-} \bar{C} = \lim_{S\to 0^+} \bar{C}, \quad \lim_{S\to 0^-} \bar{p}_a = \lim_{S\to 0^+} \bar{p}_a, \tag{9.31}$$

$$\lim_{S\to 0^-} \left(\frac{\tau D_m}{D_{Lf}} - \bar{\alpha}_L \bar{p}'_{axf}\right)\frac{\partial \bar{C}}{\partial \bar{n}} = \bar{v}_{front} S_{n0}, \quad \lim_{S\to 0^-} \frac{\partial \bar{p}_a}{\partial \bar{n}} = \bar{k}(S_{n0}) \lim_{S\to 0^+} \frac{\partial \bar{p}_a}{\partial \bar{n}}, \tag{9.32}$$

where $\bar{n} = n/l^*$; n is the normal vector of the propagating planar NAPL dissolution-front; and $\bar{v}_{front}$ is the dimensionless propagation speed of the planar NAPL dissolution-front.

When the planar NAPL dissolution-front is in a stable state, base solutions for this theoretical problem can be expressed as follows:

$$\bar{C}_b(\xi) = 1, \quad \bar{p}_{ab}(\xi) = \bar{p}'_{ax0}\xi + \bar{p}_{aC1}, \quad S_{nb} = S_{n0} \quad \text{(in the downstream region)}, \tag{9.33}$$

$$\bar{C}_b(\xi) = \exp\left(-\frac{\bar{p}'_{axf}}{\frac{\tau D_m}{D_{Lf}} - \bar{\alpha}_L \bar{p}'_{axf}}\xi\right) = \exp(\xi), \quad \bar{p}_{ab}(\xi) = \bar{p}'_{axf}\xi + \bar{p}_{aC2}, \quad S_{nb} = 0 \quad \text{(in the upstream region)}, \tag{9.34}$$

where $\bar{C}_b$, $\bar{p}_{ab}$ and S_{nb} are the base solutions for the dimensionless concentration of the NAPL species in the aqueous phase, the dimensionless aqueous phase pressure and the saturation of the NAPL, respectively; $\bar{p}_{aC1}$ and $\bar{p}_{aC2}$ are two constants to be determined.

$$\xi = \bar{x} - \bar{v}_{front}\bar{t}, \quad \bar{v}_{front} = -\frac{\bar{p}'_{axf}}{S_{no}} = \frac{1}{S_{n0}}. \tag{9.35}$$

To determine the critical condition under which a planar NAPL dissolution-front in the fluid-saturated porous medium can become unstable, a small dimensionless time-dependent perturbation is added to the planar NAPL dissolution-front. As a result, the total solutions of the NAPL dissolution system are equal to the sum of the base and perturbed solutions as follows:

$$S(\xi, \bar{y}, \bar{t}) = \xi - \delta \exp(\bar{\omega}\bar{t})\cos(\bar{m}\bar{y}), \tag{9.36}$$

$$\bar{p}_{atotal}(\xi, \bar{y}, \bar{t}) = \bar{p}_{ab}(\xi, \bar{t}) + \delta \hat{p}_a(\xi)\exp(\bar{\omega}\bar{t})\cos(\bar{m}\bar{y}), \tag{9.37}$$

$$\bar{C}_{total}(\xi, \bar{y}, \bar{t}) = \bar{C}_b(\xi, \bar{t}) + \delta \hat{C}(\xi)\exp(\bar{\omega}\bar{t})\cos(\bar{m}\bar{y}), \tag{9.38}$$

where $\bar{\omega}$ is the dimensionless growth rate of the perturbation; $\bar{m}$ is the dimensionless wavenumber of the perturbation; $\hat{C}$ and $\hat{p}_a$ are the perturbed solutions for the dimensionless concentration of the NAPL species in the aqueous phase and the dimensionless aqueous phase pressure, respectively; δ is the amplitude of the perturbation and $\delta \ll 1$ by the definition of the linear stability analysis.

Since $S(\xi, \bar{y}, \bar{t})$ is a function of coordinates ξ and $\bar{y}$, the following derivatives exist mathematically:

$$\left(\frac{\partial}{\partial \xi}\right)_{\xi} = \frac{\partial S}{\partial \xi}\frac{\partial}{\partial S} = \left(\frac{\partial}{\partial \xi}\right)_{S}, \quad \left(\frac{\partial}{\partial \bar{y}}\right)_{\xi} = \frac{\partial S}{\partial \bar{y}}\frac{\partial}{\partial S} + \left(\frac{\partial}{\partial \bar{y}}\right)_{S} = \frac{\partial S}{\partial \bar{y}}\left(\frac{\partial}{\partial \xi}\right)_{S} + \left(\frac{\partial}{\partial \bar{y}}\right)_{S}, \tag{9.39}$$

$$\left(\frac{\partial^2}{\partial \xi^2}\right)_{\xi} = \left(\frac{\partial^2}{\partial \xi^2}\right)_{S}, \quad \left(\frac{\partial^2}{\partial \bar{y}^2}\right)_{\xi} = \frac{\partial^2 S}{\partial \bar{y}^2}\frac{\partial}{\partial \xi} + \left(\frac{\partial S}{\partial \bar{y}}\right)^2 \frac{\partial^2}{\partial \xi^2} + 2\frac{\partial S}{\partial \bar{y}}\frac{\partial^2}{\partial \xi \partial \bar{y}} + \left(\frac{\partial^2}{\partial \bar{y}^2}\right)_{S}. \tag{9.40}$$

Substituting the total solutions into the governing equations (i.e. Eqs. (9.28) and (9.29)) yields the first-order perturbation equations for the NAPL dissolution system as follows:

$$\hat{C} = 0, \quad \frac{\partial^2 \hat{p}_a}{\partial \xi^2} - \bar{m}^2 \hat{p}_a + \bar{m}^2 \bar{p}'_{ax0} = 0 \quad \text{(in the downstream region)}, \tag{9.41}$$

$$\begin{aligned} &\frac{\partial^2 \hat{C}}{\partial \xi^2} - \frac{\partial \hat{C}}{\partial \xi} - \Lambda \bar{m}^2 \hat{C} + \Lambda \bar{m}^2 \exp(\xi) + (1 - \bar{\alpha}_L)\exp(\xi)\frac{\partial \hat{p}_a}{\partial \xi} - \bar{\alpha}_L \exp(\xi)\frac{\partial^2 \hat{p}_a}{\partial \xi^2} = 0\,, \\ &\frac{\partial^2 \hat{p}_a}{\partial \xi^2} - \bar{m}^2 \hat{p}_a + \bar{m}^2 \bar{p}'_{axf} = 0 \qquad \text{(in the upstream region)}, \end{aligned} \tag{9.42}$$

where

$$\Lambda = \frac{\frac{\tau D_m}{D_{Lf}} + \bar{\alpha}_T}{\frac{\tau D_m}{D_{Lf}} + \bar{\alpha}_L} = \frac{\tau D_m + \alpha_T v_{af}}{\tau D_m + \alpha_L v_{af}}. \tag{9.43}$$

It needs to be pointed out that in the process of deriving the first-order perturbation equation for solute transport in the NAPL dissolution system, the following expression has been considered:

$$\begin{aligned} &\nabla \cdot \left[\left(\frac{\tau D_m}{D_{Lf}}\mathbf{I} - \frac{\partial \bar{p}_{atotal}}{\partial \bar{x}}\bar{\mathbf{D}}_{dispersion}\right) \cdot \nabla \bar{C}_{total}\right] \\ &= \frac{\partial}{\partial \bar{x}}\left[\left(\frac{\tau D_m}{D_{Lf}} - \bar{\alpha}_L \frac{\partial \bar{p}_{atotal}}{\partial \bar{x}}\right)\frac{\partial \bar{C}_{total}}{\partial \bar{x}}\right] + \frac{\partial}{\partial \bar{y}}\left[\left(\frac{\tau D_m}{D_{Lf}} - \bar{\alpha}_T \frac{\partial \bar{p}_{atotal}}{\partial \bar{x}}\right)\frac{\partial \bar{C}_{total}}{\partial \bar{y}}\right] \\ &= \left[\left(\frac{\tau D_m}{D_{Lf}} - \bar{\alpha}_L \frac{\partial \bar{p}_{atotal}}{\partial \bar{x}}\right)\frac{\partial^2 \bar{C}_{total}}{\partial \bar{x}^2} - \bar{\alpha}_L \frac{\partial^2 \bar{p}_{atotal}}{\partial \bar{x}^2}\frac{\partial \bar{C}_{total}}{\partial \bar{x}}\right] \\ &\quad + \left[\left(\frac{\tau D_m}{D_{Lf}} - \bar{\alpha}_T \frac{\partial \bar{p}_{atotal}}{\partial \bar{x}}\right)\frac{\partial^2 \bar{C}_{total}}{\partial \bar{y}^2} - \bar{\alpha}_T \frac{\partial^2 \bar{p}_{atotal}}{\partial \bar{x}\partial \bar{y}}\frac{\partial \bar{C}_{total}}{\partial \bar{y}}\right]. \end{aligned} \tag{9.44}$$

For the sake of convenience, the following expression is used to derive the first-order perturbation equations:

$$\frac{\partial}{\partial \xi} = \frac{\partial}{\partial \bar{x}}. \tag{9.45}$$

Substituting Eqs. (9.37) and (9.38) into the first term on the right-hand side of Eq. (9.44) and considering $\partial^2 \bar{p}_{ab}/(\partial \bar{x}^2) = 0$ simultaneously yield the following equation:

$$\begin{aligned}
&\left(\frac{\tau D_m}{D_{Lf}} - \bar{\alpha}_L \frac{\partial \bar{p}_{atotal}}{\partial \bar{x}}\right) \frac{\partial^2 \bar{C}_{total}}{\partial \bar{x}^2} - \bar{\alpha}_L \frac{\partial^2 \bar{p}_{atotal}}{\partial \bar{x}^2} \frac{\partial \bar{C}_{total}}{\partial \bar{x}} \\
&= \left(\frac{\tau D_m}{D_{Lf}} - \bar{\alpha}_L \frac{\partial \bar{p}_{ab}}{\partial \bar{x}}\right) \frac{\partial^2 \bar{C}_b}{\partial \bar{x}^2} + \left(\frac{\tau D_m}{D_{Lf}} - \bar{\alpha}_L \frac{\partial \bar{p}_{ab}}{\partial \bar{x}}\right) \frac{\partial^2 \hat{C}}{\partial \bar{x}^2} \delta \exp(\bar{\omega} \bar{t}) \cos(\bar{m} \bar{y}) \\
&- \bar{\alpha}_L \frac{\partial \hat{p}_a}{\partial \bar{x}} \frac{\partial^2 \bar{C}_b}{\partial \bar{x}^2} \delta \exp(\bar{\omega} \bar{t}) \cos(\bar{m} \bar{y}) - \bar{\alpha}_L \frac{\partial \hat{p}_a}{\partial \bar{x}} \frac{\partial^2 \hat{C}}{\partial \bar{x}^2} [\delta \exp(\bar{\omega} \bar{t}) \cos(\bar{m} \bar{y})]^2 \\
&- \bar{\alpha}_L \frac{\partial^2 \hat{p}_a}{\partial \bar{x}^2} \frac{\partial \bar{C}_b}{\partial \bar{x}} \delta \exp(\bar{\omega} \bar{t}) \cos(\bar{m} \bar{y}) - \bar{\alpha}_L \frac{\partial^2 \hat{p}_a}{\partial \bar{x}^2} \frac{\partial \hat{C}}{\partial \bar{x}} [\delta \exp(\bar{\omega} \bar{t}) \cos(\bar{m} \bar{y})]^2.
\end{aligned} \tag{9.46}$$

If the second-order terms due to the perturbation are neglected, then Eq. (9.46) can be rewritten in the following form:

$$\begin{aligned}
&\left(\frac{\tau D_m}{D_{Lf}} - \bar{\alpha}_L \frac{\partial \bar{p}_{atotal}}{\partial \bar{x}}\right) \frac{\partial^2 \bar{C}_{total}}{\partial \bar{x}^2} - \bar{\alpha}_L \frac{\partial^2 \bar{p}_{atotal}}{\partial \bar{x}^2} \frac{\partial \bar{C}_{total}}{\partial \bar{x}} \\
&= \left(\frac{\tau D_m}{D_{Lf}} - \bar{\alpha}_L \frac{\partial \bar{p}_{ab}}{\partial \bar{x}}\right) \frac{\partial^2 \bar{C}_b}{\partial \bar{x}^2} + \left(\frac{\tau D_m}{D_{Lf}} - \bar{\alpha}_L \frac{\partial \bar{p}_{ab}}{\partial \bar{x}}\right) \frac{\partial^2 \hat{C}}{\partial \bar{x}^2} \delta \exp(\bar{\omega} \bar{t}) \cos(\bar{m} \bar{y}) \\
&- \bar{\alpha}_L \frac{\partial \hat{p}_a}{\partial \bar{x}} \frac{\partial^2 \bar{C}_b}{\partial \bar{x}^2} \delta \exp(\bar{\omega} \bar{t}) \cos(\bar{m} \bar{y}) - \bar{\alpha}_L \frac{\partial^2 \hat{p}_a}{\partial \bar{x}^2} \frac{\partial \bar{C}_b}{\partial \bar{x}} \delta \exp(\bar{\omega} \bar{t}) \cos(\bar{m} \bar{y}).
\end{aligned} \tag{9.47}$$

Compared with the previous case when dispersion is considered as the scalar of a constant (Chadam et al. 1986; Zhao et al. 2008b, c), two new terms, namely

$$-\bar{\alpha}_L \frac{\partial \hat{p}_a}{\partial \bar{x}} \frac{\partial^2 \overline{C}_b}{\partial \bar{x}^2} \delta \exp(\overline{\omega t}) \cos(\overline{my}) \text{ and } \quad \bar{\alpha}_L \frac{\partial^2 \hat{p}_a}{\partial \bar{x}^2} \frac{\partial \overline{C}_b}{\partial \bar{x}} \delta \exp(\overline{\omega t}) \cos(\overline{my}),$$

are added in the process of deriving the first-order perturbation equation for solute (i.e. NAPL) transport when dispersion is considered as the second-order tensor.

Similarly, if Eqs. (9.37) and (9.38) are substituted into the second term on the right-hand side of Eq. (9.44) with the consideration of $\partial^2 \bar{p}_{ab}/(\partial \bar{x} \partial \bar{y}) = 0$ and $\partial \bar{C}_b/\partial \bar{y} = 0$, then the following equation can be obtained when the second-order terms are neglected:

$$\left(\frac{\tau D_m}{D_{Lf}} - \bar{\alpha}_T \frac{\partial \bar{p}_{atotal}}{\partial \bar{x}}\right) \frac{\partial^2 \bar{C}_{total}}{\partial \bar{y}^2} - \bar{\alpha}_T \frac{\partial^2 \bar{p}_{atotal}}{\partial \bar{x} \partial \bar{y}} \frac{\partial \bar{C}_{total}}{\partial \bar{y}}$$
$$= \left(\frac{\tau D_m}{D_{Lf}} - \bar{\alpha}_T \frac{\partial \bar{p}_{ab}}{\partial \bar{x}}\right) \frac{\partial^2 \bar{C}_b}{\partial \bar{y}^2} + \left(\frac{\tau D_m}{D_{Lf}} - \bar{\alpha}_T \frac{\partial \bar{p}_{ab}}{\partial \bar{x}}\right) \frac{\partial^2 \left[\hat{C} \cos(\bar{m}\bar{y})\right]}{\partial \bar{y}^2} \delta \exp(\omega \bar{t}). \tag{9.48}$$

In this situation, compared with the previous case when dispersion is considered as the scalar of a constant, no extra term is contributed to the first-order perturbation equation for solute (i.e. NAPL) transport when dispersion is considered as the second-order tensor.

Since the dispersion tensor is a function of perturbed dimensionless pressure (i.e. the term related to $\delta \hat{p}_a(\xi) \exp(\bar{\omega} \bar{t}) \cos(\bar{m}\bar{y})$), the first-order perturbation equations when solute dispersion is considered as the second-order tensor are different from those when it is considered as the scalar of a constant (Chadam et al. 1986). Obviously, consideration of the dependence of the dispersion tensor on the perturbed dimensionless pressure results in much more complicated perturbation equations with two additional terms, so that special mathematical treatments should be used to derive the corresponding analytical solutions. Due to the neglect of this fact, the analytical solutions derived by Imhoff and Miller (1996) are conceptually questionable.

The boundary conditions for the first-order perturbation problem are:

$$\lim_{\bar{x} \to \infty} \hat{C} = 0, \quad \lim_{\bar{x} \to \infty} \frac{\partial \hat{p}_a}{\partial \xi} = 0 \quad \text{(downstream boundary)}, \tag{9.49}$$

$$\lim_{\bar{x} \to -\infty} \hat{C} = 0, \quad \lim_{\bar{x} \to -\infty} \frac{\partial \hat{p}_a}{\partial \xi} = 0 \quad \text{(upstream boundary)}. \tag{9.50}$$

Similarly, the interface conditions for this first-order perturbation problem can be expressed as follows:

$$\lim_{S \to 0^-} \hat{C} = \lim_{S \to 0^+} \hat{C}, \quad \lim_{S \to 0^-} \hat{p}_a = \lim_{S \to 0^+} \hat{p}_a, \tag{9.51}$$

$$\lim_{S \to 0^-} \left[\frac{\partial \hat{C}}{\partial \bar{n}} - \bar{\alpha}_L \frac{\partial \hat{p}_a}{\partial \bar{n}}\right] = \bar{\omega} S_{n0}, \quad \lim_{S \to 0^-} \frac{\partial \hat{p}_a}{\partial \bar{n}} = \bar{k}(S_{n0}) \lim_{S \to 0^-} \frac{\partial \hat{p}_a}{\partial \bar{n}}. \tag{9.52}$$

Solving Eqs. (9.41) and (9.42) with the boundary and interface conditions (i.e. Eqs. (9.49)–(9.52)) yields the following analytical solutions:

$$\hat{C} = 0, \quad \hat{p}_a(\xi) = \bar{p}'_{ax0} \left[1 - \frac{1 - \beta}{1 + \beta} \exp(-\bar{m}\xi)\right] \quad \text{(in the downstream region)}, \tag{9.53}$$

$$\hat{C}(\xi) = \exp(\xi) - \frac{(1+\beta) + (1-\overline{\alpha}_L\overline{m})(1-\beta)}{1+\beta}\exp(\sigma\xi) + (1-\overline{\alpha}_L\overline{m})\frac{1-\beta}{1+\beta}\exp[(\overline{m}+1)\xi],$$
$$\hat{p}_a(\xi) = \left[1 + \frac{1-\beta}{1+\beta}\exp(\overline{m}\xi)\right] \quad \text{(in the upstream region)}, \tag{9.54}$$

where

$$\beta = \overline{k}(S_{n0}), \qquad \sigma = \frac{\sqrt{1+4\overline{m}^2\Lambda}+1}{2}. \tag{9.55}$$

Inserting Eqs. (9.54) into (9.52) yields the following equation for the dimensionless growth rate of the small perturbation:

$$\overline{\omega} = \frac{1}{(1+\beta)S_{n0}}\left\{2 - [2-\overline{\alpha}_L\overline{m}(1-\beta)]\frac{\sqrt{1+4\overline{m}^2\Lambda}+1}{2} + (\overline{m}-\overline{\alpha}_L\overline{m}^2)(1-\beta)\right\}. \tag{9.56}$$

Equation (9.56) indicates that the dimensionless growth rate of the NAPL dissolution system is not only dependent on the ratio of transverse dispersion to longitudinal dispersion (i.e. Λ), but also dependent on the dimensionless longitudinal dispersivity itself (i.e. $\overline{\alpha}_L$). This demonstrates that Eq. (9.56) is conceptually different from the dimensionless growth rate derived by Imhoff and Miller (1996). Due to this conceptual difference, their model predictions are questionable so that the critical condition under which a planar NAPL dissolution-front becomes unstable cannot be appropriately derived. This is another weakness of the previous approach used by Imhoff and Miller (1996).

9.2.2 Current Approach

To overcome the weaknesses associated with the previous approach, the following dimensionless quantities are used to derive the dimensionless governing equations of the NAPL dissolution system:

$$\overline{x} = \frac{x}{L^*}, \quad \overline{y} = \frac{y}{L^*}, \quad \overline{C} = \frac{C}{C_{eq}}, \quad \overline{p}_a = \frac{p_a}{p_a^*}, \tag{9.57}$$

$$\overline{\alpha}_L = \frac{\alpha_L}{L^*}, \quad \overline{\alpha}_T = \frac{\alpha_T}{L^*}, \quad \overline{t} = \frac{t}{t^*}\varepsilon, \quad \varepsilon = \frac{C_{eq}}{\rho_n} \ll 1 \tag{9.58}$$

where

$$t^* = \frac{\phi}{\beta_0}, \quad L^* = \sqrt{\tau D_m t^*}, \quad p_a^* = \frac{\phi \tau D_m}{\psi(S_{nf})}, \tag{9.59}$$

$$\overline{\mathbf{D}}_{\mathbf{h}} = \frac{\mathbf{D}_{\mathbf{h}}}{\tau D_m}, \quad \overline{\psi}(S_n) = \frac{\psi(S_n)}{\psi(S_{nf})}, \quad \psi(S_n) = \frac{k(S_n)}{\mu_a}, \tag{9.60}$$

where $\psi(S_{nf})$ is the value of $\psi(S_n)$ at $S_n = S_{nf}$; and $S_{nf} = 0$ is the saturation of the NAPL after it is completely dissolved in the NAPL dissolution system.

Inserting Eqs. (9.57)–(9.60) into (9.9)–(9.11) yields the following dimensionless governing equations for the NAPL dissolution system:

$$\frac{\partial S_n}{\partial \bar{t}} + S_n^{\beta_1}(1 - \overline{C}) = 0, \tag{9.61}$$

$$\varepsilon \frac{\partial}{\partial \bar{t}}[(1 - S_n)\bar{C}] - \nabla \cdot \left\{[(1 - S_n)\overline{\mathbf{D}}_{\mathbf{h}}] \cdot \nabla \bar{C} + \bar{C}\overline{\psi}(S_n)\nabla \bar{p}_a\right\} + \frac{\partial S_n}{\partial \bar{t}} = 0, \tag{9.62}$$

$$\varepsilon \frac{\partial S_n}{\partial \bar{t}} + \nabla \cdot [\overline{\psi}(S_n)\nabla \bar{p}_a] + \frac{\varepsilon}{\bar{\rho}_a} S_n^{\beta_1}(1 - \bar{C}) = 0, \tag{9.63}$$

where

$$\overline{\mathbf{D}}_{\mathbf{h}} = \begin{bmatrix} 1 - \bar{\alpha}_L \frac{\overline{\psi}(S_n)}{1 - S_n} \frac{\partial \bar{p}_a}{\partial \bar{x}} & 0 \\ 0 & 1 - \bar{\alpha}_T \frac{\overline{\psi}(S_n)}{1 - S_n} \frac{\partial \bar{p}_a}{\partial \bar{x}} \end{bmatrix}. \tag{9.64}$$

The boundary conditions in this case can be expressed in the dimensionless form as follows:

$$\lim_{\bar{x}\to\infty} \bar{C} = 1, \quad \lim_{\bar{x}\to\infty} \frac{\partial \bar{p}_a}{\partial \bar{x}} = \bar{p}'_{ax0} \quad \text{(downstream boundary)}, \tag{9.65}$$

$$\lim_{\bar{x}\to-\infty} \bar{C} = 0, \quad \lim_{\bar{x}\to-\infty} \frac{\partial \bar{p}_a}{\partial \bar{x}} = \bar{p}'_{axf} \quad \text{(upstream boundary)}. \tag{9.66}$$

In the limit case of ε approaching zero, the dimensionless governing equations of the NAPL dissolution problem in both the downstream region and the upstream region can be expressed as follows:

$$\bar{C} = 1, \quad \nabla^2 \bar{p}_a = 0, \quad S_n = S_{n0} \quad \text{(in the downstream region)}, \tag{9.67}$$

$$\nabla \cdot \left\{\left[\left(\mathbf{I} - \frac{\partial \bar{p}_a}{\partial \bar{x}} \overline{\mathbf{D}}^*_{dispersion}\right) \cdot \nabla \bar{C}\right] + \bar{C}\nabla \bar{p}_a\right\} = 0, \quad \nabla^2 \bar{p}_a = 0, \quad S_n = 0 \quad \text{(in the upstream region)}, \tag{9.68}$$

where

$$\bar{\mathbf{D}}^*_{dispersion} = \begin{bmatrix} \bar{\alpha}_L & 0 \\ 0 & \bar{\alpha}_T \end{bmatrix}. \tag{9.69}$$

If the planar NAPL dissolution-front is denoted by $S(\bar{x}, \bar{t}) = 0$, then the dimensionless aqueous phase fluid pressure, dimensionless NAPL concentration and the mass fluxes of both the NAPL and the aqueous phase fluid should be continuous on $S(\bar{x}, \bar{t}) = 0$. This leads to the following interface conditions for the planar NAPL dissolution-front propagation problem:

$$\lim_{S\to 0^-} \bar{C} = \lim_{S\to 0^+} \bar{C}, \quad \lim_{S\to 0^-} \bar{p}_a = \lim_{S\to 0^+} \bar{p}_a, \tag{9.70}$$

$$\lim_{S\to 0^-} (1 - \bar{\alpha}_L \bar{p}'_{axf}) \frac{\partial \bar{C}}{\partial \bar{n}} = \bar{v}_{front} S_{n0}, \quad \lim_{S\to 0^-} \frac{\partial \bar{p}_a}{\partial \bar{n}} = \frac{\psi(S_{n0})}{\psi(S_{nf})} \lim_{S\to 0^+} \frac{\partial \bar{p}_a}{\partial \bar{n}}, \tag{9.71}$$

where $\bar{n} = n/L^*$; n is the normal vector of the propagating planar NAPL dissolution-front; and $\bar{v}_{front}$ is the dimensionless propagation speed of the planar NAPL dissolution-front.

Base solutions for this theoretical problem can be expressed as follows:

$$\bar{C}_b(\xi) = 1, \quad \bar{p}_{ab}(\xi) = \bar{p}'_{ax0}\xi + \bar{p}_{aC1}, \quad S_{nb} = S_{n0} \quad \text{(in the downstream region)}, \tag{9.72}$$

$$\bar{C}_b(\xi) = \exp\left(-\frac{\bar{p}'_{axf}}{1 - \bar{\alpha}_L \bar{p}'_{axf}}\xi\right), \quad \bar{p}_{ab}(\xi) = \bar{p}'_{axf}\xi + \bar{p}_{aC2}, \quad S_{nb} = 0 \quad \text{(in the upstream region)}, \tag{9.73}$$

where $\bar{p}_{aC1}$ and $\bar{p}_{aC2}$ are two constants.

$$\xi = \bar{x} - \bar{v}_{front}\bar{t}, \quad \bar{p}'_{ax0} = \frac{\psi(S_{nf})}{\psi(S_{n0})}\bar{p}'_{axf}, \quad \bar{v}_{front} = -\frac{\bar{p}'_{axf}}{S_{no}}. \tag{9.74}$$

By following the same procedures as those used in the above linear stability analysis, the first-order perturbation equations of the NAPL dissolution system can be expressed as

$$\hat{C} = 0, \quad \frac{\partial^2 \hat{p}_a}{\partial \xi^2} - \bar{m}^2 \hat{p}_a + \bar{m}^2 \bar{p}'_{ax0} = 0 \quad \text{(in the downstream region)}, \tag{9.75}$$

$$\left(1-\bar{\alpha}_L\bar{p}'_{axf}\right)\frac{\partial^2\hat{C}}{\partial\xi^2}+\bar{p}'_{axf}\frac{\partial\hat{C}}{\partial\xi}-\left(1-\bar{\alpha}_T\bar{p}'_{axf}\right)\bar{m}^2\hat{C}$$
$$-\frac{\left(1-\bar{\alpha}_T\bar{p}'_{axf}\right)\bar{m}^2\bar{p}'_{axf}}{1-\bar{\alpha}_L\bar{p}'_{axf}}\exp\left(-\frac{\bar{p}'_{axf}}{1-\bar{\alpha}_L\bar{p}'_{axf}}\xi\right)$$
$$-\frac{\bar{p}'_{axf}}{1-\bar{\alpha}_L\bar{p}'_{axf}}\left(1+\frac{\bar{\alpha}_L\bar{p}'_{axf}}{1-\bar{\alpha}_L\bar{p}'_{axf}}\right)\exp\left(-\frac{\bar{p}'_{axf}}{1-\bar{\alpha}_L\bar{p}'_{axf}}\xi\right)\frac{\partial\hat{p}_a}{\partial\xi} \tag{9.76}$$
$$+\frac{\bar{\alpha}_L\bar{p}'_{axf}}{1-\bar{\alpha}_L\bar{p}'_{axf}}\exp\left(-\frac{\bar{p}'_{axf}}{1-\bar{\alpha}_L\bar{p}'_{axf}}\xi\right)\frac{\partial^2\hat{p}_a}{\partial\xi^2}=0\,,$$
$$\frac{\partial^2\hat{p}_a}{\partial\xi^2}-\bar{m}^2\hat{p}_a+\bar{m}^2\bar{p}'_{axf}=0 \qquad \text{(in the upstream region)},$$

Since the dispersion tensor is a function of perturbed dimensionless pressure (i.e. the term related to $\delta\hat{p}_a(\xi)\exp(\bar{\omega}\bar{t})\cos(\bar{m}\bar{y})$), the first-order perturbation equations derived here are different from those derived by Imhoff and Miller (1996) previously.

Although the corresponding boundary conditions are the same as those expressed by Eqs. (9.49) and (9.50), the interface conditions for this first-order perturbation problem can be expressed as follows:

$$\lim_{S\to 0^-}\hat{C}=\lim_{S\to 0^+}\hat{C},\quad \lim_{S\to 0^-}\hat{p}_a=\lim_{S\to 0^+}\hat{p}_a, \tag{9.77}$$

$$\lim_{S\to 0^-}\left[\left(1-\bar{\alpha}_L\bar{p}'_{axf}\right)\frac{\partial\hat{C}}{\partial\bar{n}}+\frac{\bar{\alpha}_L\bar{p}'_{axf}}{1-\bar{\alpha}_L\bar{p}'_{axf}}\frac{\partial\hat{p}_a}{\partial\bar{n}}\right]=\bar{\omega}S_{n0},\quad \lim_{S\to 0^-}\frac{\partial\hat{p}_a}{\partial\bar{n}}=\frac{\psi(S_{n0})}{\psi(S_{nf})}\lim_{S\to 0^+}\frac{\partial\hat{p}_u}{\partial\bar{n}}. \tag{9.78}$$

Solving Eqs. (9.75) and (9.76) with the corresponding boundary and interface conditions yields the following analytical solutions:

$$\hat{C}=0,\quad \hat{p}_a(\xi)=\bar{p}'_{ax0}\left[1-\frac{1-\beta}{1+\beta}\exp(-\bar{m}\xi)\right] \quad \text{(in the downstream region)}, \tag{9.79}$$

$$\hat{C}(\xi)=-\frac{\bar{p}'_{axf}}{1-\bar{\alpha}_L\bar{p}'_{axf}}\left\{\exp\left(-\frac{\bar{p}'_{axf}}{1-\bar{\alpha}_L\bar{p}'_{axf}}\xi\right)-\frac{(1+\beta)+\left(\frac{1}{1-\bar{\alpha}_L\bar{p}'_{axf}}-\bar{\alpha}_L\bar{m}\right)(1-\beta)}{1+\beta}\exp(\sigma\xi)\right\}$$
$$-\frac{\bar{p}'_{axf}}{1-\bar{\alpha}_L\bar{p}'_{axf}}\left\{\left(\frac{1}{1-\bar{\alpha}_L\bar{p}'_{axf}}-\bar{\alpha}_L\bar{m}\right)\frac{1-\beta}{1+\beta}\exp\left[\left(\bar{m}-\frac{\bar{p}'_{axf}}{1-\bar{\alpha}_L\bar{p}'_{axf}}\right)\xi\right]\right\},$$

$$\hat{p}_a(\xi) = \bar{p}'_{axf}\left[1 + \frac{1-\beta}{1+\beta}\exp(\bar{m}\xi)\right] \quad \text{(in the upstream region)}, \tag{9.80}$$

where

$$\beta = \frac{\psi(S_{n0})}{\psi(S_{nf})} = \frac{k(S_{n0})}{k(S_{nf})}, \tag{9.81}$$

$$\sigma = \frac{\sqrt{\left(\frac{\bar{p}'_{axf}}{1-\bar{\alpha}_L\bar{p}'_{axf}}\right)^2 + \frac{4\bar{m}^2\left(1-\bar{\alpha}_T\bar{p}'_{axf}\right)}{1-\bar{\alpha}_L\bar{p}'_{axf}}} - \frac{\bar{p}'_{axf}}{1-\bar{\alpha}_L\bar{p}'_{axf}}}{2}. \tag{9.82}$$

Inserting Eqs. (9.80) into (9.78) yields the following equation for the dimensionless growth rate of the perturbed NAPL dissolution-front:

$$\begin{aligned}\bar{\omega} = {} & \frac{\bar{p}'_{axf}}{(1+\beta)S_{n0}}\left\{\frac{(1+\beta)\bar{p}'_{axf}}{1-\bar{\alpha}_L\bar{p}'_{axf}} + \left[(1+\beta) + \left(\frac{1}{1-\bar{\alpha}_L\bar{p}'_{axf}} - \bar{\alpha}_L\bar{m}\right)(1-\beta)\right]\sigma\right\} \\ & - \frac{\bar{p}'_{axf}}{(1+\beta)S_{n0}}\left[\left(\bar{m} - \frac{\bar{p}'_{axf}}{1-\bar{\alpha}_L\bar{p}'_{axf}} - \bar{\alpha}_L\bar{m}^2\left(1-\bar{\alpha}_L\bar{p}'_{axf}\right)\right)\frac{(1-\beta)}{1-\bar{\alpha}_L\bar{p}'_{axf}}\right].\end{aligned} \tag{9.83}$$

Equation (9.83) indicates that the dimensionless growth rate is not only dependent on the dimensionless wavenumber of the perturbation, but also dependent on the dimensionless pressure gradient of the aqueous phase fluid in the upstream direction of the NAPL dissolution system. This means that in order to determine the value of $\bar{\omega}$, either the dimensionless pressure gradient of the aqueous phase fluid applied to the upstream of the NAPL dissolution system ($\bar{p}'_{axf}$) or the dimensionless wavenumber of the perturbation ($\bar{m}$) may be kept constant. In the former case, the value of the dimensionless growth rate is positive when the dimensionless wavenumber of the perturbation is very small. This indicates that if the dimensionless wavelength of the perturbation is infinite, then the NAPL dissolution front is unstable, even if the absolute value of the dimensionless pressure gradient of the aqueous phase fluid in the upstream direction of the NAPL dissolution system is very small. However, to allow a perturbation of infinite dimensionless wavelength to take place, the dimensionless geometrical characteristic length of a NAPL dissolution system, which is defined as the dimensionless geometrical length of the NAPL dissolution system in the direction perpendicular to the aqueous phase liquid flow (i.e. $\bar{L}_y = L_y/L^*$), should be equal to infinity. Therefore, for the NAPL dissolution system of an infinite domain, the stability of a NAPL dissolution front can be independent of the value of the dimensionless pressure gradient of the aqueous phase fluid applied to the upstream of the NAPL dissolution system (i.e. $\bar{p}'_{axf}$). This is the situation that was considered by Chadam et al. (1986).

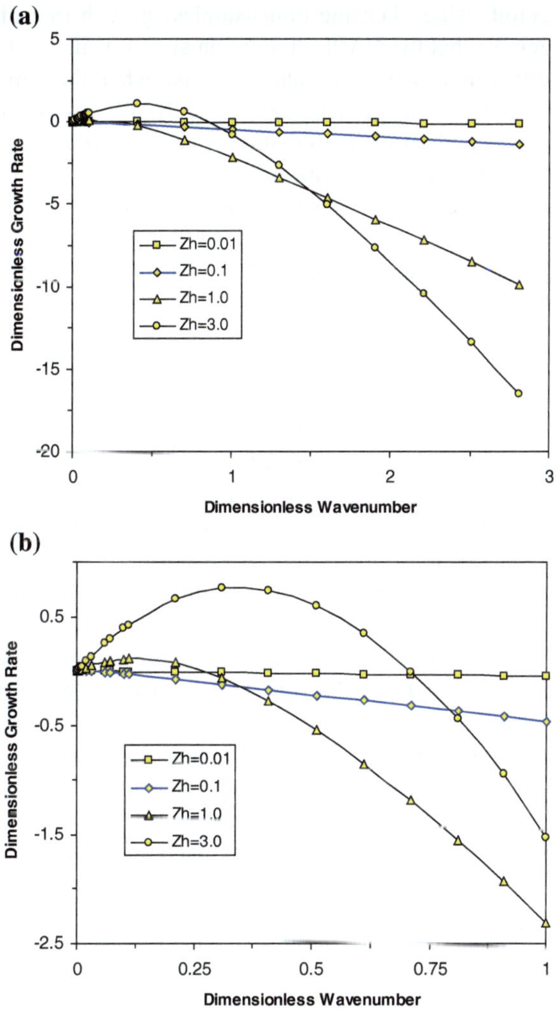

Fig. 9.1 Distributions of the dimensionless growth rate due to different Zhao numbers ($S_{n0} = 0.2$, $S_{ai} = 0.15$, $\bar{\alpha}_L = 0.2$, $\bar{\alpha}_T = 0.02$): **a** Entire view; **b** Zoomed-in view

The dynamics of a NAPL dissolution system in this particular case can be shown in Fig. 9.1, where $Zh = -\bar{p}'_{axf}$ and Zh is called the Zhao number (Zhao et al. 2009). To obtain the numerical results in this figure, the following parameters are used: the initial saturation (S_{n0}) of the NAPL is 0.2; the irreducible saturation (S_{ai}) of the aqueous phase fluid is 0.15; the dimensionless longitudinal dispersivity ($\bar{\alpha}_L$) is 0.2; and the dimensionless transverse dispersivity ($\bar{\alpha}_T$) is 0.02. Since $\bar{p}'_{axf} < 0$ when the aqueous phase fluid flow is along the positive x direction, the values of Zh are positive in the theoretical analysis. It is noted that for a given Zhao number (i.e. Zh), there is a cutoff dimensionless wavenumber corresponding to the dimensionless growth rate of zero. This indicates that if the dimensionless wavenumber of a perturbation in the NAPL dissolution system is greater than this cutoff value, then the dimensionless growth rate of the perturbation is less than zero so that the NAPL dissolution system is in a stable state. Since the value of the dimensionless growth rate is positive when the dimensionless wavenumber of the perturbation approaches zero, the NAPL dissolution system is unstable as the wavelength of the perturbation approaches infinity. As the Zhao number increases, both the cutoff dimensionless wavenumber and the dimensionless growth rate within the range of positive values increase significantly. This demonstrates that an increase in (the magnitude of) the upstream aqueous phase fluid flow can remarkably destabilize the NAPL dissolution system so that a planar dissolution front can easily evolve into complicated morphologies.

Since most NAPL dissolution systems in the real world are of finite domains, it is desirable to investigate the instability of NAPL dissolution fronts in aqueous phase fluid-saturated porous media of finite domains. If a NAPL dissolution system has a finite domain, then the dimensionless critical characteristic length (i.e. $\bar{L}_{critical}$) of the NAPL dissolution system can be represented by half the dimensionless wavelength associated with the given dimensionless wavenumber ($\bar{m}$) of a perturbation. This is the second case that is considered in this chapter. In this situation, the dimensionless geometrical characteristic length of a NAPL dissolution system is equal to the dimensionless critical characteristic length (i.e. $\bar{L}_y = \bar{L}_{critical}$) of the NAPL dissolution system, so that half the dimensionless wavelength of the perturbation can be accommodated within the NAPL dissolution system of a finite domain. This means that when the dimensionless geometrical characteristic length (i.e. $\bar{L}_y$) of a NAPL dissolution system is given, the minimum value for the dimensionless wavenumber ($\bar{m}$) of the perturbation can be determined from the dimensionless critical characteristic length of the NAPL dissolution system. Note that this minimum value for the dimensionless wavenumber of the perturbation corresponds to the maximum value of the dimensionless wavelength of the perturbation. To enables such a perturbation to grow, there is a requirement for the injected aqueous phase fluid flow in the upstream direction of the NAPL dissolution system. In other words, the injected aqueous phase fluid flow must be strong enough to allow the NAPL dissolution front to become unstable. As a result, the corresponding minimum absolute value of the dimensionless pressure gradient of the aqueous phase fluid in the upstream direction of a NAPL dissolution system can be defined as the critical Zhao number of the NAPL dissolution system having a finite domain (Zhao et al. 2009).

Figure 9.2 shows the dynamics of a NAPL dissolution system in this particular situation. To obtain the numerical solutions in this figure, the same basic parameters as those used to obtain the numerical results in Fig. 9.1 are employed here for the initial saturation (S_{n0}) of the NAPL, the irreducible saturation (S_{ai}) of the aqueous phase fluid, the dimensionless longitudinal dispersivity ($\bar{\alpha}_L$), and the dimensionless transverse dispersivity ($\bar{\alpha}_T$) in the NAPL dissolution system. For a given dimensionless wavenumber (i.e. $\bar{m}$), there is a critical Zhao number that corresponds to a change in the sign of the dimensionless growth rate from negative to positive. Thus, if the Zhao number of a NAPL dissolution system is greater than

Fig. 9.2 Distributions of the dimensionless growth rate due to different dimensionless wavenumbers ($S_{n0} = 0.2$, $S_{ai} = 0.15$, $\bar{\alpha}_L = 0.2$, $\bar{\alpha}_T = 0.02$): **a** Entire view; **b** Zoomed-in view

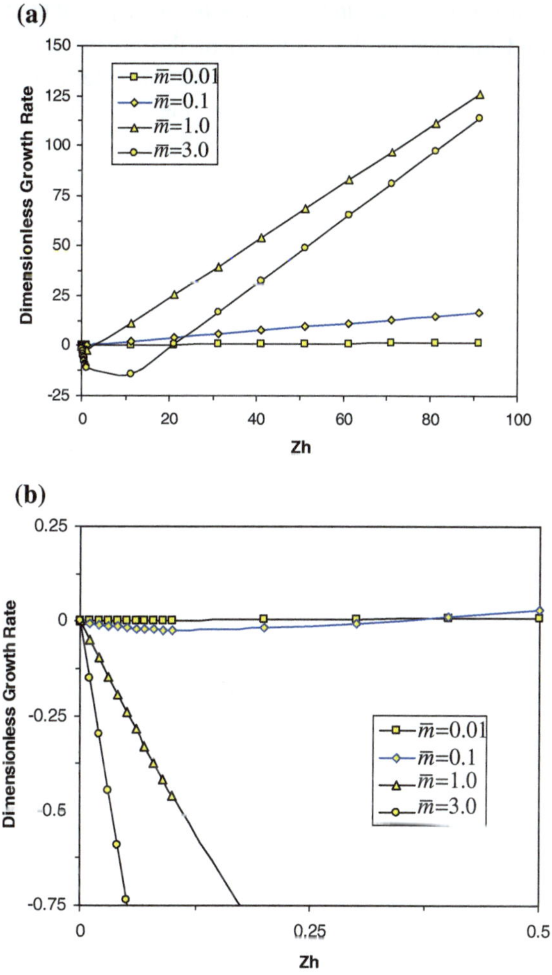

this critical value, then the dimensionless growth rate of the perturbation is positive so that the NAPL dissolution system is in an unstable state. Since the value of the dimensionless growth rate is negative when the Zhao number of a NAPL dissolution system is less than the critical Zhao number, the NAPL dissolution system is stable for the given dimensionless wavenumber (i.e. $\bar{m}$). As the dimensionless wavenumber increases, both the critical Zhao number and the dimensionless growth rate within the range of positive values increase dramatically. This indicates that for the NAPL dissolution system of a finite domain, the dimensionless growth rate associated with a higher Zhao number can be greater than that associated with the critical Zhao number of the NAPL dissolution system. As a result, fingering morphologies may become the dominant patterns of NAPL dissolution fronts in NAPL dissolution systems involving finite domains.

Since it is the dimensionless pressure gradient (i.e. $\bar{p}'_{axf}$) of the aqueous phase fluid (applied to the upstream of the NAPL dissolution system) that forms the driving force for the NAPL dissolution system of a finite domain, it is reasonable to take $\bar{m}$ as a constant so that the critical condition under which the NAPL dissolution system becomes unstable can be derived. In the work of Imhoff and Miller (1996), they set $\bar{p}'_{axf} = 1$ in the process of deriving the critical dimensionless wavenumber of the perturbation, implying that the dimensional pressure gradient (p'_{axf}) of the aqueous phase fluid applied to the upstream of the NAPL dissolution system was treated as a constant, rather than as an independent variable for the NAPL dissolution system. Thus the Darcy velocity that is equal to the product of $(-k_f/\mu_a)$ and p'_{axf} was equivalently treated as a constant, rather than an independent variable.

Letting $\bar{\omega} = 0$ yields the following characteristic equation for the critical Zhao number (i.e. $Zh_{critical}$) of the NAPL dissolution system having a finite domain.

$$\begin{aligned}&\frac{(1+\beta)Zh_{critical}}{1+\bar{\alpha}_L Zh_{critical}} - \left[(1+\beta) + \left(\frac{1}{1+\bar{\alpha}_L Zh_{critical}} - \bar{\alpha}_L\bar{m}\right)(1-\beta)\right]\sigma_1\\&+\left[\left(\bar{m} + \frac{Zh_{critical}}{1+\bar{\alpha}_L Zh_{critical}} - \bar{\alpha}_L\bar{m}(1+\bar{\alpha}_L Zh_{critical})\right)\frac{(1-\beta)}{1+\bar{\alpha}_L Zh_{critical}}\right] = 0,\end{aligned} \tag{9.84}$$

where

$$\sigma_1 = \frac{\sqrt{\left(\frac{Zh_{critical}}{1+\bar{\alpha}_L Zh_{critical}}\right)^2 + \frac{4\bar{m}(1+\bar{\alpha}_L Zh_{critical})}{1+\bar{\alpha}_L Zh_{critical}}} + \frac{Zh_{critical}}{1+\bar{\alpha}_L Zh_{critical}}}{2}. \tag{9.85}$$

With $\bar{m} = 1$ taken as an example, Eq. (9.84) can be expressed as follows:

$$\begin{aligned}&\frac{(1+\beta)Zh_{critical}}{(1+\bar{\alpha}_L)Zh_{critical}} - \left[(1+\beta) + \left(\frac{1}{(1+\bar{\alpha}_L)Zh_{critical}} - \bar{\alpha}_L\right)(1-\beta)\right]\sigma_2\\&+\left[\left(\bar{m} + \frac{Zh_{critical}}{(1+\bar{\alpha}_L)Zh_{critical}} - \bar{\alpha}_L(1+\bar{\alpha}_L Zh_{critical})\right)\frac{(1-\beta)}{(1+\bar{\alpha}_L)Zh_{critical}}\right] = 0,\end{aligned} \tag{9.86}$$

where

$$\sigma_2 = \frac{\sqrt{\left(\frac{Zh_{critical}}{1+\bar{\alpha}_L Zh_{critical}}\right)^2 + \frac{4(1+\bar{\alpha}_T Zh_{critical})}{1+\bar{\alpha}_L Zh_{critical}}} + \frac{Zh_{critical}}{1+\bar{\alpha}_L Zh_{critical}}}{2}. \tag{9.87}$$

Physically, $\bar{m} = 1$ means that the dimensional wavelength (i.e. λ) of the perturbation has the same order of magnitude as the intrinsic characteristic length (L^*) of a NAPL dissolution system because $\bar{m} = 2\pi L^*/\lambda$.

Since $(1 + \bar{\alpha}_L Zh_{critical}) > 0$, Eq. (9.86) can be rewritten into the following form:

$$(1+\beta)Zh_{critical} - \{(1+\beta)(1+\bar{\alpha}_L Zh_{critical}) + [1-\bar{\alpha}_L(1+\bar{\alpha}_L Zh_{critical})](1-\beta)\}\sigma_2$$
$$+\left(1+\frac{Zh_{critical}}{1+\bar{\alpha}_L Zh_{critical}} - \bar{\alpha}_L(1+\bar{\alpha}_L Zh_{critical})\right)(1-\beta) = 0. \tag{9.88}$$

If $\bar{\alpha}_L = \bar{\alpha}_T = 0$, then Eq. (9.88) has the following solution:

$$Zh_{critical} = \frac{(3-\beta)(1+\beta)}{2(1-\beta)}. \tag{9.89}$$

This solution is identical to the previous theoretical results when the solute dispersion effect is neglected in the mineral dissolution system (Zhao et al. 2008b, c).

Based on the characteristic quantities expressed in Eq. (9.59) for the NAPL dissolution system, the dimensionless Zhao number is defined as follows (Zhao et al. 2009):

$$Zh = -\bar{p}'_{axf} = -\frac{p'_{axf}L^*}{p^*_a} = \frac{V_{axf}}{\sqrt{\phi\tau D_m}}\sqrt{\frac{1}{\beta_0}} = F_{advection}F_{dispersion}F_{dissolution}, \tag{9.90}$$

where

$$F_{advection} = V_{axf} \quad \text{(representing the aqueous phase fluid advection)}, \tag{9.91}$$

$$F_{dispersion} = \frac{1}{\sqrt{\phi\tau D_m}} \quad \text{(representing the NAPL solute diffusion/dispersion)}, \tag{9.92}$$

$$F_{dissolution} = \sqrt{\frac{1}{\beta_0}} \quad \text{(representing the kinetics of the NAPL dissolution)}, \tag{9.93}$$

where V_{axf} is the Darcy velocity of the aqueous phase fluid after the NAPL is completely dissolved in the NAPL dissolution system.

Equation (9.90) indicates that the Zhao number is a comprehensive dimensionless number to represent the three major controlling mechanisms simultaneously taking place in the NAPL dissolution system. By means of both the Zhao number and the critical Zhao number, the instability of NAPL dissolution fronts in fluid-saturated porous media of finite domains can be assessed. If $Zh < Zh_{critical}$, then the NAPL dissolution system is subcritical, while if $Zh > Zh_{critical}$, then the NAPL dissolution system is supercritical. When $Zh = Zh_{critical}$, the NAPL dissolution system is in a critical state.

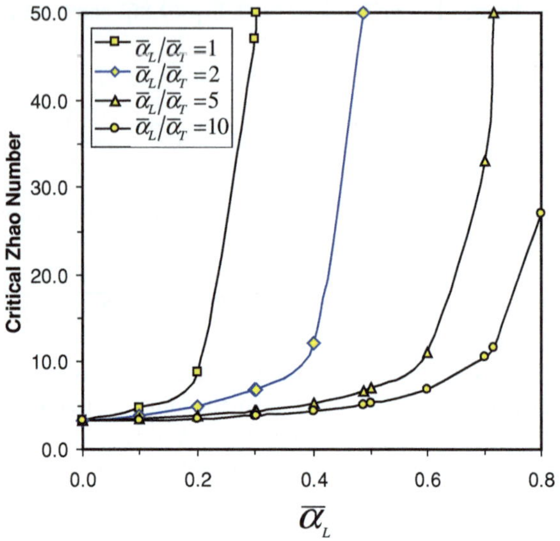

Fig. 9.3 Variations of the critical Zhao number with the dimensionless longitudinal dispersivity due to different dispersion ratios ($S_{n0} = 0.2$, $S_{ai} = 0.15$)

9.2.3 Theoretical Understanding of the Effects of Solute Dispersion on NAPL Dissolution-Front Instability in Two-Dimensional Fluid-Saturated Porous Media

The present theoretical solution can be used to investigate theoretically the effect of solute dispersion on NAPL dissolution-front instability in a fluid-saturated porous medium. For this purpose, the critical Zhao numbers associated with different dimensionless solute-dispersion coefficients in the case of $|\bar{m}| = 1$ are evaluated from the characteristic equation (i.e. Eq. (9.88)). As indicated by Eqs. (9.5) and (9.81), $\beta = (1 - S_{ai} - S_{n0})^3 \big/ (1 - S_{ai})^3$, implying that β is a function of both the initial saturation (S_{n0}) of the NAPL and the irreducible saturation (S_{ai}) of the aqueous phase fluid so that the critical Zhao number is a function of $\bar{\alpha}_L$, $\bar{\alpha}_T$, S_{n0} and S_{ai}. Therefore, for a given set of S_{n0} and S_{ai}, the critical Zhao number is only dependent on $\bar{\alpha}_L$ and $\bar{\alpha}_T$. In the case of $S_{n0} = 0.2$ and $S_{ai} = 0.15$, the distributions of the critical Zhao number due to different solute dispersion ratios are shown in Fig. 9.3, where the solute dispersion ratio (as represented by $\bar{\alpha}_L/\bar{\alpha}_T$) is defined as the ratio of the dimensionless longitudinal dispersivity ($\bar{\alpha}_L$) to the dimensionless transverse dispersivity ($\bar{\alpha}_T$). It is obvious that with an increase in the dimensionless longitudinal dispersivity, the critical Zhao number can increase significantly, especially for the smaller solute-dispersion ratios such as $\bar{\alpha}_L/\bar{\alpha}_T = 1$ and $\bar{\alpha}_L/\bar{\alpha}_T = 2$. For a given dimensionless longitudinal dispersivity, the critical Zhao number also increases significantly with an increase in the solute dispersion ratio, $\bar{\alpha}_L/\bar{\alpha}_T$. This indicates that an increase in either the longitudinal dispersivity or the dispersion ratio stabilizes the NAPL dissolution front so that it becomes more difficult for a planar NAPL dissolution-front to evolve into different morphologies.

9.3 Computational Simulation of NAPL Dissolution-Front Instability Problems in Two-Dimensional Fluid-Saturated Porous Media

Although the present theoretical results can be used to assess whether or not a NAPL dissolution system involving a finite domain is supercritical, it is very difficult, if not impossible, to simulate the morphological evolution of a NAPL dissolution front using theoretical methods. For this reason, numerical methods are often used as a complementary way to simulate the morphological evolutions of NAPL dissolution fronts in supercritical NAPL dissolution systems involving finite domains. Since numerical methods are approximate solution methods, they should be verified using some benchmark problems, the analytical solutions for which are available for comparison. To verify the accuracy of the computational simulation, a benchmark problem is constructed from the theoretical analysis derived under subcritical Zhao numbers in the previous study (Zhao et al. 2009). As a result, the numerical solution obtained from the benchmark problem can be compared with the corresponding analytical solution. To further verify whether or not the proposed numerical procedure can be used to simulate the morphological evolution of NAPL dissolution fronts in fluid-saturated porous media of supercritical Zhao numbers, a computational model that was presented by Imhoff et al. (2002) on the basis of their previous laboratory experiments (Imhoff et al. 1996) is used in this investigation. Since computational simulation results for the complicated morphological evolution patterns of NAPL dissolution fronts compared well with those observed from the corresponding laboratory experiments (Imhoff et al. 1996; Miller et al. 1998), this computational model can be used as a typical representative of the related laboratory experiments. On the other hand, because both the detailed input parameters and output results of this computational model are openly available (Imhoff et al. 2002), it can be served as a useful benchmark problem for verifying numerical procedures that are used to simulate the morphological evolution of NAPL dissolution fronts in fluid-saturated porous media of supercritical Zhao numbers. For this reason, the proposed numerical procedure will be verified by comparing the current results with those previously obtained from the same laboratory experiment-based computational model (Imhoff et al. 2002).

9.3.1 Formulation of the Proposed Numerical Procedure for Simulating the Evolution of NAPL Dissolution Fronts

The dimensionless governing equations (i.e. Eqs. (9.61)–(9.63)) of a NAPL dissolution problem in a two-dimensional fluid-saturated porous medium can be solved using a combination of the finite element and finite difference methods (Zhao et al. 2011). Toward this end, the finite element method is used to discretize the geometrical shape of the problem domain, while the finite difference method is used to discretize the dimensionless time. Since the dimensionless governing equations of a NAPL dissolution system are highly nonlinear, the segregated algorithm, in which Eqs. (9.61)–(9.63) are solved separately in a sequential manner, is used to derive the formulation of the proposed numerical procedure.

For a given dimensionless time-step, $\bar{t}+\Delta\bar{t}$, the saturation of the NAPL can be denoted by $(S_n)_{\bar{t}+\Delta\bar{t}} = (S_n)_{\bar{t}} + (\Delta S_n)_{\bar{t}+\Delta\bar{t}}$, where $(S_n)_{\bar{t}}$ is the saturation of the NAPL at the previous time-step and $(\Delta S_n)_{\bar{t}+\Delta\bar{t}}$ is the saturation increment of the NAPL at the current time-step. Since Eq. (9.61) is a nonlinear equation, it can be linearized using the following Taylor expansion:

$$(S_n)^{\beta_1}_{\bar{t}+\Delta\bar{t}} = \left[(S_n)_{\bar{t}} + (\Delta S_n)_{\bar{t}+\Delta\bar{t}}\right]^{\beta_1} = (S_n)^{\beta_1}_{\bar{t}} + \beta_1 (S_n)^{\beta_1-1}_{\bar{t}} (\Delta S_n)_{\bar{t}+\Delta\bar{t}}. \tag{9.94}$$

Using the backward difference scheme, Eq. (9.61) can be written as follows:

$$\left[\frac{1}{\Delta\bar{t}}(S_n)^{1-\beta_1}_{\bar{t}} + \beta_1(1-\bar{C}_{\bar{t}+\Delta\bar{t}})\right](\Delta S_n)_{\bar{t}+\Delta\bar{t}} = -(S_n)_{\bar{t}}(1-\bar{C}_{\bar{t}+\Delta\bar{t}}), \tag{9.95}$$

where $\bar{C}_{\bar{t}+\Delta\bar{t}}$ is the dimensionless concentration of the NAPL at the current time-step; $\Delta\bar{t}$ is the dimensionless time increment at the current time-step.

In the finite difference sense, the following relationships exist:

$$\begin{aligned} \varepsilon\frac{\partial}{\partial\bar{t}}[(1-S_n)\bar{C}] &= \varepsilon\frac{\Delta\left[\left(1-(S_n)_{\bar{t}+\Delta\bar{t}}\right)\bar{C}_{\bar{t}+\Delta\bar{t}}\right]}{\Delta\bar{t}} \\ &= -\varepsilon\bar{C}_{\bar{t}+\Delta\bar{t}}\frac{\Delta(S_n)_{\bar{t}+\Delta\bar{t}}}{\Delta\bar{t}} + \varepsilon\left[1-(S_n)_{\bar{t}+\Delta\bar{t}}\right]\frac{\Delta\bar{C}_{\bar{t}+\Delta\bar{t}}}{\Delta\bar{t}}, \end{aligned} \tag{9.96}$$

$$\frac{\partial S_n}{\partial\bar{t}} = \frac{\Delta(S_n)_{\bar{t}+\Delta\bar{t}}}{\Delta\bar{t}} = -(1-\bar{C}_{\bar{t}+\Delta\bar{t}})(S_n)^{\beta_1}_{\bar{t}+\Delta\bar{t}}, \tag{9.97}$$

$$\nabla\cdot[(1-S_n)\bar{\mathbf{D}}_{\mathbf{h}}]\cdot\nabla\bar{C} = \nabla\cdot\left[\left(1-(S_n)_{\bar{t}+\Delta\bar{t}}\right)(\bar{\mathbf{D}}_{\mathbf{h}})_{\bar{t}+\Delta\bar{t}}\right]\cdot\nabla\bar{C}_{\bar{t}+\Delta\bar{t}}, \tag{9.98}$$

$$\begin{aligned} \nabla\cdot[\bar{C}\bar{\psi}(S_n)\nabla\bar{p}_a] &= \bar{C}\nabla\cdot[\bar{\psi}(S_n)\nabla\bar{p}_a] + \nabla\bar{p}_a\cdot[\bar{\psi}(S_n)\nabla\bar{C}] \\ &= \bar{C}_{\bar{t}+\Delta\bar{t}}\nabla\cdot\left[\bar{\psi}\left((S_n)_{\bar{t}+\Delta\bar{t}}\right)\nabla(\bar{p}_a)_{\bar{t}+\Delta\bar{t}}\right] + \nabla(\bar{p}_a)_{\bar{t}+\Delta\bar{t}}\cdot\left[\bar{\psi}\left((S_n)_{\bar{t}+\Delta\bar{t}}\right)\nabla\bar{C}_{\bar{t}+\Delta\bar{t}}\right]. \end{aligned} \tag{9.99}$$

Considering Eqs. (9.96) and (9.97) simultaneously yields the following equation:

$$\varepsilon\frac{\partial}{\partial\bar{t}}[(1-S_n)\bar{C}]+\frac{\partial S_n}{\partial\bar{t}}=(1-\varepsilon\bar{C}_{\bar{t}+\Delta\bar{t}})\frac{\Delta(S_n)_{\bar{t}+\Delta\bar{t}}}{\Delta\bar{t}}+\varepsilon\left[1-(S_n)_{\bar{t}+\Delta\bar{t}}\right]\frac{\Delta\bar{C}_{\bar{t}+\Delta\bar{t}}}{\Delta\bar{t}}. \tag{9.100}$$

Since ε is much smaller than one, Eq. (9.100) can be approximately expressed as follows:

$$\begin{aligned}\varepsilon\frac{\partial}{\partial\bar{t}}[(1-S_n)\bar{C}]+\frac{\partial S_n}{\partial\bar{t}}&=\frac{\Delta(S_n)_{\bar{t}+\Delta\bar{t}}}{\Delta\bar{t}}+\varepsilon\left[1-(S_n)_{\bar{t}+\Delta\bar{t}}\right]\frac{\Delta\bar{C}_{\bar{t}+\Delta\bar{t}}}{\Delta\bar{t}}\\&=-(1-\bar{C}_{\bar{t}+\Delta\bar{t}})(S_n)^{\beta_1}_{\bar{t}+\Delta\bar{t}}+\varepsilon\left[1-(S_n)_{\bar{t}+\Delta\bar{t}}\right]\frac{\Delta\bar{C}_{\bar{t}+\Delta\bar{t}}}{\Delta\bar{t}}.\end{aligned} \tag{9.101}$$

Substituting Eqs. (9.98), (9.99) and (9.101) into (9.62) yields the following finite difference equation:

$$\begin{aligned}&\left\{\frac{\varepsilon}{\Delta\bar{t}}\left[1-(S_n)_{\bar{t}+\Delta\bar{t}}\right]+(S_n)^{\beta_1}_{\bar{t}+\Delta\bar{t}}\right\}\bar{C}_{\bar{t}+\Delta\bar{t}}-\nabla\cdot\left[\left(1-(S_n)_{\bar{t}+\Delta\bar{t}}\right)(\bar{\mathbf{D}}_{\mathbf{h}})_{\bar{t}+\Delta\bar{t}}\right]\cdot\nabla\bar{C}_{\bar{t}+\Delta\bar{t}}\\&-\nabla(\bar{p}_a)_{\bar{t}+\Delta\bar{t}}\cdot\left[\bar{\psi}\left((S_n)_{\bar{t}+\Delta\bar{t}}\right)\nabla\bar{C}_{\bar{t}+\Delta\bar{t}}\right]-\bar{C}_{\bar{t}+\Delta\bar{t}}\nabla\cdot\left[\bar{\psi}\left((S_n)_{\bar{t}+\Delta\bar{t}}\right)\nabla(\bar{p}_a)_{\bar{t}+\Delta\bar{t}}\right]\\&=\frac{\varepsilon}{\Delta\bar{t}}\left[1-(S_n)_{\bar{t}+\Delta\bar{t}}\right]\bar{C}_{\bar{t}}+(S_n)^{\beta_1}_{\bar{t}+\Delta\bar{t}}.\end{aligned} \tag{9.102}$$

Similarly, Eq. (9.63) can be rewritten in the following discretized form:

$$\nabla\cdot\left[\bar{\psi}(S_n)\nabla\bar{p}_a\right]=\nabla\cdot\left[\bar{\psi}\left((S_n)_{\bar{t}+\Delta\bar{t}}\right)\nabla(\bar{p}_a)_{\bar{t}+\Delta\bar{t}}\right]-\varepsilon\left(1-\frac{1}{\bar{\rho}_a}\right)(S_n)^{\beta_1}_{t+\Delta t}(1-\bar{C}_{\bar{t}+\Delta\bar{t}}). \tag{9.103}$$

Substituting Eqs. (9.103) into (9.102) yields the following equation:

$$\begin{aligned}&\left\{\frac{\varepsilon}{\Delta\bar{t}}\left[1-(S_n)_{\bar{t}+\Delta\bar{t}}\right]+(S_n)^{\beta_1}_{\bar{t}+\Delta\bar{t}}\right\}\bar{C}_{\bar{t}+\Delta\bar{t}}-\nabla\cdot\left[\left(1-(S_n)_{\bar{t}+\Delta\bar{t}}\right)(\bar{\mathbf{D}}_{\mathbf{h}})_{\bar{t}+\Delta\bar{t}}\right]\cdot\nabla\bar{C}_{\bar{t}+\Delta\bar{t}}\\&-\nabla(\bar{p}_a)_{\bar{t}+\Delta\bar{t}}\cdot\left[\bar{\psi}\left((S_n)_{\bar{t}+\Delta\bar{t}}\right)\nabla\bar{C}_{\bar{t}+\Delta\bar{t}}\right]-\varepsilon\bar{C}_{\bar{t}+\Delta\bar{t}}\left\{\left(1-\frac{1}{\bar{\rho}_a}\right)(S_n)^{\beta_1}_{\bar{t}+\Delta\bar{t}}(1-\bar{C}_{\bar{t}+\Delta\bar{t}})\right\}\\&=\frac{\varepsilon}{\Delta\bar{t}}\left[1-(S_n)_{\bar{t}+\Delta\bar{t}}\right]\bar{C}_{\bar{t}}+(S_n)^{\beta_1}_{\bar{t}+\Delta\bar{t}}.\end{aligned} \tag{9.104}$$

Since $\varepsilon \ll 1$, Eq. (9.104) can be approximately expressed in the following form:

$$\left\{\frac{\varepsilon}{\Delta\bar{t}}\left[1-(S_n)_{\bar{t}+\Delta\bar{t}}\right]+(S_n)_{\bar{t}+\Delta\bar{t}}^{\beta_1}\right\}\bar{C}_{\bar{t}+\Delta\bar{t}}-\nabla\cdot\left[\left(1-(S_n)_{\bar{t}+\Delta\bar{t}}\right)(\bar{\mathbf{D}}_{\mathbf{h}})_{\bar{t}+\Delta\bar{t}}\right]\cdot\nabla\bar{C}_{\bar{t}+\Delta\bar{t}}$$
$$-\nabla(\bar{p}_a)_{\bar{t}+\Delta\bar{t}}\cdot\left[\bar{\psi}\left((S_n)_{\bar{t}+\Delta\bar{t}}\right)\nabla\bar{C}_{\bar{t}+\Delta\bar{t}}\right]=\frac{\varepsilon}{\Delta\bar{t}}\left[1-(S_n)_{\bar{t}+\Delta\bar{t}}\right]\bar{C}_{\bar{t}}+(S_n)_{\bar{t}+\Delta\bar{t}}^{\beta_1}. \tag{9.105}$$

In brief, the dimensionless governing equations (i.e. Eqs. (9.61)–(9.63)) of a NAPL dissolution system under the condition of $\varepsilon \ll 1$ can be expressed in the finite difference form as follows:

$$\left[\frac{1}{\Delta\bar{t}}(S_n)_{\bar{t}}^{1-\beta_1}+\beta_1(1-\bar{C}_{\bar{t}+\Delta\bar{t}})\right](\Delta S_n)_{\bar{t}+\Delta\bar{t}}=-(S_n)_{\bar{t}}(1-\bar{C}_{\bar{t}+\Delta\bar{t}}), \tag{9.106}$$

$$\left\{\frac{\varepsilon}{\Delta\bar{t}}\left[1-(S_n)_{\bar{t}+\Delta\bar{t}}\right]+(S_n)_{\bar{t}+\Delta\bar{t}}^{\beta_1}\right\}\bar{C}_{\bar{t}+\Delta\bar{t}}-\nabla\cdot\left[\left(1-(S_n)_{\bar{t}+\Delta\bar{t}}\right)(\bar{\mathbf{D}}_{\mathbf{h}})_{\bar{t}+\Delta\bar{t}}\right]\cdot\nabla\bar{C}_{\bar{t}+\Delta\bar{t}}$$
$$-\nabla(\bar{p}_a)_{\bar{t}+\Delta\bar{t}}\cdot\left[\bar{\psi}\left((S_n)_{\bar{t}+\Delta\bar{t}}\right)\nabla\bar{C}_{\bar{t}+\Delta\bar{t}}\right]=\frac{\varepsilon}{\Delta\bar{t}}\left[1-(S_n)_{\bar{t}+\Delta\bar{t}}\right]\bar{C}_{\bar{t}}+(S_n)_{\bar{t}+\Delta\bar{t}}^{\beta_1}, \tag{9.107}$$

$$\nabla\cdot\left[\bar{\psi}(S_n)\nabla\bar{p}_a\right]=\nabla\cdot\left[\bar{\psi}\left((S_n)_{\bar{t}+\Delta\bar{t}}\right)\nabla(\bar{p}_a)_{\bar{t}+\Delta\bar{t}}\right]=\varepsilon\left(1-\frac{1}{\bar{\rho}_a}\right)(S_n)_{\bar{t}+\Delta\bar{t}}^{\beta_1}(1-\bar{C}_{\bar{t}+\Delta\bar{t}}). \tag{9.108}$$

Note that Eqs. (9.61)–(9.63) are valid for any values of ε, so that if ε is close to unity, they should be used to obtain numerical solutions instead of Eqs. (9.106)–(9.108). However, the direct use of Eqs. (9.61)–(9.63) may result in more computer efforts, compared with the use of Eqs. (9.106)–(9.108). Since most NAPL dissolution problems may fall in the range of $\varepsilon \ll 1$ and analytical solutions are only available in this range, Eqs. (9.106)–(9.108) are used to obtain numerical solutions in this chapter.

Using both the proposed segregated scheme and the finite element method (Zienkiewicz 1977; Zhao et al. 2009), Eqs. (9.106)–(9.108) are solved separately and sequentially for the saturation of the NAPL, the dimensionless concentration of the NAPL and the dimensionless pressure of the aqueous phase fluid at the current time-step. To consider the coupling effect between these three equations, an iteration scheme is also used in the process of finding the numerical solution. The following convergence criterion needs to be satisfied so that a convergent solution can be obtained.

$$E=Max\left(\sqrt{\sum_{i=1}^{N_{S_n}}\left[(S_n)_{i,\bar{t}+\Delta\bar{t}}^{k}-(S_n)_{i,\bar{t}+\Delta\bar{t}}^{k-1}\right]^2},\ \sqrt{\sum_{i=1}^{N_{\bar{C}}}\left(\bar{C}_{i,\bar{t}+\Delta\bar{t}}^{k}-\bar{C}_{i,\bar{t}+\Delta\bar{t}}^{k-1}\right)^2},\ \sqrt{\sum_{i=1}^{N_{\bar{p}_a}}\left[(\bar{p}_a)_{i,\bar{t}+\Delta\bar{t}}^{k}-(\bar{p}_a)_{i,\bar{t}+\Delta\bar{t}}^{k-1}\right]^2}\right)<\bar{E}, \tag{9.109}$$

where E and $\bar{E}$ are the maximum error at the k-th iteration step and the allowable error limit; N_{S_n}, $N_{\bar{C}}$ and $N_{\bar{p}a}$ are the total numbers of the degree-of-freedom for the saturation of the NAPL, the dimensionless concentration of the NAPL in the aqueous phase fluid and the dimensionless pressure of the aqueous phase fluid, respectively; k is the index number at the current iteration step and k - 1 is the index number at the previous iteration step; $(S_n)^k_{i,\bar{t}+\Delta\bar{t}}$, $\bar{C}^k_{i,\bar{t}+\Delta\bar{t}}$ and $(\bar{p}_a)^k_{i,\bar{t}+\Delta\bar{t}}$ are the saturation of the NAPL, the dimensionless concentration of the NAPL and the dimensionless pressure of the aqueous phase fluid for node i at both the current time-step and the current iteration step; $(S_n)^{k-1}_{i,\bar{t}+\Delta\bar{t}}$, $\bar{C}^{k-1}_{i,\bar{t}+\Delta t}$ and $(\bar{p}_a)^{k-1}_{i,\bar{t}+\Delta\bar{t}}$ are the saturation of the NAPL, the dimensionless concentration of the NAPL and the dimensionless pressure of the aqueous phase fluid for node i at the current time-step but at the previous iteration step.

9.3.2 Verification of the Proposed Numerical Procedure for Simulating the Evolution of NAPL Dissolution Fronts

Due to inevitable round-off errors in computation and discretization errors in temporal and spatial variables, it is necessary to verify the proposed numerical procedure so that meaningful numerical results can be obtained from a discretized computational model. For this reason two benchmark problems, for which either the analytical results or previous solutions are available, are considered in this subsection.

Figure 9.4 shows the geometry and boundary conditions of the first benchmark problem, in which the dimensionless pressure gradient of the aqueous phase fluid (i.e. $\bar{p}'_{axf} = -0.02$) is applied to the left boundary. This means that there is a horizontal throughflow from the left to the right of the computational model. The dimensionless height and width of the computational model are 200 and 400 respectively. Except for the left boundary, the initial residual saturation of the NAPL is 0.2, while the initial dimensionless concentration of the NAPL in the aqueous phase fluid is unity within the computational domain. The final residual saturation of a value of zero is applied to the left boundary as a boundary condition of the computational domain. Both the top and the bottom boundaries are assumed to be impermeable for the NAPL and aqueous phase fluid.

The following parameters are used in the corresponding computations: the initial saturation (S_{n0}) of the NAPL is 0.2; the irreducible saturation (S_{ai}) of the aqueous phase fluid is 0.15; the dimensionless longitudinal dispersivity ($\bar{\alpha}_L$) is 0.2; the dimensionless transverse dispersivity (α_T) is 0.02; the ratio (ε) of the equilibrium concentration of the NAPL species in the aqueous phase fluid to the density of the NAPL itself is 0.001; the density ratio ($\bar{\rho}_a$) of the aqueous phase fluid to the NAPL is 1.0/1.46; the value of β_1 is 0.87; and the Zhao number

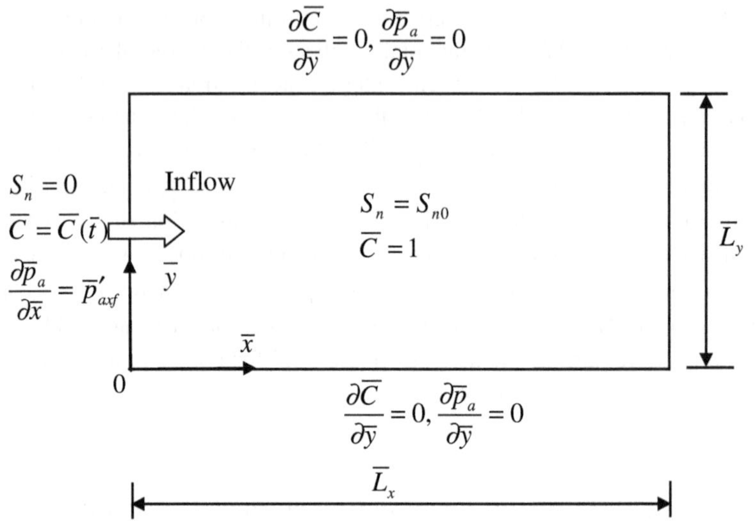

Fig. 9.4 Geometry and boundary conditions for the first benchmark problem

(i.e. $-\bar{p}'_{axf} = 0.02$) applied to the left boundary of the NAPL dissolution system is 0.02. Since the critical Zhao number of the system is approximately equal to 0.052, the NAPL dissolution system considered here is in a subcritical state. In this case, the corresponding dimensionless propagation speed of the planar NAPL dissolution-front is 0.1. To appropriately simulate the propagation of the NAPL dissolution front, the whole computational domain is simulated by 19,701 four-node square elements of 20,000 nodal points in total. The initial residual saturation field of the NAPL is randomly perturbed by a small amount of 1 % of the originally-input saturation of the NAPL (i.e. $S_{n0} = 0.2$) before running the computational model. This means that the resulting initial residual saturation is of a random distribution, which has a mean value of the homogeneous residual saturation (i.e. $S_{n0} = 0.2$) and a variation of 0.002 (i.e. 1 % of $S_{n0} = 0.2$) in the whole computational domain. The dimensionless time-step length is 20.

Since the computational domain of the benchmark problem is of finite size, a time-dependent-dimensionless-concentration boundary condition (i.e. $\bar{C}(\bar{t}) = \exp(\bar{p}'_{axf}\bar{v}_{front}\bar{t})$) needs to be applied to the left boundary so that the numerical solutions can be compared with the analytical solutions. From the previous study (Zhao et al. 2009), the analytical solutions for this benchmark problem can be expressed as follows:

$$\bar{C}(\bar{x},\bar{t}) = 1, \quad \bar{p}_a(\bar{x},\ \bar{t}) = -\bar{p}'_{ax0}(\bar{L}_x - \bar{x}) + 100, \quad S_n(\bar{x},\bar{t}) = S_{n0} \quad (\bar{x} > \bar{v}_{front}\bar{t}), \tag{9.110}$$

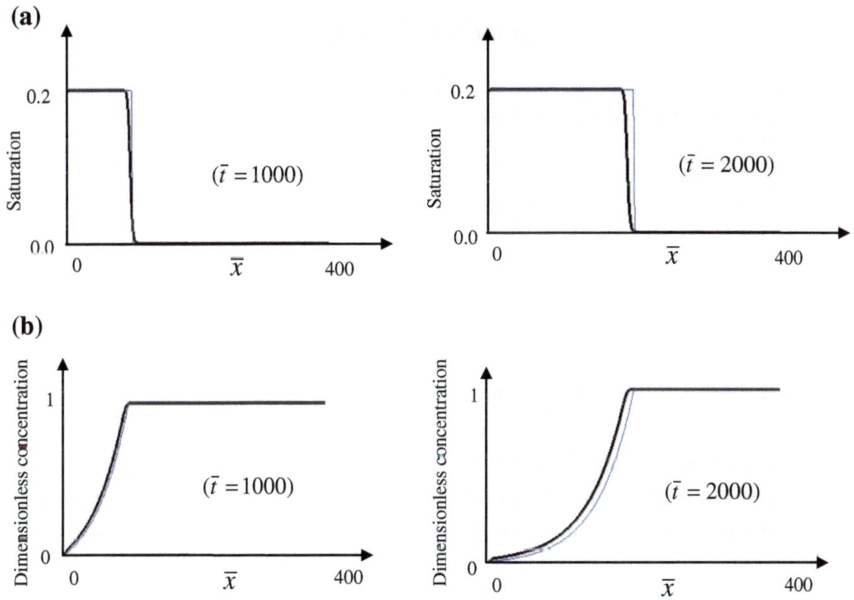

Fig. 9.5 Comparison of the numerical solutions with analytical results at different time instants: **a** Residual saturation and **b** Dimensionless concentration

$$\bar{C}(\bar{x},\bar{t}) = \exp[-\bar{p}'_{axf}(\bar{x}-\bar{v}_{front}\bar{t})], \; \bar{p}_a(\bar{x},\bar{t}) = \bar{p}'_{axf}(\bar{x}-\bar{v}_{front}\bar{t}) - \bar{p}'_{ax0}(\bar{L}_x-\bar{v}_{front}\bar{t}) + 100,$$
$$S_n(\bar{x},\bar{t}) = 0 \qquad (\bar{x} \leq \bar{v}_{front}\bar{t}). \tag{9.111}$$

Note that the dimensionless pressure ($\bar{p}_{a0}$) of the aqueous phase fluid on the downstream boundary is assumed to be 100 in the benchmark problem.

Figure 9.5 shows the comparison of numerical solutions with analytical ones for the residual saturation of the NAPL and the dimensionless concentration of the NAPL in the aqueous phase fluid at two different time instants. In this figure, thick lines show the numerical results, while thin lines show the corresponding analytical solutions. It can be observed that the overall numerical solutions agree well with the analytical solutions, except for a smooth effect on the numerically-simulated propagation front of the NAPL saturation. This fact demonstrates that the proposed numerical procedure is capable of simulating the planar dissolution-front propagation within the fluid-saturated porous medium of a finite domain under subcritical Zhao number conditions. Since a finite value for the ratio (ε) of the equilibrium concentration of the NAPL species in the aqueous phase fluid to the density of the NAPL itself is used in the computational model, there is a discrepancy between the computational model used to produce the numerical results and the theoretical model used to derive the analytical solutions. Because of this discrepancy, both the sharpness and the dimensionless propagation speed of the

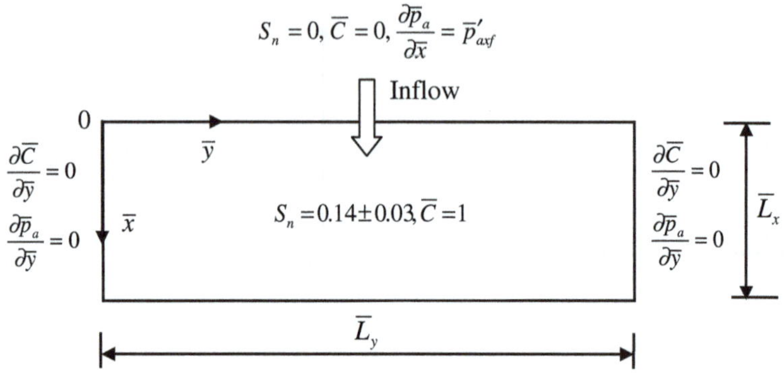

Fig. 9.6 Geometry and boundary conditions for the second benchmark problem

NAPL dissolution front obtained from the computational model differ slightly from those obtained from the theoretical analysis. To reduce this discrepancy, the value for the ratio (ε) of the equilibrium concentration of the NAPL species in the aqueous phase fluid to the density of the NAPL itself should be used as small as possible in the computational model.

As mentioned previously, the second benchmark problem under consideration is a computational model (as shown in Fig. 9.6) that is based on the laboratory experiments conducted by Imhoff et al. (1996). Miller et al. (1998) have successfully used this computational model to reproduce the same fingering phenomena as those observed in the corresponding laboratory experiments. The two primary purposes of considering this benchmark problem are: (1) to verify the proposed numerical procedure when it is used to simulate the morphological evolution of NAPL dissolution fronts in supercritical systems; and (2) to demonstrate how to use the Zhao number and the critical Zhao number to assess the instability likelihood of a NAPL dissolution system. In order to establish this computational model, the data associated with Experiment 13 in the experimental work of Imhoff et al. (1996) was used so that this computational model has a solid foundation of experiment results. The NAPL considered in the experiment is trichloroethylene (i.e. TCE). For the purpose of reproducing this experiment numerically, Imhoff et al. (2002) used the following parameters in their numerical experiments: the aqueous solubility (C_{eq}) of the NAPL is $1.27\,\text{kg/m}^3$; the irreducible saturation (S_{ai}) of the aqueous phase (i.e. water) is 0.104; the molecular diffusion coefficient (D_m) of the NAPL is $8.4 \times 10^{-10}\,\text{m}^2/\text{s}$; the porosity ($\phi$) of the porous medium is 0.355; the permeability (k_f) of the porous medium in the isotropic case is $1.0 \times 10^{-12}\,\text{m}^2$; the initial residual saturation (S_{n0}) of the NAPL is 0.14; the dynamic viscosity (μ_a) of the aqueous phase (i.e. water) is $1.0 \times 10^{-3} (\text{N} \cdot \text{s/m}^2)$; the density ($\rho_a$) of the aqueous phase (i.e. water) is $1,000\,\text{kg/m}^3$; the density (ρ_n) of the NAPL is $1,460\,\text{kg/m}^3$; the ratio (ε) of the aqueous solubility (C_{eq}) to the density (ρ_n) of the NAPL is 0.87×10^{-3}; the Darcy velocity (V_{axf}) of the injected aqueous phase fluid (i.e. water) is

$1.0/86400\,\text{m/s} = 1.16 \times 10^{-5}\,\text{m/s}$; the tortuosity ($\tau$) of the porous medium is 0.66; the longitudinal and transverse dispersivities (α_L and α_T) of the porous medium are 7.2×10^{-4} m and 1.44×10^{-4} m, respectively; the geometrical mean particle diameter (d_{50}) is 3.6×10^{-4} m; the length of the model in the injected flow direction is 0.30 m; and the length in the direction perpendicular to the injected flow is 1.0 m.

Using the above parameters, the Reynolds number of the NAPL dissolution system is as follows:

$$Re = \frac{d_{50}\rho_a V_{axf}}{\phi\mu_a(1 - S_{n0})} = \frac{3.6 \times 10^{-4} \times 1000 \times 1.16 \times 10^{-5}}{0.355 \times 10^{-3} \times (1 - 0.14)} = 1.368 \times 10^{-2}. \tag{9.112}$$

When the correlation of Imhoff et al. (1994) is used, both β_1 and β_0 involved in the mass transfer rate coefficient (K) can be evaluated using the following equations:

$$\beta_1 = 0.87, \tag{9.113}$$

$$\beta_0 = \frac{186\phi^{0.87}\text{Re}^{0.71}D_m}{d_{50}^2} = \frac{186 \times 0.355^{0.87} \times 0.01368^{0.71} \times 8.4 \times 10^{-10}}{(3.6 \times 10^{-4})^2} = 2.325 \times 10^{-2}(1/\text{s}). \tag{9.114}$$

Thus the intrinsic characteristic time, length and aqueous pressure of the NAPL dissolution system can be determined as follows:

$$t^* = \frac{\phi}{\beta_0} = \frac{0.355}{2.325 \times 10^{-2}} = 15.27\,\text{s}, \tag{9.115}$$

$$L^* = \sqrt{\tau D_m t^*} = \sqrt{0.66 \times 8.4 \times 10^{-10} \times 15.27} = 9.20 \times 10^{-5}\,\text{m}, \tag{9.116}$$

$$p_a^* = \frac{\phi\tau D_m}{\psi(S_{nf})} = \frac{\phi\tau D_m\mu_a}{k_f} = \frac{0.355 \times 0.66 \times 8.4 \times 10^{-10} \times 10^{-3}}{10^{-12}} = 0.1968\,(\text{N/m}^2). \tag{9.117}$$

As stated in a previous publication (Zhao et al. 2010a), both the Zhao number and its critical value (i.e. the critical Zhao number) of a NAPL dissolution system are needed to assess the dissolution front instability of the NAPL dissolution system. If the Zhao number of a NAPL dissolution system is greater than the corresponding critical Zhao number of the NAPL dissolution system, then the NAPL dissolution front is unstable in the NAPL dissolution system. From the previous studies (Zhao et al. 2010a), the Zhao number of the NAPL dissolution system under consideration is:

$$Zh = -\bar{p}'_{axf} = -\frac{p'_{axf}L^*}{p_a^*} = \frac{V_{axf}}{\sqrt{\phi\tau D_m}}\sqrt{\frac{1}{\beta_0}}$$
$$= \frac{1.16 \times 10^{-5}}{\sqrt{0.355 \times 0.66 \times 8.4 \times 10^{-10} \times 2.352 \times 10^{-2}}} = 5.392. \quad (9.118)$$

To establish a computational model for the NAPL dissolution system under consideration, other dimensionless quantities can be evaluated as follows:

$$\bar{L}_x = \frac{L_x}{L^*} = \frac{0.3}{9.2 \times 10^{-5}} = 3260.9, \quad \bar{L}_y = \frac{L_y}{L^*} = \frac{1.0}{9.2 \times 10^{-5}} = 10869.6, \quad (9.119)$$

$$\bar{\alpha}_L = \frac{\alpha_L}{L^*} = \frac{7.2 \times 10^{-4}}{9.2 \times 10^{-5}} = 7.826, \quad \bar{\alpha}_T = \frac{\alpha_T}{L^*} = \frac{1.44 \times 10^{-4}}{9.2 \times 10^{-5}} = 1.5652. \quad (9.120)$$

The critical Zhao number (i.e. $Zh_{critical}$) of the NAPL dissolution system can be determined from the following characteristic equation:

$$\frac{(1+\beta)Zh_{critical}}{1+\bar{\alpha}_L Zh_{critical}} - \left[(1+\beta) + \left(\frac{1}{1+\bar{\alpha}_L Zh_{critical}} - \bar{\alpha}_L \bar{m}\right)(1-\beta)\right]\sigma_1$$
$$+ \left[\left(\bar{m} + \frac{Zh_{critical}}{1+\bar{\alpha}_L Zh_{critical}} - \bar{\alpha}_L \bar{m}^2(1+\bar{\alpha}_L Zh_{critical})\right)\frac{(1-\beta)}{1+\bar{\alpha}_L Zh_{critical}}\right] = 0, \quad (9.121)$$

where

$$\sigma_1 = \frac{\sqrt{\left(\frac{Zh_{critical}}{1+\bar{\alpha}_L Zh_{critical}}\right)^2 + \frac{4\bar{m}^2(1+\bar{\alpha}_T Zh_{critical})}{1+\bar{\alpha}_L Zh_{critical}}} + \frac{Zh_{critical}}{1+\bar{\alpha}_L Zh_{critical}}}{2}, \quad (9.122)$$

$$\beta = \frac{\psi(S_{n0})}{\psi(S_{nf})} = \frac{k(S_{n0})}{k(S_{nf})}. \quad (9.123)$$

Since $\bar{m}$, $\bar{\alpha}_L$ and $\bar{\alpha}_T$ are dependent on L^*, it is necessary to establish a relationship between L^* and $Zh_{critical}$ below. Based on the dimensionless quantities expressed in Eqs. (9.59) and (9.60), the dimensionless Zhao number is defined as follows:

$$Zh = -\bar{p}'_{axf} = -\frac{p'_{axf}L^*}{p_a^*} = \frac{V_{axf}}{\sqrt{\phi\tau D_m}}\sqrt{\frac{1}{\beta_0}}, \quad (9.124)$$

where V_{axf} is the Darcy velocity of the aqueous phase fluid after the NAPL is completely dissolved in the NAPL dissolution system.

Equation (9.124) can be straightforwardly rewritten in the following form:

$$V_{axf} = Zh\sqrt{\phi\tau D_m\beta_0}. \tag{9.125}$$

Using this Equation, the Reynolds number of the NAPL dissolution system can be expressed as

$$Re = \frac{d_{50}\rho_a Zh}{\mu_a(1 - S_{n0})}\sqrt{\frac{\tau D_m\beta_0}{\phi}}. \tag{9.126}$$

This leads to the following equation:

$$\beta_0 = \frac{186\phi^{0.87}D_m}{d_{50}^2}\left(\frac{d_{50}\rho_a Zh}{\mu_a(1 - S_{n0})}\sqrt{\frac{\tau D_m}{\phi}}\right)^{0.71}\beta_0^{0.355}. \tag{9.127}$$

Note that in the critical state, $Zh = Zh_{critical}$, so that β_0 can be expressed as follows:

$$\begin{aligned}\beta_0 &= \left[\frac{186\phi^{0.87}D_m}{d_{50}^2}\left(\frac{d_{50}\rho_a Zh_{critical}}{\mu_a(1 - S_{n0})}\sqrt{\frac{\tau D_m}{\phi}}\right)^{0.71}\right]^{\frac{1}{0.645}} \\ &\approx \frac{3301\phi^{0.798}\tau^{0.55}D_m^{2.1}}{d_{50}^2}\left(\frac{\rho_a Zh_{critical}}{\mu_a(1 - S_{n0})}\right)^{1.1}.\end{aligned} \tag{9.128}$$

Finally, L^* is expressed as a function of β_0.

$$L^* = \sqrt{\frac{\phi\tau D_m}{\beta_0}}. \tag{9.129}$$

Consideration of Eqs. (9.128) and (9.129) yields the following equation:

$$Zh_{critical} = \frac{\mu_a(1 - S_{n0})}{\rho_a D_m(L^*)^{1.818}}\left(\frac{d_{50}^2\phi^{0.202}\tau^{0.45}}{3301}\right)^{\frac{1}{1.1}}. \tag{9.130}$$

For the NAPL dissolution system of a finite domain, the critical geometrical characteristic length (L_y) of the system should be equal to half the longest wavelength (λ) of the perturbation that can grow in the system, so that the following condition exists:

$$\frac{\lambda}{2} = L_y. \tag{9.131}$$

Since $m = 2\pi/\lambda = \pi/L_y$, where m is the wavenumber of the perturbation, the dimensionless wavenumber of the perturbation can be expressed as follows:

$$\bar{m} = mL^* = \frac{\pi}{L_y}L^* = \frac{\pi}{L_y^*}. \tag{9.132}$$

Consideration of Eqs. (9.130) and (9.132) simultaneously yields the following equation:

$$Zh_{critical} = \frac{\mu_a(1 - S_{n0})\pi^{1.818}}{\rho_a D_m(\bar{m}L_y)^{1.818}} \left(\frac{d_{50}^2 \phi^{0.202} \tau^{0.45}}{3301}\right)^{\frac{1}{1.1}}. \tag{9.133}$$

Equation (9.133) indicates that for the NAPL dissolution system of a finite domain, the critical Zhao number of the system is inversely proportional to the critical geometrical characteristic length (L_y) of the system. Thus, with an increase in the critical geometrical characteristic length (L_y) of the system, there is a decrease in the critical Zhao number of the system. This implies that for the perturbation of a given wavelength, the NAPL dissolution system becomes more unstable with the increase of the critical geometrical characteristic length (L_y) of the system.

Substituting Eq. (9.128) into Eq. (9.124) yields the Zhao number of a NAPL dissolution system when the correlation of Imhoff et al. (1994) can be used for describing the mass transfer rate coefficient of the NAPL dissolution system.

$$Zh = \left\{\frac{d_{50}V_{axf}}{57.45\phi^{0.899}\tau^{0.775}D_m^{1.55}}\left(\frac{\mu_a(1 - S_{n0})}{\rho_a}\right)^{0.55}\right\}^{\frac{1}{1.55}}. \tag{9.134}$$

Again, Eq. (9.134) indicates that the Zhao number of a NAPL dissolution system is directly proportional to the injected Darcy velocity of the aqueous phase fluid in the NAPL dissolution system. As a result, with an increase in the injected Darcy velocity of the aqueous phase fluid in a NAPL dissolution system, there is an increase in the Zhao number of the NAPL dissolution system. This means that the NAPL dissolution system becomes more unstable with an increase in the injected Darcy velocity of the aqueous phase fluid within the NAPL dissolution system. On the other hand, for a given Zhao number, the injected Darcy velocity of the aqueous phase fluid within the NAPL dissolution system can be straightforwardly determined from either Eqs. (9.124) or (9.134). Thus, the Zhao number of a NAPL dissolution system, rather than the injected Darcy velocity of the aqueous phase fluid within the NAPL dissolution system, is used to investigate three kinds of potential fingering modes in the next section later.

If Eq. (9.128) is substituted into Eq. (9.129), then L^* is expressed as a function of $Zh_{critical}$ so that the critical Zhao number of the NAPL dissolution system can be determined by solving the corresponding characteristic equation (i.e. Eq. (9.121)) of the system. As a result, the solution for the critical Zhao number of the NAPL dissolution system is equal to 0.011 approximately. Since the Zhao number of the NAPL dissolution system is about 5.4 (see Eq. (9.118)), which is much greater than the critical Zhao number, the system under consideration is in an unstable state.

Note that although both the NAPL morphological profile evolution and the NAPL morphological interface roughness evolution can be used to compare the current simulation results with the previous ones (Imhoff et al. 2002), the former is adopted in this investigation because it is the NAPL morphological profile evolution that can be used to show the three different kinds of fingering modes, as demonstrated later in Sect. 9.4. However, for the simulation of NAPL dissolution-front instability in heterogeneous porous media, the latter may be the best way to represent the simulation results, since the finger locations are not necessarily in the same locations for the random heterogeneous porous medium.

Figure 9.7 shows the comparison of the current simulation results with the previous solutions that can be viewed as a representative of the related laboratory experimental observations (Imhoff et al. 2002). Note that in this figure, the previous solutions are presented in a black-white form, so that the color code is only applicable to the current simulation results. Two cases are considered for comparison. In the first case, the longitudinal and transverse dispersivities are 7.2×10^{-4} m and 1.44×10^{-4} m, while in the second case, both of them are equal to 7.2×10^{-4} m. To examine the possible effect of the Darcy velocity on the fingering patterns of the NAPL dissolution front, two different Darcy velocities, namely 1.16×10^{-5} m/s (equivalent to 1 m/day) and 2.52×10^{-5} m/s (equivalent to 2.18 m/day), are used to produce the related simulation results. Since the current simulation results compare well with the previous solutions, it demonstrates that the proposed numerical procedure is suitable and useful for simulating the morphological evolution of NAPL dissolution fronts in supercritical systems.

9.4 Three Different Kinds of Modes Associated with Morphological Evolution of NAPL Dissolution Fronts in Supercritical Systems

After the proposed numerical procedure is verified through two benchmark problems, it is used to investigate three potential modes of NAPL dissolution fronts in the supercritical systems of finite domains. Compared with previous publications (Miller et al. 1998; Zhao et al. 2010b), this is the major contribution of this chapter. The geometrical and boundary conditions of the computational model are the same as those used in the first benchmark problem (see Fig. 9.4). To examine three different modes of the NAPL dissolution fronts in supercritical systems, three cases with two different dimensionless wavenumbers, namely 0.005 and 0.001, are considered in the corresponding computations. In the first two cases,

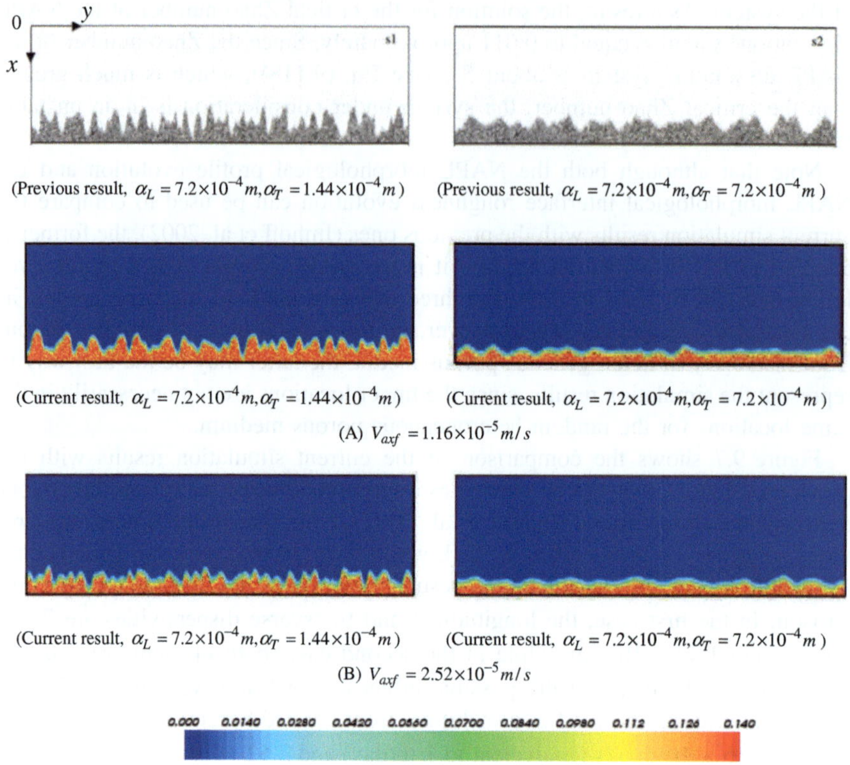

Fig. 9.7 Comparison of current results with the previous ones due to different longitudinal and transversal dispersivities

the dimensionless wavenumber is 0.005, while in the third case, the dimensionless wavenumber is 0.001. Common parameters used in three cases are as follows: the initial saturation (S_{n0}) of the NAPL is 0.2; the irreducible saturation (S_{ai}) of the aqueous phase fluid is 0.15; the dimensionless longitudinal and transverse dispersivities ($\bar{\alpha}_L$ and $\bar{\alpha}_T$) are 0.2 and 0.02, respectively; the ratio (ε) of the equilibrium concentration of the NAPL species in the aqueous phase fluid to the density of the NAPL itself is 0.001; the density ratio ($\bar{\rho}_a$) of the aqueous phase fluid to the NAPL is 1.0/1.46; the value of β_1 is 0.87. To simulate the propagation of NAPL dissolution fronts appropriately, the whole computational domain is simulated by 99,301 four-node square elements of 100,000 nodal points in total. The initial residual saturation field of the NAPL is randomly perturbed by a small amount of 1 % of the originally-input saturation of the NAPL (i.e. $S_{n0} = 0.2$) before running the computational model. This means that the resulting initial residual saturation is of a random distribution, which has a mean value of the homogeneous residual saturation (i.e. $S_{n0} = 0.2$) and a variation of 0.002 (i.e. 1 % of $S_{n0} = 0.2$) in the whole computational domain. Due to the consideration of three different ratios of the Zhao number to the critical Zhao number, three different kinds of modes, namely the fundamental mode, the fingering mode and the fractal mode, have been obtained from the corresponding computational models.

9.4.1 The Fundamental Mode

The fundamental mode is obtained in the first case. The characteristic of this case is that the Zhao number of the NAPL dissolution system is in the lower range of the supercritical Zhao numbers so that the Zhao number and its critical value have the same order of magnitude. The Zhao number used in this case is 0.06, while the critical Zhao number is 0.017. This means that the ratio of the Zhao number to its critical value is equal to 3.53. The dimensionless lengths of the computational domain are 628 and 1,570 in the $\bar{x}$ and $\bar{y}$ directions, respectively. In this case, the dimensionless time step length is 11.0 in the computational simulation.

Figure 9.8 shows the evolution process of the fundamental mode associated with the NAPL dissolution front in a lower supercritical Zhao number system. This mode is characterized by half the longest wavelength of the perturbation that the computational model can accommodate in the $\bar{y}$ direction. Note that the fundamental mode of the NAPL dissolution front is similar to that of temperature in a convective instability system within the crust of the Earth (Nield and Bejan 1992; Zhao et al. 2008a). Since the Zhao number of the NAPL dissolution system is in the lower range of supercritical Zhao numbers, the amplitude of the fundamental mode grows very slowly so that it takes a considerable time for the mode to be viewed with naked eyes. For example, this mode can be clearly identified after the dimensionless time is equal to 3850. With an increase in the dimensionless time, the amplitude of the fundamental mode grows accordingly.

9.4.2 The Fingering Mode

The (normal) fingering mode is obtained in the second case. The characteristic of this case is that the Zhao number of the NAPL dissolution system is in the middle range of the supercritical Zhao numbers so that the Zhao number is one order of magnitude higher than its critical value. The Zhao number used in this case is 0.5, while the critical Zhao number remains 0.017 because the same dimensionless wavenumber is used for both the first and the second cases. In this situation, the ratio of the Zhao number to its critical value is equal to 29.4. The dimensionless lengths of the computational domain are the same as those used in the first case. In this situation, the dimensionless time step length is 1.3 in the computational simulation.

Figure 9.9 shows the evolution process of the fingering mode for the NAPL dissolution front in a moderate supercritical Zhao number system. This kind of mode is characterized by the pattern of a few fingers in the NAPL dissolution

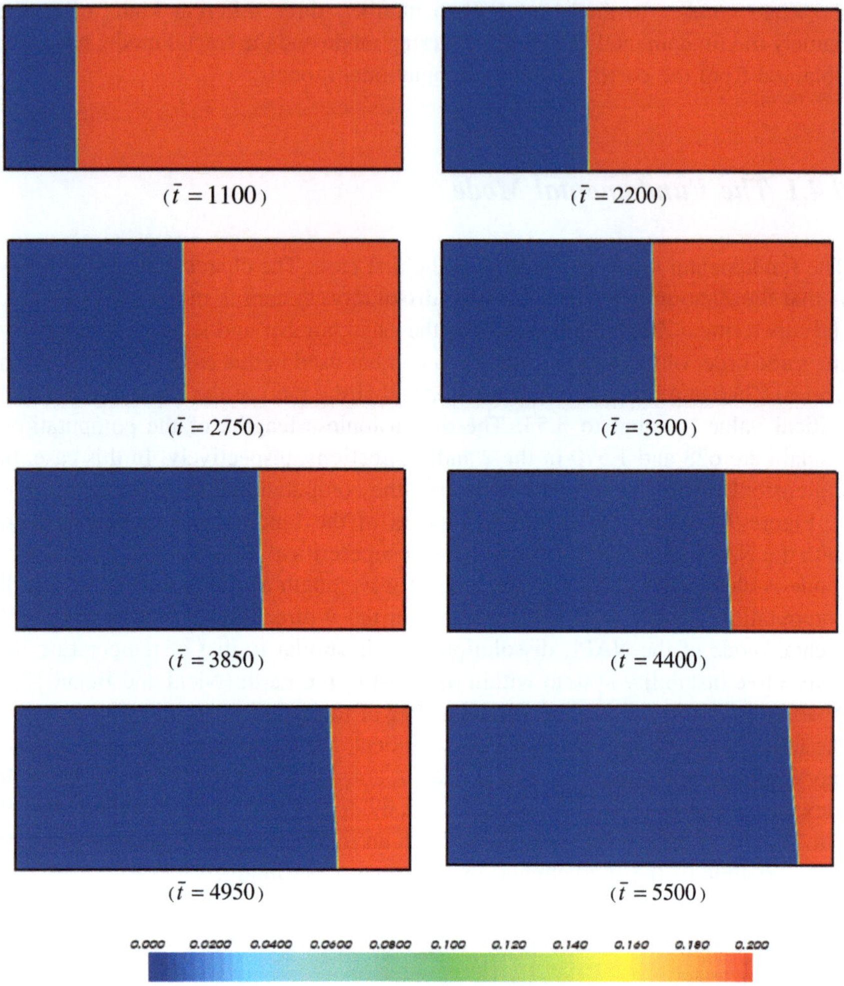

Fig. 9.8 The evolution processes of the fundamental mode (the first kind of mode) for a NAPL dissolution front in the supercritical system

system. Since the Zhao number of the NAPL dissolution system is in the middle range of supercritical Zhao numbers, the length of the fingers (i.e. the amplitude of the fingering mode) grows relatively faster, compared with that of the fundamental mode. However, the width of the finger still grows very slowly. This kind of mode can be clearly observed after the dimensionless time is equal to 260 in Fig. 9.9. Similarly, with an increase in the dimensionless time, the length of the finger grows gradually.

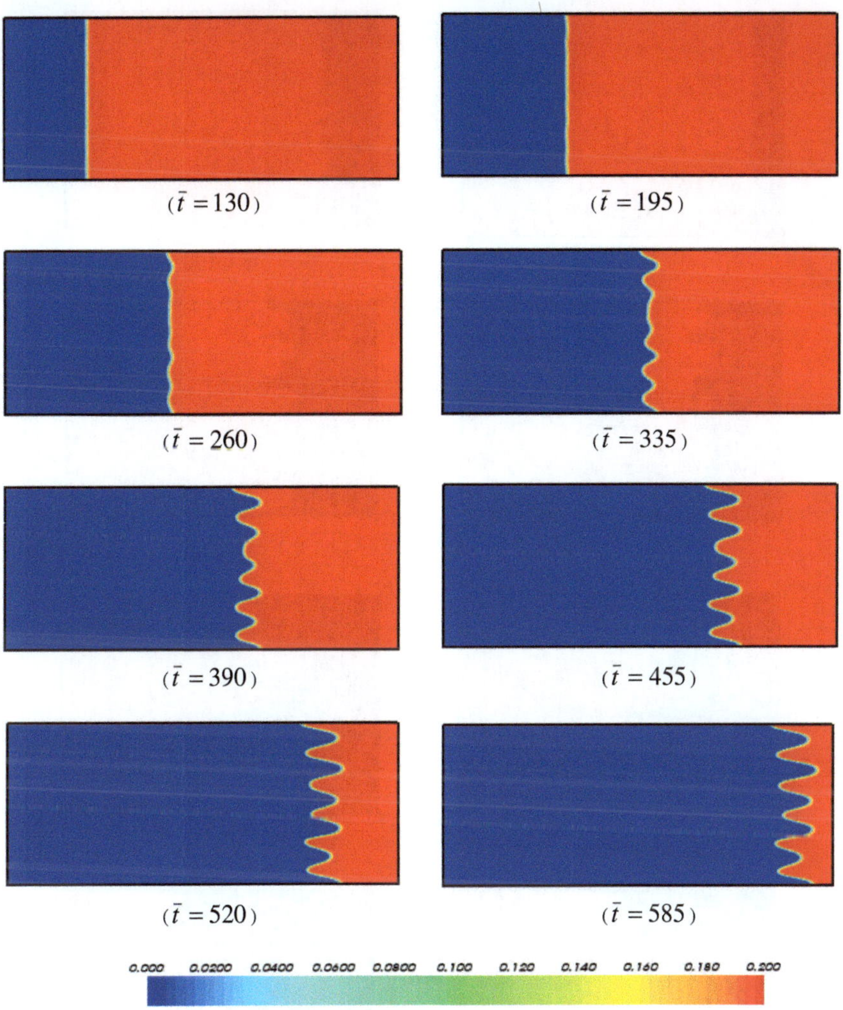

Fig. 9.9 The evolution processes of the fingering mode (the second kind of mode) for a NAPL dissolution front in the supercritical system

9.4.3 The Fractal Mode

The fractal mode is obtained in the third case. The characteristic of this case is that the Zhao number of the NAPL dissolution system is in the higher range of the supercritical Zhao numbers of the system so that the ratio of the Zhao number to its critical value is at least one order of magnitude higher than that associated with the second case (i.e. the normal fingering mode case). The Zhao number used in this case is 1.0, while the critical Zhao number is 3.34×10^{-3}. This implies that the ratio of the Zhao number to its critical value is equal to 283.3. The dimensionless

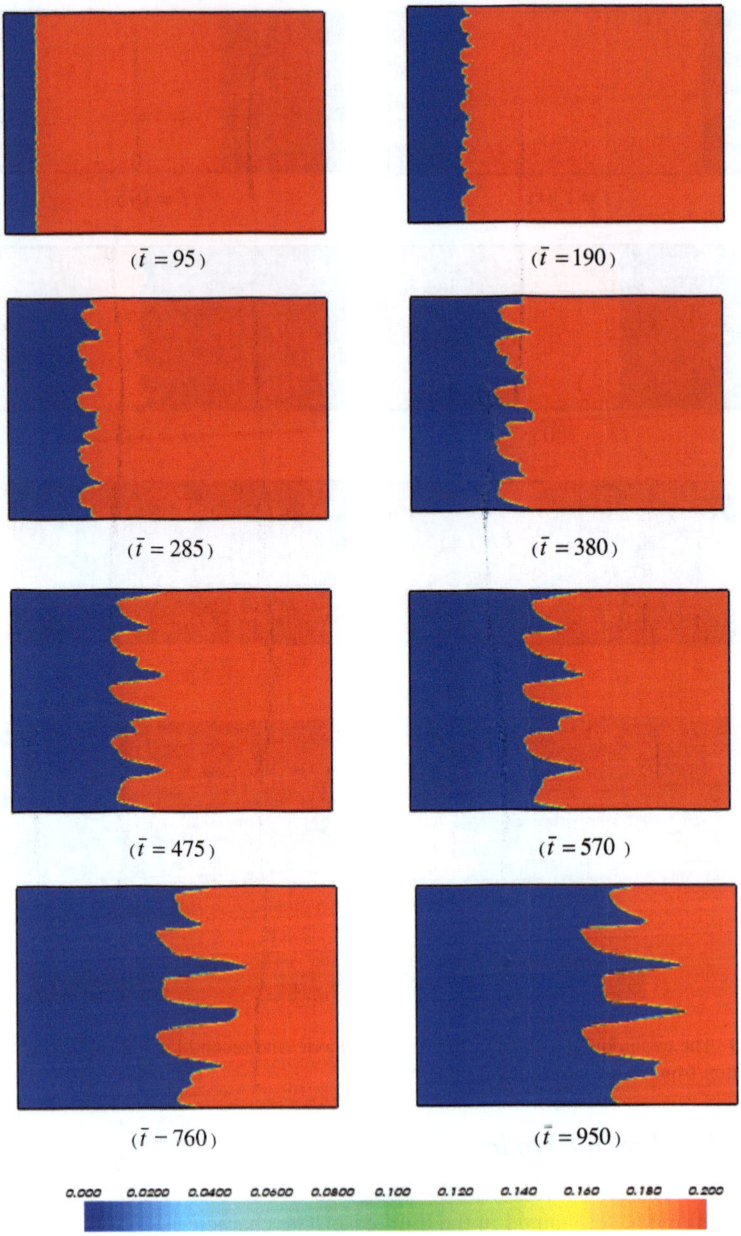

Fig. 9.10 The evolution processes of the fractal mode (the third kind of mode) for a NAPL dissolution front in the supercritical system

lengths of the computational domain are 3142 and 4713 in the $\bar{x}$ and $\bar{y}$ directions, respectively. In this case, the dimensionless time step length is 1.9 in the computational simulation.

Figure 9.10 shows the evolution process of the fractal mode associated with the NAPL dissolution front in a higher supercritical Zhao number system. This kind of mode has been commonly observed in the previous laboratory experiments and numerical simulations (Imhoff et al. 1996; Miller et al. 1998). It is characterized by the fractal pattern consisting of several irregular fingers. As the Zhao number of the NAPL dissolution system is in the higher range of supercritical Zhao numbers, the amplitude of the fractal mode grows very fast, compared with that of the normal fingering mode that is comprised of a few smooth fingers. At the early stage of the simulation, many small fingers of shorter wavelengths are generated in the computational model (see $\bar{t} = 95$). As time goes on, these small fingers grow in length and width, sometimes merging to form longer and wider fingers of irregular fractal shapes. As shown in Fig. 9.10, this kind of mode can be observed after the dimensionless time is equal to 190 in the computational simulation.

In summary, a theoretical criterion that can be used to assess the instability of planar NAPL dissolution-fronts in two-dimensional fluid-saturated porous media of finite domains has been established. Not only can the present theoretical results be used for theoretical understanding of the effect of solute dispersion on the instability of a NAPL dissolution-front in the fluid-saturated porous medium of either a finite domain or an infinite domain, but also they can be used as benchmark solutions for verifying numerical methods employed to simulate detailed morphological evolution processes of NAPL dissolution fronts in two-dimensional fluid-saturated porous media. The related simulation results have demonstrated that: (1) the proposed numerical procedure is useful and applicable for simulating the morphological evolution of NAPL dissolution fronts in two-dimensional fluid-saturated porous media of finite domains; (2) if the Zhao number of a NAPL dissolution system is in the lower range of the supercritical Zhao numbers, the fundamental mode is predominant; (3) if the Zhao number is in the middle range of the supercritical Zhao numbers, the (normal) fingering mode is the predominant pattern of the NAPL dissolution front; and (4) if the Zhao number is in the higher range of the supercritical Zhao numbers, the fractal mode is predominant for the NAPL dissolution front.

References

Alt-Epping P, Smith L (2001) Computing geochemical mass transfer and water/rock ratios in submarine hydrothermal systems: implications for estimating the vigour of convection. Geofluids 1:163–181

Bear J (1972) Dynamics of fluids in porous media. American Elsevier Publishing Company, New York

Chadam J, Hoff D, Merino E, Ortoleva P, Sen A (1986) Reactive infiltration instabilities. IMA J Appl Math 36:207–221

Chadam J, Ortoleva P, Sen A (1988) A weekly nonlinear stability analysis of the reactive infiltration interface. IMA J Appl Math 48:1362–1378

Chen JS, Liu CW (2002) Numerical simulation of the evolution of aquifer porosity and species concentrations during reactive transport. Comput Geosci 28:485–499

Chen JS, Liu CW, Lai GX, Ni CF (2009) Effects of mechanical dispersion on the morphological evolution of a chemical dissolution front in a fluid-saturated porous medium. J Hydrol 373:96–102

Geller JT, Hunt JR (1993) Mass transfer from nonaqueous phase organic liquids in water-saturated porous media. Water Resour Res 29:833–845

Imhoff PT, Miller CT (1996) Dissolution fingering during the solubilization of nonaqueous phase liquids in saturated porous media: 1. Model predictions. Water Resour Res 32:1919–1928

Imhoff PT, Jaffe PR, Pinder GF (1994) An experimental study of complete dissolution of a nonaqueous phase liquid in saturated porous media. Water Resour Res 30:307–320

Imhoff PT, Thyrum GP, Miller CT (1996) Dissolution fingering during the solubilization of nonaqueous phase liquids in saturated porous media: 2. Experimental observations. Water Resour Res 32:1929–1942

Imhoff PT, Farthing MW, Gleyzer SN, Miller CT (2002) Evolving interface between clean and nonaqueous phase liquid (NAPL)-contaminated regions in two-dimensional porous media. Water Resour Res 38:1093–1106

Imhoff PT, Farthing MW, Miller CT (2003) Modeling NAPL dissolution fingering with upscaled mass transfer rate coefficients. Adv Water Resour 26:1097–1111

Miller CT, Poirier-McNeil MM, Mayer AS (1990) Dissolution of trapped nonaqueous phase liquids: mass transfer characteristics. Water Resour Res 26:2783–2796

Miller CT, Christakos TG, Imhoff PT, McBride JF, Pedit JA, Trangenstein JA (1998) Multiphase flow and transport modeling in heterogeneous porous media: challenges and approaches. Adv Water Resour 21:77–120

Nield DA, Bejan A (1992) Convection in porous media. Springer, New York

Ormond A, Ortoleva P (2000) Numerical modeling of reaction-induced cavities in a porous rock. J Geophys Res 105:16737–16747

Ortoleva P, Chadam J, Merino E, Sen A (1987) Geochemical self-organization II: the reactive-infiltration instability. Am J Sci 287:1008–1040

Powers SE, Abriola LM, Weber WJ Jr (1994) An experimental investigation of nonaqueous phase liquid dissolution in saturated subsurface systems: transient mass transfer rates. Water Resour Res 30:321–332

Raffensperger JP, Garven G (1995) The formation of unconformity-type uranium ore deposits: coupled hydrochemical modelling. Am J Sci 295:639–696

Renard F, Gratier JP, Ortoleva P, Brosse E, Bazin B (1998) Self-organization during reactive fluid flow in a porous medium. Geophys Res Lett 25:385–388

Schafer D, Schafer W, Kinzelbach W (1998a) Simulation of reactive processes related to biodegradation in aquifers: 1. Structure of the three-dimensional reactive transport model. J Contam Hydrol 31:167–186

Schafer D, Schafer W, Kinzelbach W (1998b) Simulation of reactive processes related to biodegradation in aquifers: 2. Model application to a column study on organic carbon degradation. J Contam Hydrol 31:187–209

Seyedabbasi MA, Farthing MW, Imhoff PT, Miller CT (2008) The influence of wettability on NAPL dissolution fingering. Adv Water Resour 31:1687–1696

Soerens TS, Sabatini DA, Harwell JH (1998) Effects of flow bypassing and nonuniform NAPL distribution on the mass transfer characteristics of NAPL dissolution. Water Resour Res 34:1657–1673

Steefel CI, Lasaga AC (1990) Evolution of dissolution patterns: permeability change due to coupled flow and reaction. In: Melchior DC, Basset RL (eds) Chemical modeling in aqueous systems II, vol 416. American chemistry society symposium series, pp 213–225

Steefel CI, Lasaga AC (1994) A coupled model for transport of multiple chemical species and kinetic precipitation/dissolution reactions with application to reactive flow in single phase hydrothermal systems. Am J Sci 294:529–592

Willson CS, Hall JL, Miller CT, Imhoff PT (1999) Factors affecting bank formation during surfactant-enhanced mobilization of residual NAPL. Environ Sci Technol 33:2440–2446

Xu TF, Samper J, Ayora C, Manzano M, Custodio E (1999) Modelling of non-isothermal multi-component reactive transport in field scale porous media flow systems. J Hydrol 214:144–164

Yeh GT, Tripathi VS (1991) A model for simulating transport of reactive multispecies components: model development and demonstration. Water Resour Res 27:3075–3094

Zhao C, Hobbs BE, Mühlhaus HB (1998) Finite element modelling of temperature gradient driven rock alteration and mineralization in porous rock masses. Comput Meth Appl Mech Eng 165:175–187

Zhao C, Hobbs BE, Mühlhaus HB, Ord A, Lin G (2003) Finite element modeling of three-dimensional steady-state convection and lead/zinc mineralization in fluid-saturated rocks. J Comput Meth Sci Eng 3:73–89

Zhao C, Hobbs BE, Ord A, Peng S, Mühlhaus HB, Liu L (2005) Numerical modeling of chemical effects of magma solidification problems in porous rocks. Int J Numer Meth Eng 64:709–728

Zhao C, Hobbs BE, Hornby P, Ord A, Peng S (2006) Numerical modelling of fluids mixing, heat transfer and non-equilibrium redox chemical reactions in fluid-saturated porous rocks. Int J Numer Meth Eng 66:1061–1078

Zhao C, Hobbs BE, Ord A (2008a) Convective and advective heat transfer in geological systems. Springer, Berlin

Zhao C, Hobbs BE, Hornby P, Ord A, Peng S, Liu L (2008b) Theoretical and numerical analyses of chemical-dissolution front instability in fluid-saturated porous rocks. Int J Numer Anal Meth Geomech 32:1107–1130

Zhao C, Hobbs BE, Ord A, Hornby P, Peng S (2008c) Effect of reactive surface areas associated with different particle shapes on chemical-dissolution front instability in fluid-saturated porous rocks. Transp Porous Media 73:75–94

Zhao C, Hobbs BE, Ord A (2009) Fundamentals of computational geoscience: numerical methods and algorithms. Springer, Berlin

Zhao C, Hobbs BE, Ord A, Peng S (2010a) Effects of mineral dissolution ratios on chemical-dissolution front instability in fluid-saturated porous media. Transp Porous Media 82:317–335

Zhao C, Hobbs BE, Ord A (2010b) Theoretical analyses of nonaqueous-phase-liquid dissolution induced instability in two-dimensional fluid-saturated porous media. Int J Numer Anal Meth Geomech 34:1767–1796

Zhao C, Hobbs BE, Regenauer-Lieb K, Ord A (2011) Computational simulation for the morphological evolution of nonaqueous-phase-liquid dissolution fronts in two-dimensional fluid-saturated porous media. Comput Geosci 15:167–183

Zienkiewicz OC (1977) The finite element method. McGraw-Hill, London

Chapter 10
Effects of Domain Shapes and Mesh Discretization Error on the Morphological Evolution of Nonaqueous-Phase-Liquid Dissolution Fronts in Fluid-Saturated Porous Media

In the field of contaminant hydrology, both land contamination and land remediation problems are often encountered. Land contamination is known as the distribution of chemical and pollutants on land sites, while land remediation is known as the cleanup of chemical and pollutants on land sites that causes health concerns to the humans and the environment. When nonaqueous phase liquids (NAPLs), such as trichloroethylene, ethylene dibromide, benzene, toluene and so forth (Miller et al. 1990), are released to groundwater, they can reside in the form of disconnected ganglia or blobs as residual saturations within the pores of porous media. This process belongs to the land contamination problem. Some NAPLs (e.g. trichloroethylene and ethylene dibromide) are heavier than water, but others (e.g. benzene and toluene) are lighter than water. Although their solubilities in groundwater are very low, the effect of such NAPLs on the quality of groundwater resources is severe because of their relatively high toxicity. Thus, it is necessary to remove such NAPLs from the contaminated land site. This process belongs to the land remediation problem, which is the main focus of this chapter.

To develop effective and efficient methods for removing the residual NAPLs from contaminated land sites, the detailed transport mechanism of NAPLs in fluid-saturated porous media has been studied, both experimentally and analytically, during the past two decades (Miller et al. 1990, 1998; Geller and Hunt 1993; Powers et al. 1994; Imhoff et al. 1994, 1996, 2002, 2003a; Soerens et al. 1998; Willson et al. 1999; Seyedabbasi et al. 2008). Notable achievements from existing laboratory experiments are as follows: (1) mass transfer rates between a NAPL and an aqueous phase liquid can be determined in a quantitative manner; (2) the fingering phenomena of NAPL dissolution fronts have been observed at the laboratory (i.e. centimeter) scale; and (3) the linear stability theory has been employed to derive the critical condition that can be used to assess the instability of NAPL dissolution fronts in fluid-saturated porous media. For example, the Zhao number (Zhao et al. 2010c), which is a comprehensive dimensionless number, was proposed to represent the three major controlling mechanisms simultaneously

C. Zhao, *Physical and Chemical Dissolution Front Instability in Porous Media*, Lecture Notes in Earth System Sciences,
DOI: 10.1007/978-3-319-08461-9_10

taking place in a NAPL dissolution system. Based on the linear stability theory, a NAPL dissolution system may physically have three different kinds of states (Imhoff and Miller 1996; Zhao et al. 2008c, 2010b). In the supercritical state NAPL dissolution fingering occurs, but it does not occur in the subcritical state. The neutral condition (or state) is just the interface between the two. Correspondingly, three kinds of the Zhao numbers, namely the subcritical Zhao number, the critical Zhao number and the supercritical Zhao number, can be used to represent these three different kinds of states in the NAPL dissolution system. As a direct result of these achievements, both mathematical and computational models (Imhoff and Miller 1996; Miller et al. 1998; Zhao et al. 2010b) were developed to simulate the morphological evolution of NAPL dissolution fronts in fluid-saturated porous media. Nevertheless, the existing computational models are mainly limited to either a square domain or a rectangular domain, so that it is necessary to investigate the effects of different domain shapes on the morphological evolution of NAPL dissolution fronts in fluid-saturated porous media.

It needs to be pointed out that NAPL dissolution fingering requires regions of continuous NAPL saturation distribution (in the form of disconnected ganglia or blobs as residual within the pores of the porous medium) and has been observed in experiments with length scales of 7 cm and larger in the mean flow direction (Imhoff et al. 2003b). This requirement is unlikely to be satisfied for most two-dimensional experimental systems considered in the laboratory where a small amount of NAPL is spilled (Chen and Jawitz 2008; DiFilippo et al. 2010). However, as demonstrated previously (Parker and Park 2004; Christ et al. 2006, 2009; Gerhard et al. 2007), for large NAPL spill sites, continuous regions of residual NAPL occurred so that the above-mentioned requirement can be met. For example, in the work of Parker and Park (2004) a NAPL spillage event resulted in vertical fingers that on average were 30 cm in diameter. These fingers primarily contained residual NAPL. On the other hand, since most grid blocks used in current field-scale simulations were on the order of 30–50 cm (in the x and y dimensions), by necessity NAPL-contaminated grid blocks only represent continuous regions of NAPL (often residual) that exceed the 7 cm scale. As a result, these simulations might ignore NAPL dissolution fingering. It is a sub grid-block process that is not accounted for. The local (grid-block scale) rate of NAPL dissolution may not be important for many field problems, since the bypassing of water around NAPL-contaminated zones (Parker and Park 2004) is the slowest process limiting NAPL dissolution into surrounding groundwater. However, recent simulations indicate that for some systems local NAPL dissolution rates are important in heterogeneous media (Maji and Sudicky 2008). For such systems, NAPL dissolution fingering may be important. Thus, an understanding of NAPL dissolution fingering may be important for developing innovative remediation strategies and technologies to some NAPL contaminated groundwater systems.

In addition to NAPL dissolution fingering, preferential flow within NAPL contaminated zones can be also caused by either medium heterogeneity (Maji and Sudicky 2008) or variations in NAPL saturation, which in nature are not uniform in space (Grant and Gerhard 2007; Zhang et al. 2007). Medium heterogeneity within NAPL-contaminated zones will result in a variation in aqueous-phase permeability. Some of this variation is associated with the variation in intrinsic permeability caused by the medium heterogeneity alone, while some is associated with the variation in NAPL saturation, which alters the relative permeability. Nevertheless, the mechanism of the preferential flow caused by NAPL dissolution fingering is different from that caused by medium heterogeneity. From the physical point of view, the former is considered as an emerging phenomenon due to the instability of a nonlinear system (Chadam et al. 1986, 1988; Ortoleva et al. 1987; Renard et al. 1998; Chen and Liu 2002, 2004; Chen et al. 2009; Zhao et al. 2008a, b, c, 2009, 2010a), while the latter is considered as the conventional phenomenon of a nonlinear system (Steefel and Lasaga 1990, 1994; Yeh and Tripathi 1991; Raffensperger and Garven 1995; Ormond and Ortoleva 2000; Alt-Epping and Smith 2001; Maji and Sudicky 2008).

Since the domain of a NAPL dissolution system in the real world may have many different shapes, it is difficult, if not impossible, to use a typical domain shape to represent all computational domains of NAPL dissolution systems encountered in the real world. However, for the purpose of investigating the effect of a domain shape on the interesting features associated with NAPL dissolution fingering, it is feasible to use a generic model of a specific shape (that is, to some extent, an artificial system) in the computational simulation, as long as some fundamental flow characteristics associated with irregular domains can be reasonably reflected in the generic model. On the other hand, due to the versatility and robustness of computational methods, any complicated domain shapes can be realistically simulated if their details can be precisely given. Compared with rectangular and square domains that are widely used in the previous studies (Imhoff and Miller 1996; Miller et al. 1998; Zhao et al. 2010b), some important flow characteristics associated with irregular domains are as follows (Zhang et al. 2007; Maji and Sudicky 2008). First, the pore-fluid flow in an irregular domain of a subcritical Zhao number is multi-directional (i.e. two-dimensional for a two-dimensional problem domain and three-dimensional for a three-dimensional problem domain) rather than unidirectional, just as what was observed in a rectangular or square domain of a subcritical Zhao. Second, the Darcy velocities of pore-fluid flow in an irregular domain of a subcritical Zhao number are not constant, but they are constant in a rectangular or square domain of a subcritical Zhao number. To select the domain shape of a generic model, a trapezoidal domain may be the reasonable choice from the following three points of view: (1) the above-mentioned flow characteristics such as the multi-directional flow and non-constant Darcy velocities can be simulated in a trapezoidal domain of a subcritical Zhao number; (2) due to the relatively simple shape, it is convenient to carry out a parameter study and theoretical estimation of mesh discretization error in the generic model of a trapezoidal domain; (3) the simulations are intended to investigate NAPL dissolution fingering in hypothetical systems to better understand the fingering process with converging and diverging flow. If complicated systems are used to represent the real world more realistically, there would be competing processes (e.g., NAPL dissolution fingering and flow bypass) that would make it more difficult to understand NAPL dissolution fingering, which is the focus of this work. For these reasons, different trapezoidal computational domains are considered, in this chapter, to investigate the effects of domain shapes on the morphological evolution of NAPL dissolution fronts in two-dimensional fluid-saturated porous media.

To facilitate mathematical treatments in the process of deriving analytical solutions, it is commonly assumed that the pore-fluid flow within a two-dimensional fluid-saturated porous medium, which may be considered as an approximate representation of a horizontal cross-section (plane) in an aquifer, is parallel to the inflow at the entrance of the analytical domain, so that the second-order dispersion tensor can be considered as a function of the averaged linear velocity component in the inflow direction. This assumption may be valid for either a square domain or a rectangular domain where the inflow is parallel to two lateral boundaries of the domain. However, when the computational domain of a NAPL dissolution problem has a complicated shape, the pore-fluid flow in the computational domain is no longer parallel to the inflow, so that two components of the averaged linear velocity need to be considered in the second-order dispersion tensor. By using the general second-order dispersion tensor of two velocity components, it is also possible to validate the assumption that was used to derive analytical solutions for the NAPL dissolution problem in the fluid-saturated porous medium of a rectangular domain.

10.1 Governing Equations of NAPL Dissolution Problems in Two-Dimensional Fluid-Saturated Porous Media

From the previous studies (Imhoff and Miller 1996; Zhao et al. 2010c), a NAPL dissolution problem can be treated as a mass transport problem. As a result, the governing equations of the NAPL dissolution problem in a two-dimensional fluid-saturated porous medium can be expressed as follows:

$$\phi \frac{\partial S_n}{\partial t} = -\frac{K}{\rho_n}(C_{eq} - C), \tag{10.1}$$

$$\phi \frac{\partial}{\partial t}[(1 - S_n)C] = \nabla \cdot [\phi(1 - S_n)\mathbf{D_h} \cdot \nabla C] + \nabla \cdot \left[\frac{k(S_n)}{\mu_a} C \nabla p_a\right] + K(C_{eq} - C), \tag{10.2}$$

$$\phi \frac{\partial S_n}{\partial t} = -\nabla \cdot \left[\frac{k(S_n)}{\mu_a} \nabla p_a\right] - \frac{K}{\rho_a}(C_{eq} - C), \tag{10.3}$$

where ϕ is the porosity; S_n is the NAPL saturation (i.e. the fraction of the void space occupied by the NAPL); ρ_a and ρ_n are the aqueous phase and nonaqueous phase densities; K is the mass transfer rate coefficient to express the exchange of the NAPL species from the nonaqueous phase to the aqueous phase; C is the solute concentration of the NAPL species in the bulk aqueous phase and is equal to the ratio of the NAPL mass in the solution to the unit volume of the solution; C_{eq} is the equilibrium concentration of the NAPL species in the aqueous phase; $\mathbf{D_h}$ is the general dispersion tensor; $k(S_n)$ is the saturation-dependent permeability of the porous medium to aqueous phase flow; μ_a is the dynamic viscosity of the aqueous phase; and p_a is the aqueous phase pressure.

Note that although many NAPLs at contaminated sites are mixtures of compounds, only a single-species NAPL is considered in this chapter. This means that the equations used here describe a special class of NAPL spills. As a result, S_n, C and p_a are three independent variables in these three equations. As demonstrated previously (Imhoff and Miller 1996; Zhao et al. 2010c), Eq. (10.1) represents the mass conservation of the nonaqueous phase, while Eq. (10.3) represents the mass conservation of the aqueous phase. Since the nonaqueous phase is assumed to be immobile, an advective term does not appear in Eq. (10.1). However, since the aqueous phase is assumed to be mobile, an advective term must appear in Eq. (10.3). In addition, Eq. (10.2) represents the mass conservation of the NAPL species in the fluid-saturated porous medium.

Based on Darcy's law, the averaged linear velocity vector of the aqueous phase can be expressed as follows (Bear 1972):

$$\mathbf{v_a} = \begin{Bmatrix} v_{ax} \\ v_{ay} \end{Bmatrix} = \frac{-k(S_n)}{\phi(1 - S_n)\mu_a} \nabla p_a, \tag{10.4}$$

where $\mathbf{v_a}$ is the averaged linear velocity vector of the aqueous phase; v_{ax} and v_{ay} are the averaged linear velocity components in the x and y directions, respectively; other quantities have the same meanings as defined previously.

The following expressions are used for the saturation-dependent permeability (Imhoff and Miller 1996; Zhao et al. 2010c) and the dispersion tensor (Scheidegger 1961; Holzbecher 1998), respectively:

$$k(S_n) = k_f \left(\frac{1 - S_n - S_{ai}}{1 - S_{ai}} \right)^3, \tag{10.5}$$

$$\mathbf{D_h} = \begin{bmatrix} \tau D_m + \alpha_L \frac{v_{ax}^2}{v_a} + \alpha_T \frac{v_{ay}^2}{v_a} & (\alpha_L - \alpha_T) \frac{v_{ax} v_{ay}}{v_a} \\ (\alpha_L - \alpha_T) \frac{v_{ax} v_{ay}}{v_a} & \tau D_m + \alpha_L \frac{v_{ay}^2}{v_a} + \alpha_T \frac{v_{ax}^2}{v_a} \end{bmatrix}, \tag{10.6}$$

where k_f is the intrinsic permeability of the porous medium after the NAPL is completely dissolved; S_{ai} is the irreducible saturation of the aqueous phase; τ is the tortuosity of the porous medium; D_m is the molecular diffusivity of the NAPL species in the aqueous phase; α_T and α_L are the transversal and longitudinal dispersivities of the NAPL species in the aqueous phase; and v_a is the amplitude of the averaged linear velocity vector ($\mathbf{v_a}$) of the aqueous phase as follows:

$$v_a = \sqrt{v_{ax}^2 + v_{ay}^2}. \tag{10.7}$$

Note that the motivation of choosing a general form of the second-order dispersion tensor (expressed in Eq. (10.6)) is to consider the effect of two-dimensional pore-fluid flow on the dispersion in a trapezoidal domain.

The previous experimental results (Imhoff and Miller 1996) indicate that the mass transfer rate coefficient (K) can be expressed in the following form:

$$K = \beta_0 S_n^{\beta_1}, \tag{10.8}$$

where β_0 is a function of the porous medium, the NAPL, and the velocity, viscosity and density of the aqueous phase fluid; β_1 is a constant.

If the computational domain of a NAPL dissolution system has a trapezoidal shape (as shown in Fig. 10.1), which may be considered either as the experimental specimen used in a laboratory experiment or as the approximate representation of a pie slice in a horizontal cross-section (plane) for radial flow in an aquifer, then the boundary conditions of the problem can be expressed as follows:

$$C = 0, \quad S_n = 0, \quad \frac{\partial p_a}{\partial x} = p'_{axf0} \quad (x = 0), \tag{10.9}$$

$$p_a = p_{a0} \quad (x = L_x), \tag{10.10}$$

$$\frac{\partial C}{\partial n} = 0, \quad \frac{\partial p_a}{\partial n} = 0 \quad (\text{at two lateral boundaries}), \tag{10.11}$$

where S_{n0} is the initial saturation of the NAPL; p'_{axf0} is the pressure gradient of the aqueous phase on the upstream boundary; p_{a0} is the pressure of the aqueous phase on the downstream boundary; L_x is the length of the problem domain in the x direction; and n is the normal vector of the lateral boundary. Since p'_{axf} drives the aqueous phase fluid flow continuously along the positive x direction, it has a negative algebraic value (i.e. $p'_{axf} < 0$) for the problem under consideration.

Except for the upstream boundary, the initial conditions of the problem for the rest of the computational domain are as follows:

$$C = C_{eq}, \quad S_n = S_{n0} \quad (0 < x \leq L_x). \tag{10.12}$$

Since the use of dimensionless governing equations has some advantages in dealing with problems of multi-scales and multi-processes (Zhao et al. 2009), it is useful to transform the above-mentioned governing equations of the problem into the following dimensionless form:

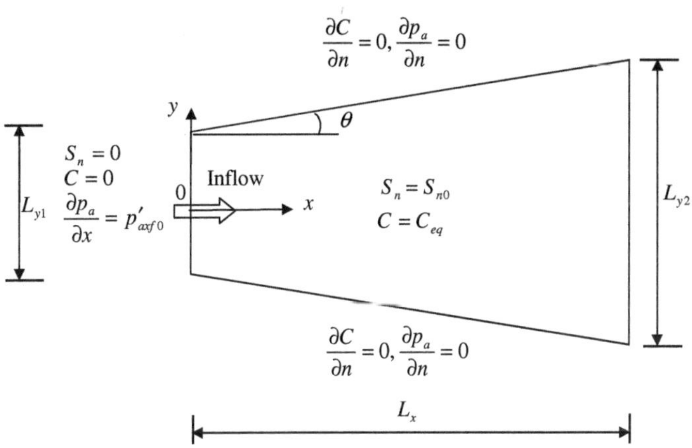

Fig. 10.1 Geometry and boundary conditions for the NAPL dissolution problem on the basis of dimensional quantities

$$\frac{\partial S_n}{\partial \bar{t}} + S_n^{\beta_1}(1-\bar{C}) = 0, \tag{10.13}$$

$$\varepsilon\frac{\partial}{\partial \bar{t}}[(1-S_n)\bar{C}] - \nabla\cdot\{[(1-S_n)\bar{\mathbf{D}}_{\mathbf{h}}]\cdot\nabla\bar{C} + \bar{C}\bar{\psi}(S_n)\nabla\bar{p}_a\} + \frac{\partial S_n}{\partial \bar{t}} = 0, \tag{10.14}$$

$$\varepsilon\frac{\partial S_n}{\partial \bar{t}} + \nabla\cdot[\bar{\psi}(S_n)\nabla\bar{p}_a] + \frac{\varepsilon}{\bar{\rho}_a}S_n^{\beta_1}(1-\bar{C}) = 0, \tag{10.15}$$

where

$$\bar{\mathbf{D}}_{\mathbf{h}} = \begin{bmatrix} \bar{d}_{11} & \bar{d}_{12} \\ \bar{d}_{21} & \bar{d}_{22} \end{bmatrix}, \tag{10.16}$$

where

$$\bar{d}_{11} = 1 + \frac{\bar{\psi}(S_n)}{1-S_n}\frac{\bar{\alpha}_L\left(\frac{\partial\bar{p}_a}{\partial\bar{x}}\right)^2 + \bar{\alpha}_T\left(\frac{\partial\bar{p}_a}{\partial\bar{y}}\right)^2}{\sqrt{\left(\frac{\partial\bar{p}_a}{\partial\bar{x}}\right)^2 + \left(\frac{\partial\bar{p}_a}{\partial\bar{y}}\right)^2}}, \quad \bar{d}_{22} = 1 + \frac{\bar{\psi}(S_n)}{1-S_n}\frac{\bar{\alpha}_L\left(\frac{\partial\bar{p}_a}{\partial\bar{y}}\right)^2 + \bar{\alpha}_T\left(\frac{\partial\bar{p}_a}{\partial\bar{x}}\right)^2}{\sqrt{\left(\frac{\partial\bar{p}_a}{\partial\bar{x}}\right)^2 + \left(\frac{\partial\bar{p}_a}{\partial\bar{y}}\right)^2}},$$

$$\bar{d}_{21} = \bar{d}_{21} = \frac{\bar{\psi}(S_n)}{1-S_n}\frac{(\bar{\alpha}_L - \bar{\alpha}_T)\frac{\partial\bar{p}_a}{\partial\bar{x}}\frac{\partial\bar{p}_a}{\partial\bar{y}}}{\sqrt{\left(\frac{\partial\bar{p}_a}{\partial\bar{x}}\right)^2 + \left(\frac{\partial\bar{p}_a}{\partial\bar{y}}\right)^2}}.$$

Note that the following dimensionless quantities are used for deriving the above dimensionless governing equations of the problem:

$$\bar{x}=\frac{x}{L^*},\quad \bar{y}=\frac{y}{L^*},\quad \bar{L}_x=\frac{L_x}{L^*},\quad \bar{L}_{y1}=\frac{L_{y1}}{L^*},\quad \bar{L}_{y2}=\frac{L_{y2}}{L^*},\quad \bar{C}=\frac{C}{C_{eq}},\tag{10.17}$$

$$\bar{p}_a=\frac{p_a}{p_a^*},\quad \bar{\alpha}_L=\frac{\alpha_L}{L^*},\quad \bar{\alpha}_T=\frac{\alpha_T}{L^*},\quad \bar{n}=\frac{n}{L^*},\quad \bar{t}=\frac{t}{t^*}\varepsilon,\quad \varepsilon=\frac{C_{eq}}{\rho_n}\ll 1,\tag{10.18}$$

where

$$t^*=\frac{\phi}{\beta_0},\quad L^*=\sqrt{\tau D_m t^*},\quad p_a^*=\frac{\phi\tau D_m}{\psi(S_{nf})},\tag{10.19}$$

$$\bar{\mathbf{D}}_\mathbf{h}=\frac{\mathbf{D}_\mathbf{h}}{\tau D_m},\quad \bar{\psi}(S_n)=\frac{\psi(S_n)}{\psi(S_{nf})},\quad \psi(S_n)=\frac{k(S_n)}{\mu_a},\tag{10.20}$$

where $\psi(S_{nf})$ is the value of $\psi(S_n)$ at $S_n=0$. Physically, the dimensionless time (i.e. $\bar{t}$) represents some scaling between the real time and intrinsic time (i.e. t^*) of the NAPL dissolution system (Zhao et al. 2010c).

It needs to be pointed out that a small dimensionless quantity (i.e. ε) is defined in Eq. (10.18), for representing the dissolution ratio of the NAPL equilibrium concentration to the mass density of the NAPL. Since this small dimensionless quantity, known as the NAPL dissolution ratio, can be used not only to mathematically determine the instability conditions of NAPL dissolution fronts in supercritical dissolution systems, but also to computationally simplify the numerical algorithm in the related computational simulation (Zhao et al. 2010a), it is kept as a parameter for the normalization of time.

Although the NAPL dissolution ratio (i.e. ε) is equal to C_e/ρ_n and ρ_n is a constant, it is the NAPL dissolution ratio (i.e. ε), rather than the equilibrium concentration (i.e. C_e) alone, that can have a significant effect on both the propagation velocity and the morphological shape of a NAPL dissolution front. To explain this point, it is necessary to briefly review the propagation mechanism of a NAPL dissolution front in the fluid-saturated porous medium as follows. In a NAPL dissolution system, the residual NAPL is resided in the form of disconnected ganglia or blobs within the pores of the porous medium. The amount of the residual NAPL per unit pore volume of the porous medium is equal to the product of the residual saturation (i.e. S_n) and the density of the residual NAPL. To remove the residual NAPL from the porous medium, the fluid (e.g. water) is injected, with a given velocity, into the region where the residual NAPL resides. Since the equilibrium concentration (i.e. C_e) represents the maximum limit that the residual NAPL can be dissolved in the fluid, the residual NAPL amount carried by the flowing fluid when it passes the NAPL dissolution front is therefore limited. After the previous flowing fluid passes the NAPL dissolution front, the fresh flowing fluid reaches the NAPL dissolution front so that some of the undissolved residual NAPL at the NAPL dissolution front can be dissolved into the fresh flowing fluid and carried away from the NAPL dissolution front by the fresh flowing fluid. Such processes continue until all the residual NAPL at the NAPL dissolution front is completely dissolved and carried away by the flowing fluid. At this stage, the NAPL dissolution front will propagate forwards. This means that in a NAPL dissolution system, there are two velocities (or speeds): one is the flowing fluid velocity, another is the propagation velocity of the NAPL dissolution front. Generally, the propagation velocity of a NAPL dissolution front is slower than that of the flowing fluid. According to the above analysis, it is recognized that the propagation velocity of a NAPL dissolution front depends directly on the NAPL dissolution ratio (i.e. ε), which can be used to determine how much flowing fluid should pass the NAPL dissolution front to enable it to propagate, rather than the equilibrium concentration (i.e. C_e) alone.

The boundary conditions can be also expressed in the following dimensionless form:

$$\bar{C} = 0, \quad S_n = 0, \quad \frac{\partial \bar{p}_a}{\partial \bar{x}} = \bar{p}'_{axf0} \quad (x = 0), \tag{10.21}$$

$$\bar{p}_a = \bar{p}_{a0} \quad (\bar{x} = \bar{L}_x), \tag{10.22}$$

$$\frac{\partial \bar{C}}{\partial \bar{n}} = 0, \quad \frac{\partial \bar{p}_a}{\partial \bar{n}} = 0 \quad \text{(at two lateral boundaries)}, \tag{10.23}$$

where $\bar{p}'_{axf0}$ is the dimensionless pressure gradient of the aqueous phase on the upstream boundary; $\bar{p}_{a0}$ is the dimensionless pressure of the aqueous phase on the downstream boundary.

Similarly, the initial conditions of the problem can be rewritten in a dimensionless form as follows:

$$\bar{C} = 1, \quad S_n = S_{n0} \quad (0 < \bar{x} \le \bar{L}_x). \tag{10.24}$$

Note that Eq. (10.24) assumes a uniform distribution field of residual NAPL saturation, which may be achieved either in the experimental specimen (of trapezoidal shape) on the laboratory scale or in large NAPL spill sites where the migration of the spilled NAPL has ceased in the system (Imhoff et al. 2003b; Gerhard et al. 2007).

To solve the dimensionless governing equations of a NAPL dissolution problem in a two-dimensional fluid-saturated porous medium, a numerical procedure consisting of a combination of the finite difference and finite element methods (Zienkiewicz 1977; Zhao et al. 2009) has been proposed in a previous study (Zhao et al. 2010a). In the proposed numerical procedure, the finite difference method is used to discretize time, while the finite element method is used to discretize space. To ensure the correctness and accuracy of the resulting numerical simulations, the proposed numerical procedure has been verified by some benchmark problems for which analytical solutions are available for comparison. Since the dimensionless governing equations of a NAPL dissolution system are highly nonlinear, the segregated algorithm, in which Eqs. (10.13)–(10.15) are solved separately and iteratively in a sequential manner, was used to derive the formulation of the proposed numerical procedure (Zhao et al. 2011). For the sake of completeness of this chapter, only the final discretized equations of the NAPL dissolution system are given below. If readers are interested in the detailed derivation processes of these equations, please refer to Chap. 9 or a previous publication (Zhao et al. 2011).

By following the numerical procedure used in Chap. 9 or a previous study (Zhao et al. 2011), the dimensionless governing equations (i.e. Eqs. (10.13)–(10.15)) of a NAPL dissolution system under the condition of $\varepsilon \ll 1$ can be expressed as follows:

$$\left[\frac{1}{\Delta\bar{t}}(S_n)_{\bar{t}}^{1-\beta_1} + \beta_1(1 - \bar{C}_{\bar{t}+\Delta\bar{t}})\right](\Delta S_n)_{\bar{t}+\Delta\bar{t}} = -(S_n)_{\bar{t}}(1 - \bar{C}_{\bar{t}+\Delta\bar{t}}), \tag{10.25}$$

$$\begin{aligned}\left\{\frac{\varepsilon}{\Delta\bar{t}}[1 - (S_n)_{\bar{t}+\Delta\bar{t}}] + (S_n)_{\bar{t}+\Delta\bar{t}}^{\beta_1}\right\}\bar{C}_{\bar{t}+\Delta\bar{t}} - \nabla\cdot[(1 - (S_n)_{\bar{t}+\Delta\bar{t}})(\bar{\mathbf{D}}_{\mathbf{h}})_{\bar{t}+\Delta\bar{t}}]\cdot\nabla\bar{C}_{\bar{t}+\Delta\bar{t}} \\ - \nabla(\bar{p}_a)_{\bar{t}+\Delta\bar{t}}\cdot[\bar{\psi}((S_n)_{\bar{t}+\Delta\bar{t}})\nabla\bar{C}_{\bar{t}+\Delta\bar{t}}] = \frac{\varepsilon}{\Delta\bar{t}}[1 - (S_n)_{\bar{t}+\Delta\bar{t}}]\bar{C}_{\bar{t}} + (S_n)_{\bar{t}+\Delta\bar{t}}^{\beta_1},\end{aligned} \tag{10.26}$$

$$\nabla\cdot\left[\bar{\psi}((S_n)_{\bar{t}+\Delta\bar{t}})\nabla(\bar{p}_a)_{\bar{t}+\Delta\bar{t}}\right] = \varepsilon\left(1 - \frac{1}{\bar{\rho}_a}\right)(S_n)_{\bar{t}+\Delta\bar{t}}^{\beta_1}(1 - \bar{C}_{\bar{t}+\Delta\bar{t}}), \tag{10.27}$$

where $(S_n)_{\bar{t}}$ and $(S_n)_{\bar{t}+\Delta\bar{t}}$ are the saturations of the NAPL at the previous and current time-steps; $(\Delta S_n)_{\bar{t}+\Delta\bar{t}}$ is the saturation increment of the NAPL at the current time-step; $\bar{C}_{\bar{t}}$ and $\bar{C}_{\bar{t}+\Delta\bar{t}}$ are the dimensionless concentrations of the NAPL at the previous and current time-steps respectively; $(\bar{p}_a)_{\bar{t}+\Delta\bar{t}}$ is the dimensionless pressure of the aqueous phase at the current time-step; $\Delta\bar{t}$ is the dimensionless time increment at the current time-step.

It should be pointed out that the main advantages in using dimensionless variables and governing equations are as follows (Zhao et al. 2008c): (1) solutions for dimensionless variables describe the behaviour of a family of problems rather than that of a particular problem. This makes the solutions more generally applicable. (2) Dimensionless variables can be used to measure the relative importance of various terms in governing equations, so that the dominant physical phenomenon can be identified for the problem. This provides a clear focus for the effective and efficient modelling of the problem. (3) Dimensionless equations can result in a significant reduction in the large differences between orders of magnitude for some terms in the corresponding dimensional equations, just like the partial differential equations that are used to describe the NAPL dissolution instability problem in this chapter. This generally makes the numerical solution more accurate and stable. For these reasons, dimensionless variables and governing equations are used in the following computational simulations. Nevertheless, the obtained dimensionless solutions can be easily transferred, if necessary, into dimensional solutions, as demonstrated in a previous study (Zhao et al. 2008c). As a result, the parametric study of a NAPL dissolution system can be directly carried out through the related dimensionless solutions. For instance, from the Zhao number (Zhao et al. 2010c), we can immediately know that an increase in the Darcy velocity of the dissolved region (i.e. V_{axf}) will cause an increase in the Zhao number of the NAPL dissolution system, so that the NAPL dissolution system becomes more unstable. On the other hand, an increase in the NAPL dissolution rate (i.e. β_0) will cause a decrease in the Zhao number of the NAPL dissolution system, so that the NAPL dissolution system becomes more stable.

10.2 Effects of Domain Shapes on the Morphological Evolution of NAPL Dissolution Fronts in Supercritical Systems

As mentioned previously, trapezoidal domains can be used to appropriately represent some fundamental flow characteristics that occur in irregular domains. For this reason, different trapezoidal computational domains are considered to investigate the effects of domain shapes on the morphological evolution of NAPL dissolution fronts in two-dimensional fluid-saturated porous media of supercritical Zhao numbers (Zhao et al. 2010c). Figure 10.1 shows the geometrical and boundary conditions for the NAPL dissolution problem of a typical trapezoidal domain, which is geometrically symmetrical to the x axis. Since the geometrical shape of such a trapezoidal computational domain can be represented by the divergent angle (i.e. θ) between a horizontal line and a lateral boundary, this angle is defined as a geometrical parameter in the corresponding computations. Note that if θ is equal to zero, then the trapezoidal shape degenerates to a rectangular one.

Based on the previous experimental measurement results that were calibrated by several laboratory tests (Imhoff and Miller 1996; Imhoff et al. 1996; Miller et al. 1998), the following parameters are selectively used in the corresponding computational models: the initial saturation (S_{n0}) of the NAPL (i.e. trichloroethylene (TCE)) is 0.2; the irreducible saturation (S_{ai}) of the aqueous phase fluid is 0.15; the dimensionless longitudinal and transverse dispersivities ($\bar{\alpha}_L$ and $\bar{\alpha}_T$) are 0.2 and 0.02, respectively; the ratio (ε) of the equilibrium concentration of the NAPL species in the aqueous phase fluid to the density of the NAPL itself is 0.001; the density ratio ($\bar{\rho}_a$) of the aqueous phase fluid to the NAPL is 1.0/1.46; the value of β_1 is 0.87. Since the dimensionless governing equations (i.e. Eqs. (10.13)–(10.15)) are used in the numerical simulation, we do not need to use specific values of the quantities such as β_0, medium porosity and tortuosity, molecular diffusivity of the NAPL and the Darcy flux at the inlet in this investigation. These quantities are represented by a comprehensive dimensionless number, known as the Zhao number (Zhao et al. 2012), in the corresponding numerical simulation.

To simulate the propagation of NAPL dissolution fronts appropriately, the finite element size has been varied to ensure that the numerical dispersion does not affect the numerical simulation results in a rectangular domain, for which analytical solutions are available for comparison with the numerical solution. Through the mesh size sensitivity analysis, it is confirmed that as long as the finite element size satisfies the mesh Peclet criterion (Daus et al. 1985), the numerical dispersion can be minimized in the computational simulation. As a result, the whole computational domain is simulated by 120,000 four-node quadrilateral elements of 120,701 nodal points in total. For the purpose of investigating the instability of a NAPL dissolution system, it is common practice to perturb the homogeneous distribution field of the initial NAPL saturation (S_{n0}) with a small amount (Zhao et al. 2010c). If the NAPL dissolution system is in a stable state, then such a small perturbation does not affect the propagation behaviour of the NAPL dissolution front in the homogeneous distribution field of the initial NAPL saturation (S_{n0}), so that an initial planar NAPL dissolution-front remains the planar shape. However, If the NAPL dissolution system is in an unstable state, then such a small perturbation can significantly affect the propagation behaviour of the NAPL dissolution front in the homogeneous distribution field of the initial NAPL saturation (S_{n0}), so that an initial planar NAPL dissolution-front can evolve into different irregular shapes. For this reason, the initial residual saturation field of the NAPL is randomly perturbed by a small amount of 1 % of the originally-input saturation of the NAPL (i.e. $S_{n0} = 0.2$) before running the computational model. This means that the resulting initial residual saturation is of a random distribution, which has a mean value of the homogeneous residual saturation (i.e. $S_{n0} = 0.2$) and a variation of 0.002 (i.e. 1 % of $S_{n0} = 0.2$) in the whole computational domain. Thus, the initial homogeneous distribution field of the NAPL saturation (S_{n0}) is replaced and reassigned by a slightly perturbed non-homogeneous distribution field of the NAPL saturation before running the computational model. Using the characteristic length (i.e. L^*) as the length scaling factor, the dimensionless length (i.e. $\bar{L}_x = L_x/L^*$) of the computational domain is 6284 in the $\bar{x}$ direction, while the dimensionless length (i.e. $\bar{L}_{y1} = L_{y1}/L^*$) of the left boundary is fixed to 3142, 2734, 2329 and 2046 in the $\bar{y}$ direction, respectively, when four different computational domains, namely $\theta = 0°$, 4°,11° and 20°, are used to investigate the effects of domain shapes on the morphological evolution of NAPL dissolution fronts in supercritical systems. The Zhao number used for all the four computational models (at the entrance of the flow) is 1.0, while the dimensionless time-step length is 3.2. From the previous theoretical study (Zhao et al. 2010c), the critical Zhao number is 3.34×10^{-3} in the case of $\theta = 0°$. As the Zhao number is much greater than its critical value, it is expected that the NAPL dissolution system under consideration is in a supercritical state. Although different values of the Zhao number can have significant effects on the morphological evolution of NAPL dissolution fronts, this issue is not considered here because it has been addressed in a rectangular domain (Zhao et al. 2011) and the main focus of this chapter is to investigate the effect of domain shapes on the morphological evolution of NAPL dissolution fronts in the fluid-saturated porous medium.

Figures 10.2, 10.3, 10.4 and 10.5 show the effects of domain shapes on the morphological evolution of NAPL dissolution fronts in the fluid-saturated porous medium at four different time instants, namely $\bar{t} = 160$, 640, 1120 and 1600,

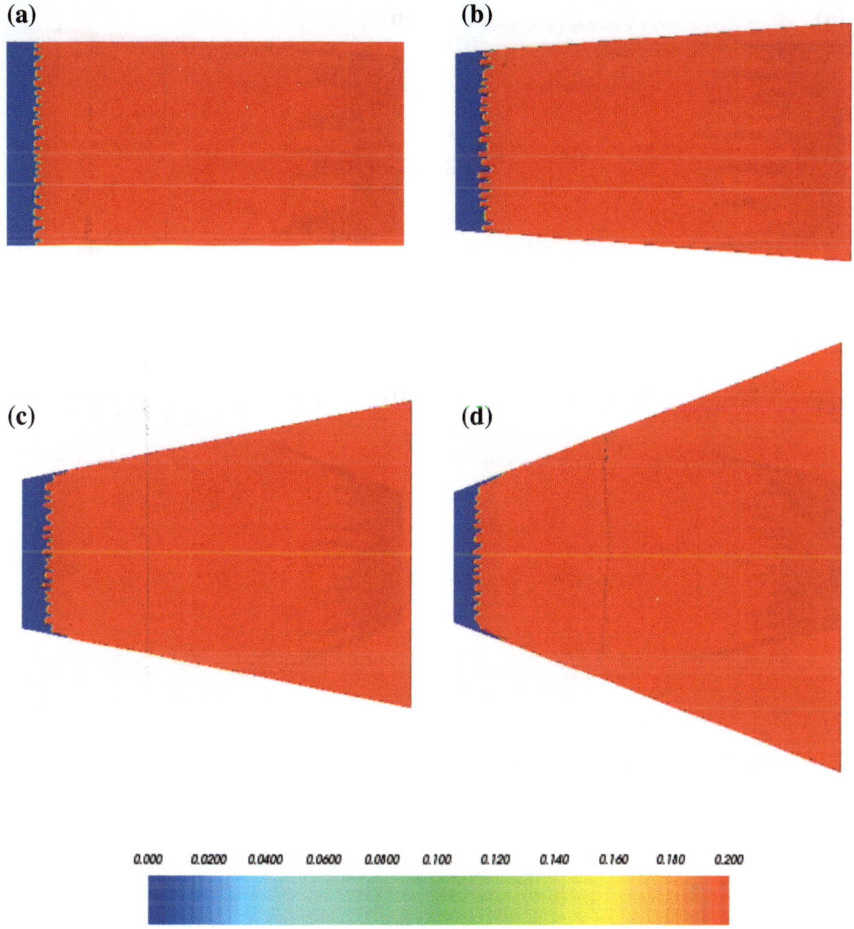

Fig. 10.2 Effects of domain shapes on the evolution of NAPL dissolution fronts represented by residual saturation in the fluid-saturated porous medium ($\bar{t} = 160$): **a** $\theta = 0$; **b** $\theta = 4°$; **c** $\theta = 11°$; **d** $\theta = 20°$

respectively. In these figures, the residual saturation of a NAPL is used to represent the NAPL dissolution front. It is observed that domain shapes can significantly affect not only the propagating speed of a NAPL dissolution front, but the morphological evolution pattern of the NAPL dissolution front as well. At the early stage of the computational simulation, the NAPL dissolution front evolves from the injected planar shape at the left boundary of the computational model into an irregular shape. For a planar NAPL dissolution-front propagating in an infinite domain, the previous theoretical analysis demonstrated that the propagating speed of the planar NAPL dissolution-front is directly proportional to the Darcy velocity of the aqueous phase fluid within the fluid-saturated porous medium. In the case of a trapezoidal domain, an increase in the divergent angle (i.e. θ) of the trapezoidal

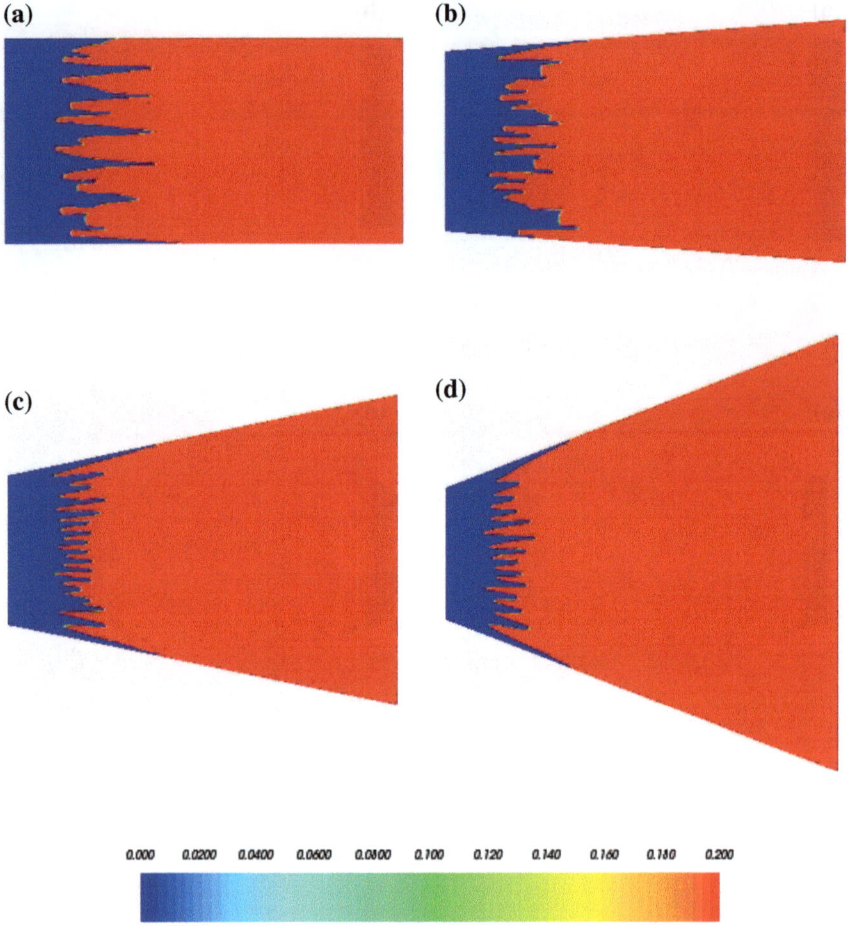

Fig. 10.3 Effects of domain shapes on the evolution of NAPL dissolution fronts represented by residual saturation in the fluid-saturated porous medium ($\bar{t} = 640$): **a** $\theta = 0$; **b** $\theta = 4°$; **c** $\theta = 11°$; **d** $\theta = 20°$

domain can lead to an increase in the area (or the length in the two-dimensional case) of a vertical cross-section that is perpendicular to the inflow direction of the aqueous phase fluid within the trapezoidal domain. From the mass conservation point of view, such an increase in the divergent angle of the trapezoidal domain can cause a decrease in the Darcy velocity of the aqueous phase fluid on the area of the vertical cross-section within the trapezoidal domain. Thus, with an increase in the divergent angle of a trapezoidal domain, the propagating speed of a planar NAPL dissolution-front decreases accordingly within the fluid-saturated porous medium. Since this theoretical prediction has good agreement with the numerical results (as shown in Fig. 10.2) at the early stage of the computational simulation, it has demonstrated that the computational model used in this investigation can

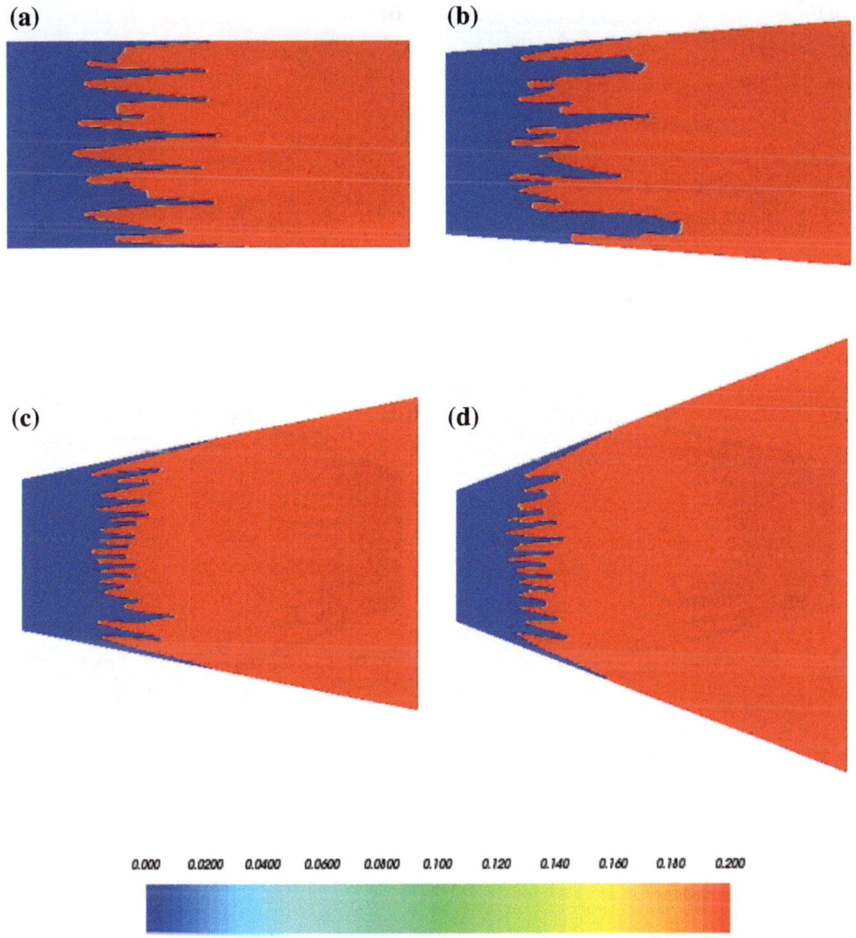

Fig. 10.4 Effects of domain shapes on the evolution of NAPL dissolution fronts represented by residual saturation in the fluid-saturated porous medium ($\bar{t} = 1120$): **a** $\theta = 0$; **b** $\theta = 4°$; **c** $\theta = 11°$; **d** $\theta = 20°$

produce reliable numerical results for simulating the evolution of a planar NAPL dissolution-front in the fluid-saturated porous medium of a trapezoidal shape.

However, it is very difficult, if not impossible, to predict theoretically the propagating speed of an irregular NAPL dissolution-front within the fluid-saturated porous medium of supercritical Zhao numbers. In this situation, the average Darcy velocity on the area of a vertical cross-section can be used to investigate the effect of domain shapes on the average propagating speed of the irregular NAPL dissolution-front. Since the aqueous phase fluid should be conservative on a vertical cross-section perpendicular to the inflow direction of the aqueous phase fluid within the trapezoidal domain, the average Darcy velocity on this vertical cross-section can be expressed in the weighted form as follows:

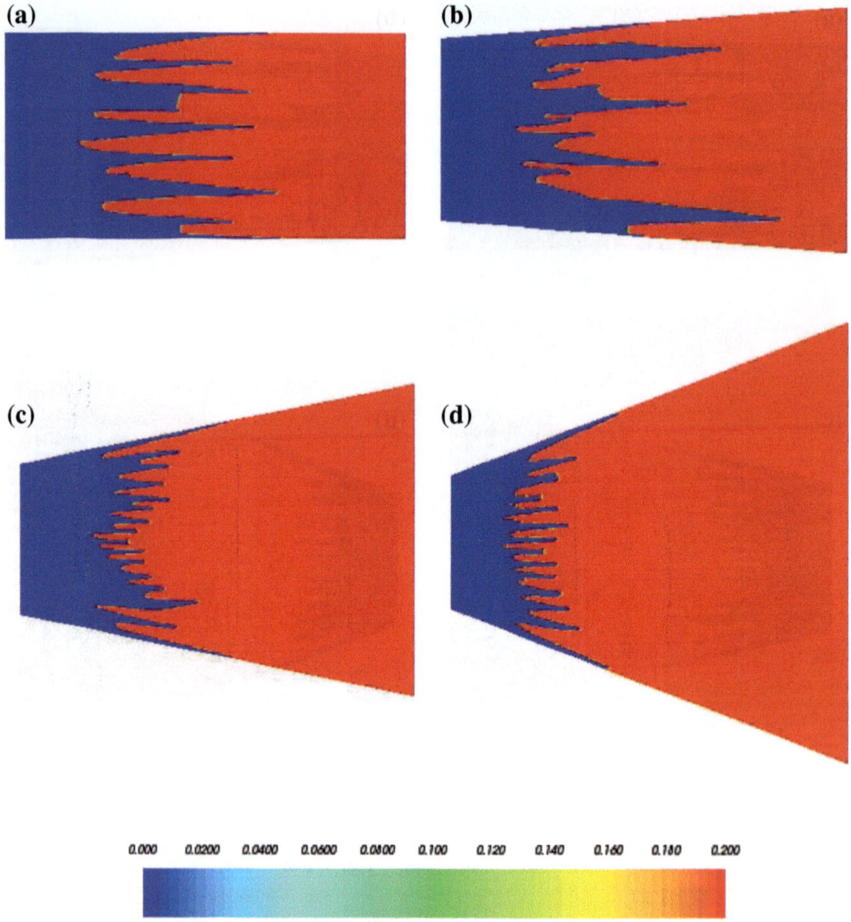

Fig. 10.5 Effects of domain shapes on the evolution of NAPL dissolution fronts represented by residual saturation in the fluid-saturated porous medium ($\bar{t} = 1600$): **a** $\theta = 0$; **b** $\theta = 4°$; **c** $\theta = 11°$; **d** $\theta = 20°$

$$v_{Darcy} = \frac{\int_{-\frac{L_y}{2}}^{+\frac{L_y}{2}} v_{Darcy} dy}{L_y}, \tag{10.28}$$

where L_y is the length of the vertical cross-section; v_{Darcy} is the Darcy velocity at a point of the vertical cross-section; $\bar{v}_{Darcy}$ is the average Darcy velocity on the vertical cross-section.

Based on the average Darcy velocity concept, it is possible to examine how the domain shape of a computational model affects the average propagating speed of an irregular NAPL dissolution-front within the fluid-saturated porous medium. For the given location of a vertical cross-section in a trapezoidal domain, an increase in the divergent angle of the trapezoidal domain leads to an increase in the length of the vertical cross-section parallel to the y direction, so that there is a decrease in the average Darcy velocity on this vertical cross-section in the trapezoidal domain. Generally, the average propagating speed of an irregular NAPL dissolution-front decreases gradually as a result of an increase in the divergent angle of the trapezoidal domain. This phenomenon can be observed from the numerical simulation results shown in Figs. 10.3, 10.4 and 10.5. However, in the case of $\theta = 4°$, the strong mergence of several irregular fingering fronts takes place within the computational domain. Due to this mergence, the fluid flow in the merged fingers of wider flow channels becomes much stronger than that in the unmerged small fingers, so that the tip of the dissolution front represented by the strongly merged finger in the case of $\theta = 4°$ has propagated faster than that represented by the weakly merged finger in the rectangular case.

In terms of the morphological evolution pattern of an irregular NAPL dissolution front, the domain shape of a computational model can affect the total numbers of irregular fingers in the fluid-saturated porous medium of supercritical Zhao numbers. For a given position at the x axis, the trapezoidal domain of a large divergent angle (i.e. θ) can provide more space in the vertical direction, compared with the trapezoidal domain of a small divergent angle. As a result, the total number of irregular fingers obtained from the trapezoidal domain of a large divergent angle is usually greater than that obtained from the trapezoidal domain of a small divergent angle. For example, in the case of $\bar{t} = 1600$, the total number of irregular fingers in the rectangular domain is equal to 8, while it is equal to 14 in the trapezoidal domain with a divergent angle of 20°. Nevertheless, as shown in Figs. 10.3, 10.4 and 10.5, the irregular fingers in the trapezoidal domain of a small divergent angle can grow much wider than those in the trapezoidal domain of a large divergent angle.

Figure 10.6 shows the effects of domain shapes on the distributions of the NAPL dimensionless concentration in the fluid-saturated porous medium at $\bar{t} = 1600$. Since the small perturbation grows with time for a supercritical NAPL dissolution system, the initial planar front of the NAPL dimensionless concentration evolves gradually into an irregular fingering one. With an increase in the dimensionless time, the amplitude of the irregular fingering NAPL concentration front increases significantly, indicating that the NAPL concentration front is morphologically unstable during its propagation within the computational domain. Compared with the computational simulation results shown in Fig. 10.5, both the NAPL residual saturation and the NAPL dimensionless concentration have similar propagation fronts because the dimensionless concentration of the NAPL approaches zero when the NAPL is completely dissolved in the upstream direction of the NAPL dissolution front. Once the NAPL is completely dissolved, the NAPL residual saturation also becomes zero in the upstream direction of the NAPL dissolution front.

It is interesting to examine NAPL dissolution fingering in a convergent trapezoidal domain where the divergent angle (as defined in Fig. 10.1) has a negative

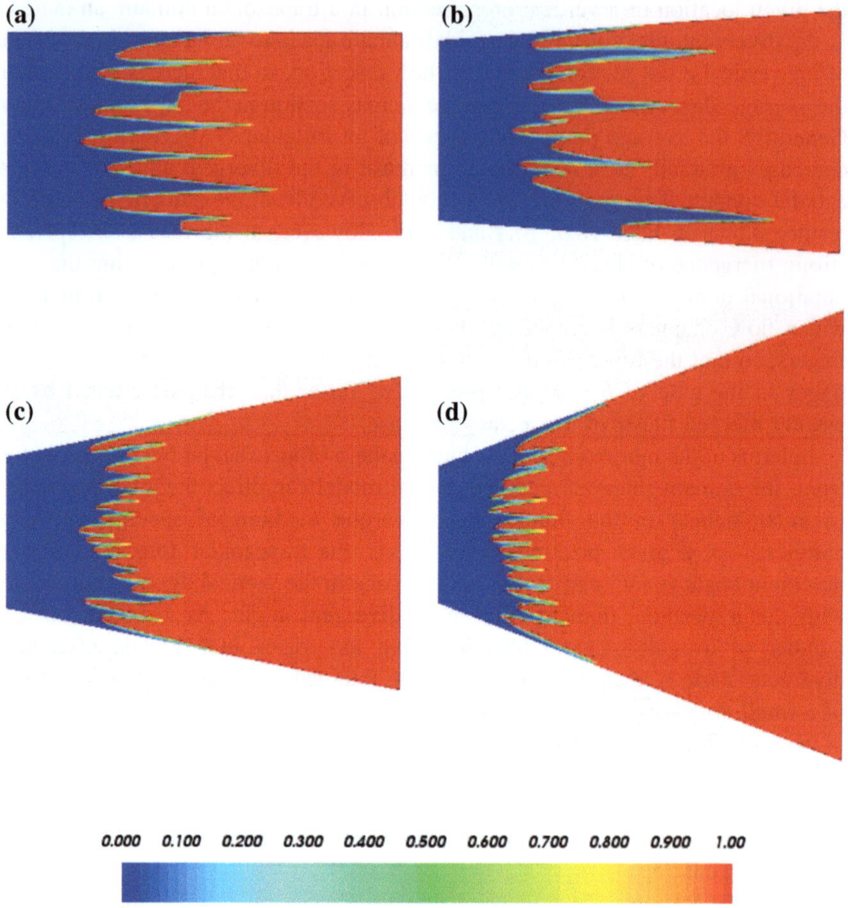

Fig. 10.6 Effects of domain shapes on the distributions of the NAPL dimensionless concentration in the fluid-saturated porous medium ($\bar{t} = 1600$): **a** $\theta = 0$; **b** $\theta = 4°$; **c** $\theta = 11°$; **d** $\theta = 20°$

value for θ. Since convergent trapezoidal domains can result in Darcy velocities increasing with an increase in the distance from the entrance of the injected flow, they may produce some interesting effects on NAPL dissolution fingering. For this purpose, a negative value of θ (i.e. $\theta = -20°$) is used to run the corresponding computational simulation of NAPL dissolution fingering in a convergent trapezoidal domain. Figure 10.7 shows the effects of the convergent domain shape (in the case of $\theta = -20°$) on the evolution of NAPL dissolution fronts (represented by residual saturation) in the fluid-saturated porous medium at four different time instants, namely $\bar{t} = 160$, 640, 1120 and 1440, respectively. Compared with the computational simulation results in the divergent trapezoidal domain of $\theta = 20°$ (see Figs. 10.2, 10.3, 10.4 and 10.5), NAPL fingers grow much faster in the convergent trapezoidal domain of $\theta = -20°$. For example, in the case of $\bar{t} = 1440$,

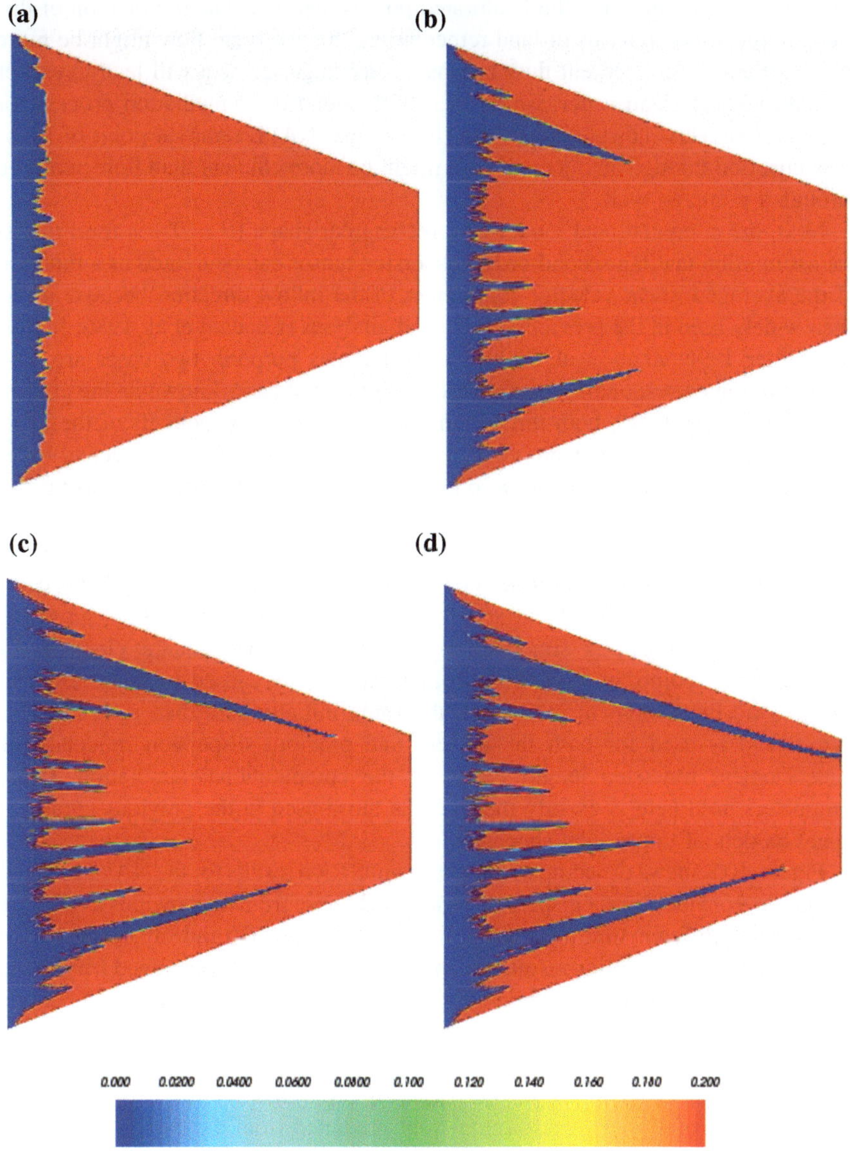

Fig. 10.7 Effects of convergent domain shape on the evolution of NAPL dissolution fronts represented by residual saturation in the fluid-saturated porous medium ($\theta = -20°$): **a** $\bar{t} = 160$; **b** $\bar{t} = 640$; **c** $\bar{t} = 1120$; **d** $\bar{t} = 1440$

one major finger has penetrated through the whole domain in the horizontal direction of the convergent trapezoidal domain in the case of $\theta = -20°$, while it only propagates about one third of the whole domain in the horizontal direction of the divergent trapezoidal domain in the case of $\theta = 20°$. This further demonstrates that domain shapes can have significant effects on the formation of NAPL dissolution fingering in the fluid-saturated porous medium. The implication of this recognition is that in terms of land remediation, the divergent flow might be more efficient than the convergent flow because faster finger growth will lead to earlier breakthrough of clean water, which then will extend the remediation process for some period. This indicates that from the residual NAPL removal point of view, flow injection through an injection well might be more efficient than flow pumping through a pumping well.

Next, the computational model of a rectangular shape ($\theta = 0°$) is used to test the assumption that the second-order dispersion tensor can be treated as a function of the averaged linear velocity component in the inflow direction because it has been widely used in the previous theoretical analyses (Chadam et al. 1986; Imhoff and Miller 1996; Zhao et al. 2008a, b, c). For this purpose, two cases are considered in the corresponding computations. In the first case (known as the current dispersion model here), both the horizontal and vertical components of the averaged linear velocity vector are considered in the evaluation of the second-order dispersion tensor, while in the second case (known as the previous dispersion model in this investigation), only the horizontal component of the averaged linear velocity vector is considered in the evaluation of the second-order dispersion tensor. In this situation, the dimensionless lengths (i.e. $\bar{L}_x$ and $\bar{L}_{y1}$) of the computational domain are 4,713 and 3,142 in the $\bar{x}$ and $\bar{y}$ directions, while the whole computational domain is simulated by 99,301 four-node square elements of 100,000 nodal points in total. To eliminate any errors caused by finite element meshes, the finite element mesh with the same initially-perturbed residual saturation field is used for both the current and previous dispersion models. The dimensionless time-step length is 1.9 in the computational simulation. Other parameters used here is exactly the same as those used in the previous computational models of trapezoidal shapes.

Figure 10.8 shows the comparison of the simulation results for the NAPL residual saturation from the current dispersion model with those from the previous dispersion model at four different time instants, namely $\bar{t} = 190, 380, 570$ and 950, respectively. In this figure, the simulation results shown in the left column are obtained from using the current dispersion model, while the simulation results shown in the right column are obtained from using the previous dispersion model. Even though the fluid flow is, strictly speaking, no longer one-dimensional at the NAPL dissolution front (see Fig. 10.9), it can be clearly observed (from Fig. 10.8) that the morphological evolution patterns of a NAPL dissolution front predicted from the current dispersion model is almost the same as those predicted from the previous dispersion model. This demonstrates that for the NAPL dissolution-front instability problem of a rectangular domain, the previous dispersion model can produce accurate theoretical and numerical solutions for the prediction of morphological evolution patterns of NAPL dissolution fronts in fluid-saturated porous media of supercritical Zhao numbers. Except for remarkable savings in computational efforts, the major benefit of using the previous dispersion model is that it enables the theoretical analysis of this kind of instability problems to become possible (Zhao et al. 2010a).

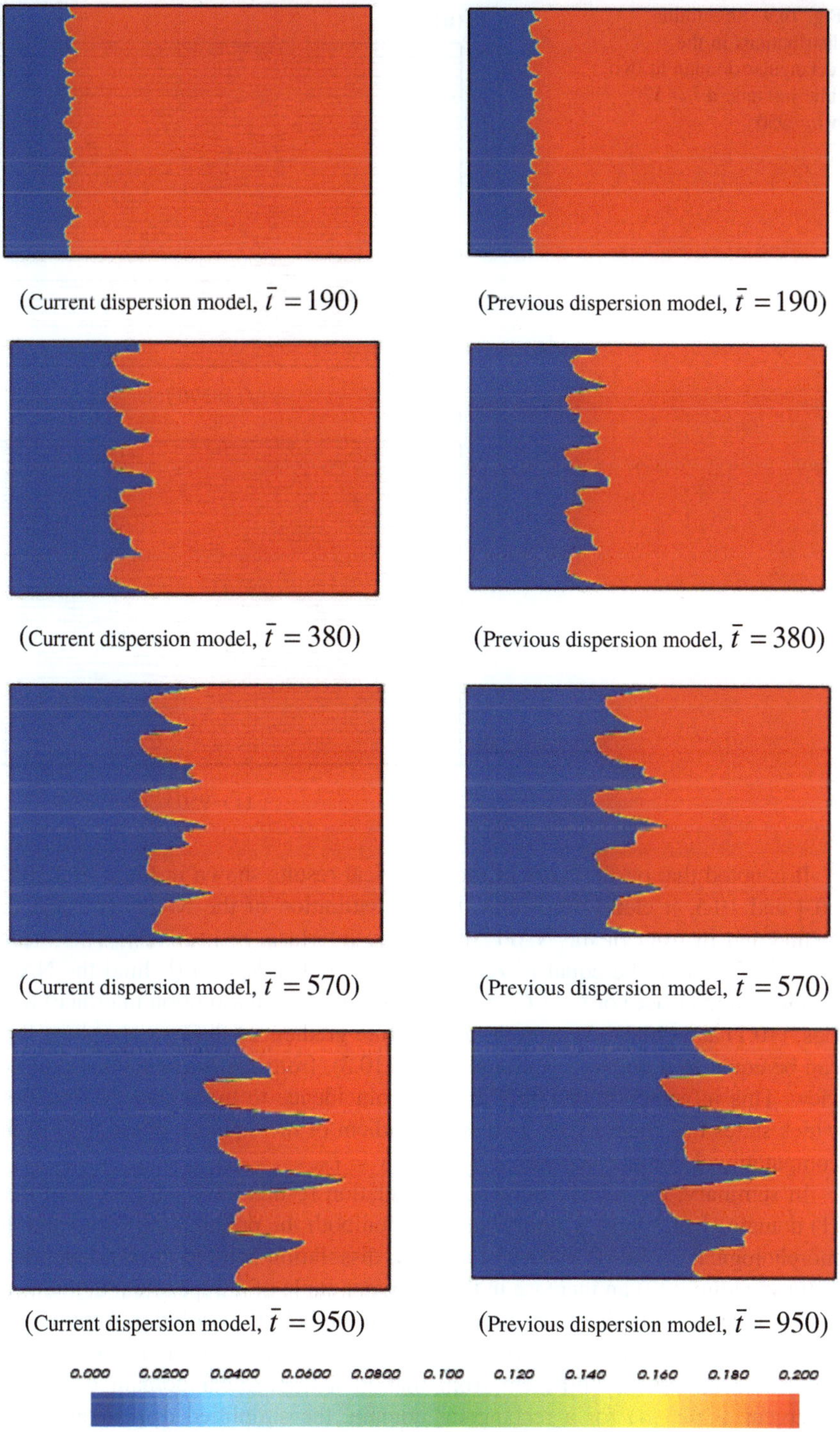

Fig. 10.8 Comparison of the simulation results from the current dispersion model with those from the previous dispersion model (NAPL residual saturation)

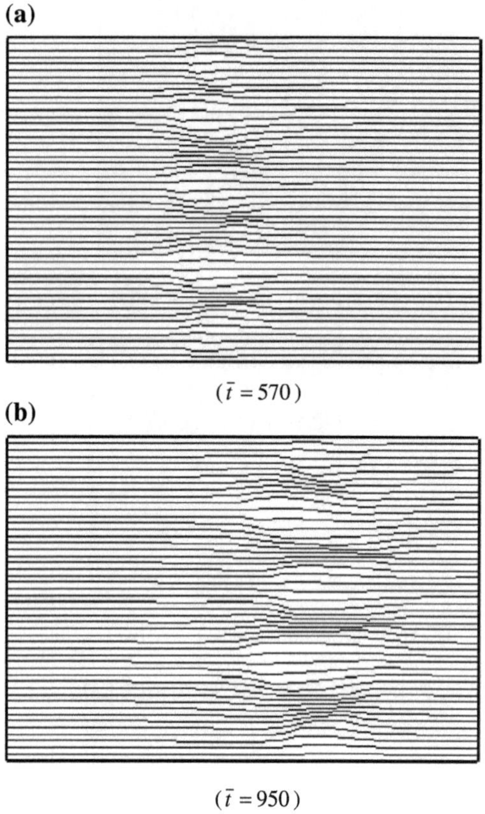

Fig. 10.9 Streamline distributions in the rectangular domain at two time instants: **a** $\bar{t} = 570$; **b** $\bar{t} = 950$

It is noted that on the basis of the numerical results shown in Figs. 10.2, 10.3, 10.4 and 10.5, it clearly indicates that the saturation of the NAPL is 0 and 0.2 behind and in front of the NAPL dissolution interface, respectively. This means that $\partial S_n/\partial t$ should be equal to zero, leading to either $S_n = 0$ (behind the NAPL dissolution interface) or $C = C_{eq}$ (in front of the NAPL dissolution interface) from Eqs. (10.1) and (10.3). As a result, the pressure gradient of the aqueous phase fluid can be equal to a nonzero constant in Eq. (10.3), from the mathematical point of view. This inference is certainly in good coincidence to the boundary condition, which states that there exists a pressure gradient of the aqueous phase fluid in the computational domain between $x = 0$ and $x = L_x$.

In summary, the related numerical simulation results have demonstrated that: (1) domain shapes have a significant effect on both the propagating speed and the morphological evolution pattern of a NAPL dissolution-front in the fluid-saturated porous medium; (2) an increase in the divergent angle of a trapezoidal domain can lead to a decrease in the propagating speed of the NAPL dissolution front; (3) the morphological evolution pattern of the NAPL dissolution-front in a rectangular domain is remarkably different from that in a trapezoidal domain of a large divergent angle; (4) for a rectangular domain, the simplified dispersion model, which is commonly used in the theoretical analysis and numerical simulation, is valid for solving NAPL dissolution-front instability problems in fluid-saturated porous media; and (5) compared with diverging flow (when the trapezoidal domain is inclined outward), converging flow (when the trapezoidal domain is inclined inward) can enhance the growth of NAPL fingers, indicating that pump-and-treat systems by extracting contaminated groundwater might enhance NAPL dissolution fingering and lead to less uniform dissolution fronts.

10.3 Effects of Mesh Discretization Error on the Morphological Evolution of NAPL Dissolution Fronts in Supercritical Systems

Compared with the simulation of a rectangular domain, the simulation of a trapezoidal domain may involve the following two effects: (1) since the two neighbour boundaries are not perpendicular each other, the inflow perpendicular to the entrance (i.e. one side of the trapezoidal domain) may change the flow direction just when it passes the entrance of the domain. This phenomenon can be called the entrance corner effect (or just the corner effect for short). (2) Since irregular quadrilateral elements, in which two neighbour sides are not perpendicular each other, must be used to simulate the trapezoidal domain, mesh discretization including mesh inclination (i.e. grid orientation) is inevitable in the computational model. This phenomenon can be called the mesh discretization effect. The main purpose of this section is to investigate both the corner effect and mesh discretization effect on the morphological evolution of NAPL dissolution fronts in trapezoidal domains consisting of fluid-saturated porous media through conducting detailed theoretical analysis and running several computational models.

For a rectangular domain, the dimensionless pressure gradient, $\bar{p}'_{axf}$, of the aqueous phase liquid on a vertical cross-section is constant in the NAPL completely dissolved region, while for a trapezoidal domain shown in Fig. 10.10, it is a function of the location of the vertical cross-section in the NAPL completely dissolved region. From the mass conservation of the aqueous phase liquid, this function can be determined and expressed in the following form:

$$\bar{p}'_{axf} = \frac{\bar{L}_{y1}}{\bar{L}_{y1} + 2\bar{x}tg\theta}\bar{p}'_{axf0}, \tag{10.29}$$

where $\bar{p}'_{axf0}$ is the dimensionless pressure gradient of the aqueous phase on the upstream boundary; $\bar{x}$ is the location of the vertical cross-section in the NAPL completely dissolved region within the trapezoidal domain.

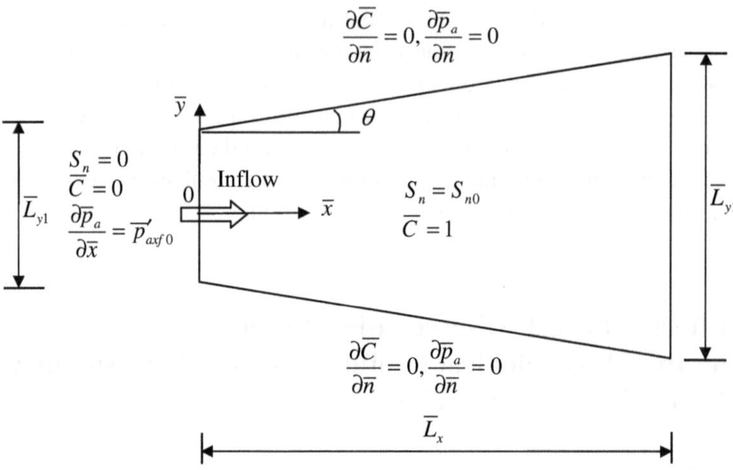

Fig. 10.10 Geometry and boundary conditions for the NAPL dissolution problem on the basis of dimensionless quantities

10.3.1 The Theoretical Basis of Mesh Discretization Error Estimation for NAPL Dissolution Problems

10.3.1.1 The Dynamic Behaviour of a NAPL Dissolution System

To facilitate the discussion of the mesh discretization error associated with the computational simulation results in the forthcoming subsections, it is necessary to briefly describe the dynamic behaviour of a NAPL dissolution system in a rectangular domain. Since the governing equations of a NAPL dissolution problem in a rectangular domain is exactly the same as those of a NAPL dissolution problem in a trapezoidal domain, the dynamic behaviour for both the rectangular domain and the trapezoidal domain should be similar so that the previous theoretical results from a rectangular domain may have some reference values when the mesh discretization error associated with the computational simulation results from a trapezoidal domain are discussed.

For the purpose of describing the dynamic behaviour of a NAPL dissolution system, the dimensionless Zhao number is defined as follows (Zhao et al. 2010c):

$$Zh = -\bar{p}'_{axf} = \frac{V_{axf}}{\sqrt{\phi \tau D_m}} \sqrt{\frac{1}{\beta_0}} = F_{advection} F_{dispersion} F_{dissolution}, \tag{10.30}$$

where V_{axf} is the Darcy velocity of the aqueous phase fluid after the NAPL is completely dissolved in the NAPL dissolution system; ϕ and τ are the porosity and tortuosity of the porous medium, respectively; D_m is the molecular diffusivity of the NAPL species in the aqueous phase; β_0 is the dissolution rate of the NAPL. It needs to be pointed out that in Eq. (10.30), $F_{advection} = V_{axf}$, representing the aqueous phase fluid advection in the NAPL dissolution system; $F_{dispersion} = 1/\sqrt{\phi\tau D_m}$, representing the NAPL solute diffusion/dispersion in the NAPL dissolution system; and $F_{dissolution} = \sqrt{1/\beta_0}$, representing the kinetics of the NAPL dissolution. Thus the Zhao number is a comprehensive dimensionless number to represent the three major controlling mechanisms (i.e. advection, dispersion and NAPL dissolution) simultaneously taking place in the NAPL dissolution system.

When the domain of a NAPL dissolution system is rectangular, then the critical Zhao number of the system has been mathematically derived in a previous study (Zhao et al. 2010c) as follows:

$$Zh_{critical} = \frac{\left(3 - \frac{k(S_{n0})}{k(S_{nf})}\right)\left(1 + \frac{k(S_{n0})}{k(S_{nf})}\right)}{2\left(1 - \frac{k(S_{n0})}{k(S_{nf})}\right)}, \tag{10.31}$$

where $Zh_{critical}$ is the critical Zhao number of the NAPL dissolution system of a rectangular shape; $k(S_{n0})$ is the saturation-dependent permeability of the porous medium to aqueous phase flow at $S_n = S_{n0}$; $k(S_{nf})$ is the saturation-dependent permeability of the porous medium to aqueous phase flow at $S_n = S_{nf}$; $S_{nf} = 0$ is the saturation of the NAPL after it is completely dissolved in the NAPL dissolution system.

By means of both the Zhao number and the critical Zhao number, the instability of NAPL dissolution fronts in fluid-saturated porous media of rectangular domains can be assessed. Note that if the domain of a NAPL dissolution system is of a rectangular shape, then the Darcy velocity of the aqueous phase fluid after the NAPL is completely dissolved (i.e. V_{axf}) is constant, while if the domain of an NAPL dissolution system is of a trapezoidal shape, then the Darcy velocity of the aqueous phase fluid after the NAPL is completely dissolved (i.e. V_{axf}) is no longer constant. In the latter case, the instability criterion derived from the rectangular shape can be applicable, provided that an appropriate Darcy velocity of the aqueous phase fluid is selected in the NAPL dissolved region.

It is interesting to note that the Zhao number can be expressed by a combination of the Peclet number and the Thiele modulus (Hong et al. 1999). If the effective diffusion/dispersion coefficient is defined as $D_{effective} = \phi\tau D_m$, then the Thiele modulus of the dissolution system can be expressed as $Tm = L^*\sqrt{\beta_0/D_{effective}}$, where L^* is the intrinsic characteristic length of the dissolution system. Similarly, the Peclet number of the dissolution system can be expressed as $Pe = (V_{axf}L^*)/D_{effective}$. Consequently, the Zhao number of the dissolution system can be expressed as follows:

$$Zh = \left(\frac{V_{axf}L^*}{D_{effective}}\right)\left(\frac{1}{L^*}\sqrt{\frac{D_{effective}}{\beta_0}}\right) = \frac{Pe}{Tm}. \tag{10.32}$$

This indicates that the Zhao number of a NAPL dissolution system is equal to the ratio of the Peclet number to the Thiele modulus of the dissolution system. Clearly, the Peclet number can be only used to describe the interaction between the advection and diffusion/dispersion processes, while the Thiele modulus can be only used to describe the interaction between the NAPL dissolution and diffusion/dispersion processes. On the contrary, the Zhao number can be used to simultaneously describe all three processes, namely the advection, NAPL dissolution and diffusion/dispersion processes, in a NAPL dissolution system. This is the main reason why the Zhao number needs to be used to describe the overall characteristic of a NAPL dissolution system.

From the linear stability theory, the NAPL dissolution front can be expressed as (Zhao et al. 2010c):

$$S(\xi, \bar{y}, \bar{t}) = \xi - \bar{\delta}_A \exp(\bar{\omega}\bar{t}) \cos(\bar{m}\bar{y}), \tag{10.33}$$

where $\bar{\omega}$ is the dimensionless growth rate; δ_A is the amplitude of an initial small perturbation; $\bar{t}$ is the dimensionless time after the perturbation is applied to the NAPL dissolution system; $\bar{m}$ is the dimensionless wavenumber of the perturbation; ξ is the coordinate of the moving coordinate system in the $\bar{x}$ direction and can be expressed as follows:

$$\xi = \bar{x} - \bar{v}_{front}\bar{t}, \quad \bar{v}_{front} = -\frac{\bar{p}'_{axf}}{S_{n0}} = \frac{Zh}{S_{n0}}. \tag{10.34}$$

Since Eq. (10.34) indicates that the initial planar NAPL dissolution-front is always represented by $\xi = 0$, the amplitude of the finger when a NAPL dissolution system is in an unstable state (in the case of $Zh > Zh_{critical}$), can be expressed as:

$$A_{Finger} = \delta_A e^{\bar{\omega}\bar{t}}, \tag{10.35}$$

where A_{Finger} is amplitude of the finger when the NAPL dissolution system is in an unstable state.

Equation (10.35) clearly indicates that for a given time, the greater the amplitude of an initial small perturbation, the greater the amplitude of the finger. Also, for a given small perturbation, the earlier the perturbation is applied to the NAPL dissolution system, the greater the amplitude of the finger.

10.3.1.2 The Propagation Characteristic of Mesh Discretization Error and the Concept of Equivalent Initial Perturbation

As demonstrated by the previous study (Zhao et al. 2010c), the basic characteristic of a NAPL dissolution-front instability problem in a fluid-saturated porous medium is that the NAPL dissolution front divides the whole problem domain into the following two sub-domains: a NAPL un-dissolved sub-domain (in the downstream direction of the NAPL dissolution front) and a NAPL completely-dissolved subdomain (in the upstream direction of the NAPL dissolution front). Thus, the instability of the NAPL dissolution system can only take place on the NAPL dissolution front. This is the main reason why the instability of the NAPL dissolution system is commonly called the NAPL dissolution-front instability problem. For this reason, the mesh discretization error associated with the NAPL dissolution-front instability problem can only propagate with the propagation of the NAPL dissolution front in the fluid-saturated porous medium.

To understand the propagation characteristic of the mesh discretization error associated with a NAPL dissolution-front instability problem, we consider a common boundary between several four-node rectangular elements (that are used to simulate a part of a rectangular domain) and assume that this common boundary contains N nodal points. For the sake of facilitating the analysis, we can assume that $N = 6$, as shown in Fig. 10.11. Suppose the distance between any two adjacent nodal points is equal, we can represent this distance by $\Delta\bar{x}$, so that the mesh discretization error associated with each nodal point is equal and can be expressed by δ_d. If the NAPL dissolution front is located at node 1 and the initial dimensionless time is considered as $\bar{t}_0 = 0$, then the dimensionless times can be expressed as $\bar{t}_1$, $\bar{t}_2$, $\bar{t}_3$, $\bar{t}_4$ and $\bar{t}_5$ when the NAPL dissolution front propagates from node 1 to nodes 2, 3, 4, 5, and 6 respectively. For a rectangular domain, the propagation speed (i.e. $\bar{v}_{front} = -\bar{p}'_{axf}\big/S_{n0} = Zh/S_{n0}$) of the NAPL dissolution front is constant (Zhao et al. 2010c), so that the time period when the NAPL dissolution front propagates between any two adjacent nodal points is also constant and can be expressed as $\Delta\bar{t}$. Thus, when the NAPL dissolution front propagates from node 1 to node 2, the mesh discretization error associated with $\Delta\bar{x}$ between nodes 1 and 2 grows and the resulting error at node 2 can be expressed as follows:

$$\delta_{d2} = \delta_d e^{\bar{\omega}\bar{t}_1} = \delta_d e^{\bar{\omega}\Delta\bar{t}}. \tag{10.36}$$

However, when the NAPL dissolution front propagates from node 2 to node 3, the total mesh discretization error associated with node 2 is equal to the sum of δ_{d2} and δ_d (i.e. the mesh discretization error associated with $\Delta\bar{x}$ between nodes 2 and 3).

$$\delta_{d2}^{total} = \delta_{d2} + \delta_d = \delta_d e^{\bar{\omega}\Delta\bar{t}} + \delta_d = \delta_d(1 + e^{\bar{\omega}\Delta\bar{t}}). \tag{10.37}$$

This means that when the NAPL dissolution front propagates from node 2 to node 3, the resulting error at node 3 can be expressed as follows:

$$\delta_{d3} = \delta_{d2}^{total} e^{\bar{\omega}\Delta\bar{t}} = \delta_d(1 + e^{\bar{\omega}\Delta\bar{t}})e^{\bar{\omega}\Delta\bar{t}} = \delta_d(e^{\bar{\omega}\Delta\bar{t}} + e^{2\bar{\omega}\Delta\bar{t}}). \tag{10.38}$$

Similarly, the resulting mesh discretization errors at nodes 4, 5 and 6 can be expressed in the following equations:

Fig. 10.11 Propagation characteristic of the mesh discretization error associated with a NAPL dissolution front propagating along the line containing 6 nodal points

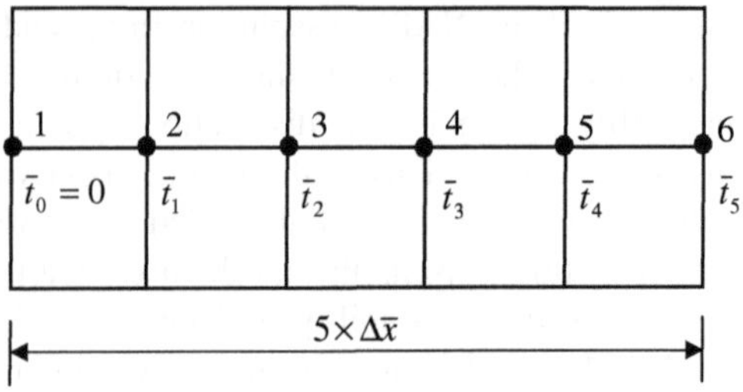

$$\delta_{d4} = \delta_d(e^{\bar{\omega}\Delta\bar{t}} + e^{2\bar{\omega}\Delta\bar{t}} + e^{3\bar{\omega}\Delta\bar{t}}), \tag{10.39}$$

$$\delta_{d5} = \delta_d(e^{\bar{\omega}\Delta\bar{t}} + e^{2\bar{\omega}\Delta\bar{t}} + e^{3\bar{\omega}\Delta\bar{t}} + e^{4\bar{\omega}\Delta\bar{t}}), \tag{10.40}$$

$$\delta_{d6} = \delta_d(e^{\bar{\omega}\Delta\bar{t}} + e^{2\bar{\omega}\Delta\bar{t}} + e^{3\bar{\omega}\Delta\bar{t}} + e^{4\bar{\omega}\Delta\bar{t}} + e^{5\bar{\omega}\Delta\bar{t}}). \tag{10.41}$$

Generally, when the NAPL dissolution front propagates from node 1 to node N, the resulting mesh discretization error at node N can be expressed as follows:

$$\begin{aligned}\delta_{dN} &= \delta_d \sum_{i=1}^{N-1} e^{(N-i)\bar{\omega}\Delta\bar{t}} = \delta_d\left(\frac{e^{\bar{\omega}\Delta\bar{t}}(e^{(N-1)\bar{\omega}\Delta\bar{t}}-1)}{e^{\bar{\omega}\Delta\bar{t}}-1}\right)\\ &\approx \delta_d\left(\frac{e^{\bar{\omega}\Delta\bar{t}}(e^{(N-1)\bar{\omega}\Delta\bar{t}})}{e^{\bar{\omega}\Delta\bar{t}}-1}\right) = \delta_d\left(\frac{e^{\bar{\omega}\Delta\bar{t}}}{e^{\bar{\omega}\Delta\bar{t}}-1}\right)e^{\bar{\omega}\bar{t}} \quad \text{(for large } N)\end{aligned} \tag{10.42}$$

Note that for a fine mesh of finite elements, δ_d can be controlled to be much less than unity (i.e. $\delta_d \ll 1$), so that $(e^{(N-1)\bar{\omega}\Delta\bar{t}} - 1)$ should be much greater than unity when the finger grows to a visible size. As a result, $(e^{(N-1)\bar{\omega}\Delta\bar{t}} \gg 1)$, so that $(e^{(N-1)\bar{\omega}\Delta\bar{t}} - 1) \approx e^{(N-1)\bar{\omega}\Delta\bar{t}}$ in the process of deriving Eq. (10.42). Thus, when the NAPL dissolution front propagates from node 1 to node N, the amplitude of the finger caused by the mesh discretization error (in a rectangular domain) can be determined by the following equation:

$$A^{mesh}_{Finger} = \delta_{dN} = \delta_d \sum_{i=1}^{N-1} e^{(N-i)\bar{\omega}\Delta\bar{t}} \approx \delta_d\left(\frac{e^{\bar{\omega}\Delta\bar{t}}}{e^{\bar{\omega}\Delta\bar{t}}-1}\right)e^{\bar{\omega}\bar{t}}, \tag{10.43}$$

where A^{mesh}_{Finger} is the amplitude of the finger caused by the mesh discretization error in the NAPL dissolution system of a rectangular domain.

Note that for a NAPL dissolution-front instability problem in a rectangular domain, the dimensionless growth rate of perturbation can be theoretically determined as follows (Zhao et al. 2010c):

$$\bar{\omega} = \frac{\bar{p}'_{axf}}{(1+\beta)S_{n0}} \left\{ \frac{(1+\beta)\bar{p}'_{axf}}{1-\bar{\alpha}_L \bar{p}'_{axf}} + \left[(1+\beta) + \left(\frac{1}{1-\bar{\alpha}_L \bar{p}'_{axf}} - \bar{\alpha}_L \bar{m} \right)(1-\beta) \right] \sigma \right\}$$
$$- \frac{\bar{p}'_{axf}}{(1+\beta)S_{n0}} \left[\left(\bar{m} - \frac{\bar{p}'_{axf}}{1-\bar{\alpha}_L \bar{p}'_{axf}} - \bar{\alpha}_L \bar{m}^2 \left(1 - \bar{\alpha}_L \bar{p}'_{axf}\right) \right) \frac{(1-\beta)}{1-\bar{\alpha}_L \bar{p}'_{axf}} \right], \tag{10.44}$$

$$\sigma = \frac{\sqrt{\left(\frac{\bar{p}'_{axf}}{1-\bar{\alpha}_L \bar{p}'_{axf}}\right)^2 + \frac{4\bar{m}^2(1-\bar{\alpha}_T \bar{p}'_{axf})}{1-\bar{\alpha}_L \bar{p}'_{axf}}} - \frac{\bar{p}'_{axf}}{1-\bar{\alpha}_L \bar{p}'_{axf}}}{2}, \tag{10.45}$$

where $\bar{p}'_{axf}$ is the dimensionless pressure gradient of the aqueous phase liquid on a vertical cross-section in the NAPL completely dissolved region; S_{n0} is the initial saturation of the NAPL; $\bar{\alpha}_T$ and $\bar{\alpha}_L$ are the dimensionless transversal and longitudinal dispersivities of the NAPL species in the aqueous phase; β is the ratio of the permeability of the NAPL un-dissolved region to that of the NAPL completely-dissolved region; $\bar{m}$ is the dimensionless wavenumber of the perturbation.

Equations (10.43)–(10.45) clearly indicate that if the mesh discretization error associated with the NAPL dissolution instability system of a rectangular domain is known, then the amplitude of the finger caused by the mesh discretization error (in a rectangular domain) can be theoretically determined. Unfortunately, the main purpose of this study is to find the mesh discretization error associated with the NAPL dissolution instability system, so that an inverse problem needs to be solved. This means that if the amplitude of the finger caused by the mesh discretization error (in a rectangular domain) is numerically evaluated, then the mesh discretization error associated with the NAPL dissolution instability system of a rectangular domain can be determined using the following equation:

$$\delta_d = \frac{A^{mesh}_{Finger}}{\sum_{i=1}^{N-1} e^{(N-i)\bar{\omega}\Delta\bar{t}}} \approx A^{mesh}_{Finger} \left(\frac{e^{\bar{\omega}\Delta\bar{t}} - 1}{e^{\bar{\omega}\Delta\bar{t}}} \right) e^{-\bar{\omega}\bar{t}}. \tag{10.46}$$

Although Eq. (10.46) is not easy to be evaluated, it provides a theoretical basis for estimating the mesh discretization error associated with the NAPL dissolution instability system of a rectangular domain.

To solve this inverse problem better, the concept of equivalent initial perturbation associated with the mesh discretization error of a NAPL dissolution system is presented below. The basic idea behind the concept of equivalent initial perturbation associated with the mesh discretization of a NAPL dissolution system is that the amplitude of the finger caused by the mesh discretization error can be approximately represented by the following equation:

$$A_{Finger}^{mesh} = \delta_d^{equivalent} e^{\bar{\omega}\bar{t}}, \tag{10.47}$$

where $\delta_d^{equivalent}$ is the equivalent initial perturbation associated with the mesh discretization error of a NAPL dissolution system.

Considering Eqs. (10.43) and (10.47) simultaneously yields the following equation:

$$\delta_d^{equivalent} = \frac{\delta_d}{e^{\bar{\omega}\bar{t}}} \sum_{i=1}^{N-1} e^{(N-i)\bar{\omega}\Delta\bar{t}} \approx \delta_d \left(\frac{e^{\bar{\omega}\Delta\bar{t}}}{e^{\bar{\omega}\Delta\bar{t}} - 1} \right) = \alpha\delta_d, \quad \alpha = \frac{e^{\bar{\omega}\Delta\bar{t}}}{e^{\bar{\omega}\Delta\bar{t}} - 1}. \tag{10.48}$$

Equation (10.48) indicates that since $\bar{\omega}\Delta\bar{t}$ is a constant (as demonstrated later), the equivalent initial perturbation associated with the mesh discretization error of a NAPL dissolution system is also a constant and theoretically computable. Since δ_d remains unknown in this equation, Eq. (10.47) is directly used to calculate the equivalent initial perturbation associated with the mesh discretization error of a NAPL dissolution system in Sect. 10.3.2. Therefore, Eq. (10.47) may be considered as the core of the proposed finger-amplitude growing theory associated with the mesh discretization error in the NAPL dissolution system.

10.3.1.3 The Product of $\bar{\omega}\Delta\bar{t}$ Associated with the NAPL Dissolution System of a Trapezoidal Domain

The proposed propagation theory of mesh discretization error associated with the NAPL dissolution system of a rectangular domain can be also extended to the mesh discretization error estimation associated with the NAPL dissolution system of a trapezoidal domain. For this purpose, we can approximately divide the trapezoidal domain into a series of rectangular sub-domains. Obviously, if the number of these rectangular sub-domains approaches infinity, then the shape of the trapezoidal domain could be accurately represented. This is equivalently to use the limit concept in mathematics. However, for each rectangular sub-domain of the same dimensionless width (i.e. $\Delta\bar{x}$) in the horizontal direction, as shown in Fig. 10.12, the dimensionless pressure gradient, $\bar{p'}_{axf}$, of the aqueous phase liquid on its left-hand-side vertical boundary, which is assumed to be the current location of the NAPL dissolution front in the trapezoidal domain, is no longer constant, as indicated by Eq. (10.29) in the previous section. This means that the propagation speed (i.e. $\bar{v}_{front} = -\bar{p}'_{axf} / S_{n0}$) of the NAPL dissolution front in each rectangular sub-domain is also no longer constant. This is one of the main differences between the NAPL dissolution system of a rectangular domain and that of a trapezoidal domain.

For the purpose of determining the product of $\bar{\omega}\Delta\bar{t}$ associated with the NAPL dissolution system of a trapezoidal domain, we need to consider a typical rectangular sub-domain shown in Fig. 10.12. For this rectangular sub-domain, the time

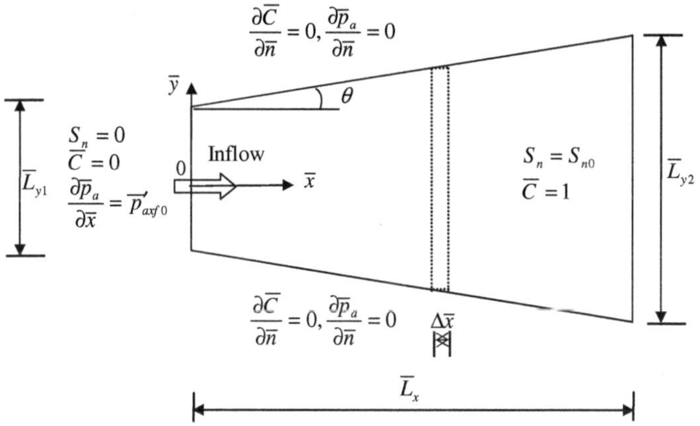

Fig. 10.12 A typical sub-domain of rectangular shape in the trapezoidal domain

period (i.e. $\Delta \bar{t}$), when the NAPL dissolution front propagates from its left-hand-side vertical boundary to its right-hand-side vertical boundary, can be expressed as follows:

$$\Delta \bar{t} = \frac{\Delta \bar{x}}{\bar{v}_{front}} = -\frac{\Delta \bar{x} S_{n0}}{\bar{p}'_{axf}} = -\frac{(\bar{L}_{y1} + 2\bar{x} tg\theta)\Delta \bar{x} S_{n0}}{\bar{L}_{y1}\bar{p}'_{axf0}}. \tag{10.49}$$

To determine the dimensionless growth rate (i.e. $\bar{\omega}$) in the typical rectangular sub-domain shown in Fig. 10.12, we need to use Eq. (10.44) to investigate how the dimensionless growth rate varies with the dimensionless pressure gradient, $\bar{p}'_{axf}$, of the aqueous phase liquid on its left-hand-side vertical boundary. Figure 10.13 shows the variation of the dimensionless growth rate with the Zhao number due to different dimensionless wavenumbers in the NAPL dissolution problem. In this figure, $Zh = -\bar{p}'_{axf}$ by definition. It is obvious that for a given dimensionless wavenumber, the dimensionless growth rate varies linearly with the Zhao number of the NAPL dissolution system.

Thus, Eq. (10.44) can be approximately rewritten in the following simplified form:

$$\bar{\omega} \approx -\bar{p}'_{axf} f(S_{n0}, \bar{\alpha}_L, \bar{\alpha}_T, \beta) = -\frac{\bar{L}_{y1}}{\bar{L}_{y1} + 2\bar{x} tg\theta}\bar{p}'_{axf0} f(S_{n0}, \bar{\alpha}_L, \bar{\alpha}_T, \beta), \tag{10.50}$$

where f is a function of S_{n0}, $\bar{\alpha}_L$, $\bar{\alpha}_T$ and β. This function represents the constant slope of the straight line. Thus, $f(S_{n0}, \bar{\alpha}_L, \bar{\alpha}_T, \beta)$ should be independent of the geometrical shape of the problem domain.

Note that for the NAPL dissolution system of a rectangular domain, the value of θ is equal to zero, so that the product of $\bar{\omega}\Delta \bar{t}$ in the case of the NAPL dissolution front propagating in a rectangular domain can be expressed as follows:

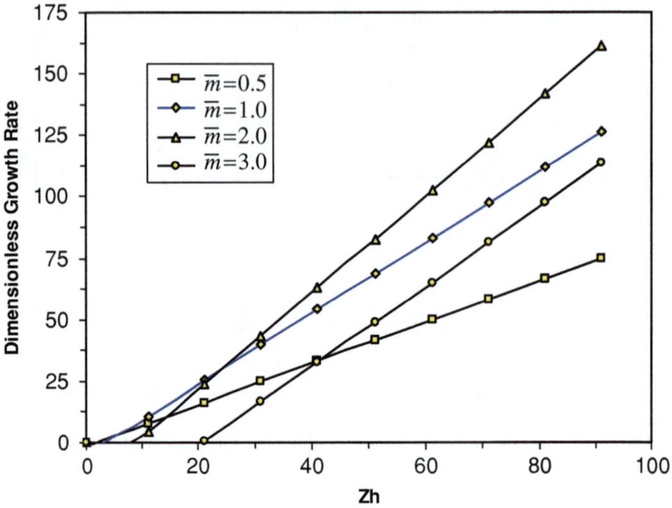

Fig. 10.13 Variation of the dimensionless growth rate with the Zhao number due to different dimensionless wavenumbers in the NAPL dissolution problem

$$(\bar{\omega}\Delta\bar{t})_{rectangular} = \Delta\bar{x}S_{n0}f(S_{n0},\ \bar{\alpha}_L,\ \bar{\alpha}_T,\ \beta). \tag{10.51}$$

Similarly, considering Eqs. (10.49) and (10.50) simultaneously yields the following product for the NAPL dissolution front propagating in a trapezoidal domain.

$$\begin{aligned}(\bar{\omega}\Delta\bar{t})_{trapezoidal} &= \left(-\frac{\bar{L}_{y1}}{\bar{L}_{y1}+2\bar{x}tg\theta}\bar{p}'_{axf0}f(S_{n0},\ \bar{\alpha}_L,\ \bar{\alpha}_T,\ \beta)\right)\times\left(-\frac{(\bar{L}_{y1}+2\bar{x}tg\theta)\Delta\bar{x}S_{n0}}{\bar{L}_{y1}\bar{p}'_{axf0}}\right)\\ &= \Delta\bar{x}S_{n0}f(S_{n0},\ \bar{\alpha}_L,\ \bar{\alpha}_T,\ \beta)\end{aligned} \tag{10.52}$$

Equations (10.51) and (10.52) indicate that although both the dimensionless growth rate (i.e. $\bar{\omega}$) and the time period (i.e. $\Delta\bar{t}$) when the NAPL dissolution front propagates the same horizontal distance (i.e. $\Delta\bar{x}$) are totally different, their products are exactly the same for the NAPL dissolution-front propagation in both the rectangular and trapezoidal domains. When the NAPL dissolution front propagates through N elements with the equal dimensionless width of $\Delta\bar{x}$, it can be easily proven that the product of $N\bar{\omega}\Delta\bar{t}$, where N is a positive integer of any value, is also exactly the same for the NAPL dissolution front propagation in both the rectangular and trapezoidal domains. This theoretical finding demonstrates that if the horizontal width of the finite elements that is used to simulate a trapezoidal domain is exactly the same as that is used to simulate a rectangular domain, the mesh discretization error propagation theory presented for the rectangular domain (in Sect. 10.3.1.2) can be, in principle, used to estimate the mesh discretization error associated with the NAPL dissolution-front propagation in the trapezoidal domain.

Thus, Eq. (10.47) is also valid for estimating the mesh discretization error associated with the NAPL dissolution-front propagation in the trapezoidal domain. This is another reason why we consider a trapezoidal domain in this investigation.

10.3.1.4 The Corner Effect at the Entrance of a Trapezoidal Domain

For a trapezoidal domain, as shown in Fig. 10.10, the two neighbour boundaries are not perpendicular each other, so that the inflow perpendicular to the entrance (i.e. one side of the trapezoidal domain) may change the flow direction just when it passes the entrance of the domain. Due to this special kind of shape, the overall flow pattern turns into a radial flow pattern at some distance away from the entrance of the flow. This means that such a corner effect can be treated as an equivalent perturbation that is just applied at the corner points of the flow entrance in the trapezoidal domain. Following the same procedures as those used in Sect. 10.3.1.2, the amplitude of the finger caused by the corner effect can be approximately represented by the following equation:

$$A_{Finger}^{corner} = \delta_c^{equivalent} e^{\bar{\omega}\bar{t}}, \tag{10.53}$$

where $\delta_c^{equivalent}$ is the equivalent initial perturbation associated with the corner effect at the entrance of a NAPL dissolution system; A_{Finger}^{corner} is the amplitude of the finger caused by the corner effect.

As a counterpart of Eq. (10.47), Eq. (10.53) may be considered as the core of the proposed finger-amplitude growing theory associated with the corner effect in the NAPL dissolution system of a trapezoidal domain. Consequently, the total amplitude of the finger (caused by both the corner effect and the mesh discretization error), which grows at either the top or the bottom boundary of the trapezoidal domain, should be equal to the sum of Eqs. (10.47) and (10.53) as follows:

$$A_{Finger}^{corner+mesh} = (\delta_c^{equivalent} + \delta_d^{equivalent}) e^{\bar{\omega}\bar{t}}. \tag{10.54}$$

where $A_{Finger}^{corner+mesh}$ is the total amplitude of the finger caused by both the corner effect and the mesh discretization error at either the top or the bottom boundary of the trapezoidal domain.

10.3.2 Corner and Mesh Discretization Effects on the Morphological Evolution of NAPL Dissolution Fronts in Supercritical Systems of Trapezoidal Domains

As shown in Fig. 10.10 in the previous section, the geometrical shape of a trapezoidal computational domain, which is symmetrical to the $\bar{x}$ axis, can be represented by the divergent angle (i.e. θ) between a horizontal line and the top boundary. This angle is defined as a geometrical parameter in the corresponding computations. If θ is equal to zero, then the trapezoidal domain degenerates to a rectangular one.

On the basis of the previous experimental measurement results that were calibrated by several laboratory tests (Imhoff and Miller 1996; Imhoff et al. 1996; Miller et al. 1998), the following parameters are selectively used in the corresponding computational models: the initial saturation (S_{n0}) of the NAPL is 0.2; the irreducible saturation (S_{ai}) of the aqueous phase fluid is 0.15; the dimensionless longitudinal and transverse dispersivities ($\bar{\alpha}_L$ and $\bar{\alpha}_T$) are 0.2 and 0.02, respectively; the ratio (ε) of the equilibrium concentration of the NAPL species in the aqueous phase fluid to the density of the NAPL itself is 0.001; the density ratio ($\bar{\rho}_a$) of the aqueous phase fluid to the NAPL is 1.0/1.46; the value of β_1 is 0.87. To simulate the propagation of NAPL dissolution fronts appropriately, the whole computational domain is simulated by 120,000 four-node quadrilateral elements of 120,701 nodal points in total. Figure 10.14 shows the meshes of finite elements used for modeling both the rectangular domain (i.e. $\theta = 0°$) and the trapezoidal domain (i.e. $\theta = 20°$). Note that each element in this figure is further divided into 100 finer elements before the computation.

For the purpose of investigating the instability of a NAPL dissolution system, it is common practice to perturb the homogeneous distribution field of the initial NAPL saturation (S_{n0}) with a small amount (Zhao et al. 2010c). For this reason, the initial residual saturation field of the NAPL is randomly perturbed by a small amount of 1 % of the originally-input saturation of the NAPL (i.e. $S_{n0} = 0.2$) before running the computational model. This means that the initial homogeneous distribution field of the NAPL saturation (S_{n0}) is replaced and reassigned by a slightly perturbed non-homogeneous distribution field of the NAPL saturation before running the computational model. The dimensionless length (i.e. $\bar{L}_x$) of the computational domain is 6,284 in the $\bar{x}$ direction, while the dimensionless length (i.e. $\bar{L}_{y1}$) of the left boundary is fixed to 3,142 and 2,046 in the $\bar{y}$ direction for two different computational domains, namely $\theta = 0°$ and 20°, respectively. The Zhao number used for the two computational models (at the entrance of the flow) is 1.0, while the dimensionless time-step length is 3.2.

10.3.2.1 Corner and No-Flow Boundary Effects of Computational Domains

Figure 10.15 shows the morphological evolution of NAPL dissolution fronts in both the rectangular ($\theta = 0°$) and the trapezoidal ($\theta = 20°$) domains consisting of the fluid-saturated porous media at two different time instants, namely $\bar{t} = 160$ and 1600, respectively. In this figure, the residual saturation of a NAPL is used to represent the NAPL dissolution front. It is observed that the NAPL dissolution front propagates much faster along the top and bottom boundaries, especially in the trapezoidal domain. This may be caused by both the no-flow boundary and

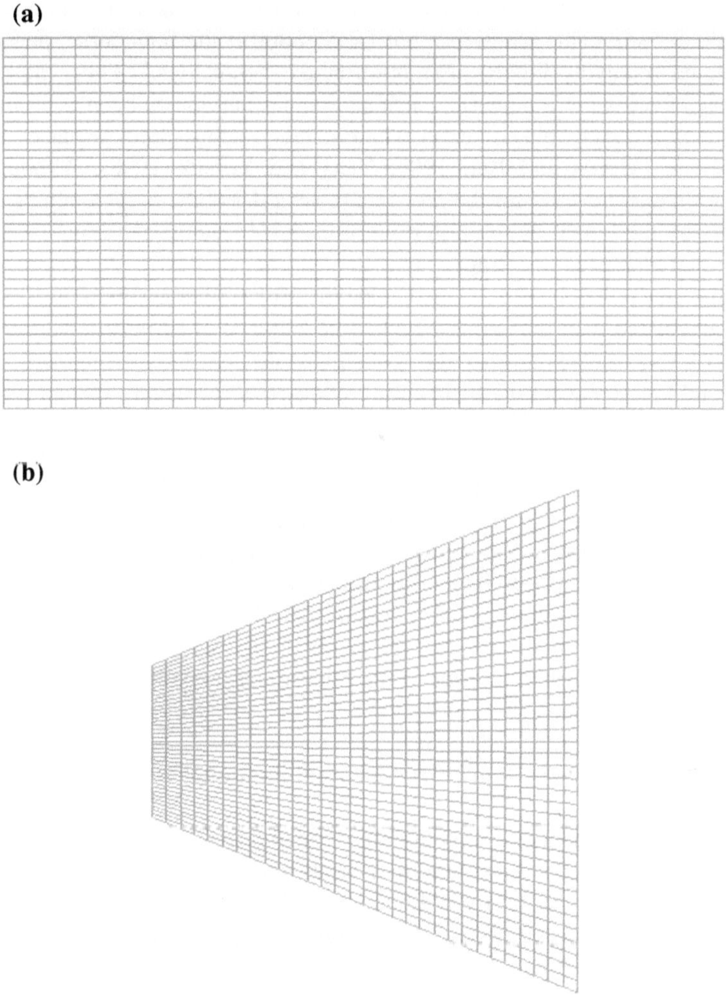

Fig. 10.14 Finite element meshes used for simulating the NAPL dissolution problem: a Rectangular domain; b Trapezoidal domain

corner effects. In the case of a rectangular domain, Kalia and Balakotaiah (2009) observed the same phenomenon in the numerical simulation of acid dissolution induced wormhole formation in carbonates. Regarding the physical reason to cause this phenomenon, they made the following explanation: "When no-flow boundaries are present in a system, an inconsistency may occur at these boundaries because of Darcy's law. As the fluid is injected in the medium, it is transported in both the axial and transverse directions. Darcy's law allows for streamwise or axial flow at the boundaries although the transverse velocity component is zero because of the boundary conditions. This leads to a deflection of the fluid carried by the perpendicular velocity component at the boundaries to the axial direction resulting in a velocity slip. This leads to more fluid going to the boundaries and in turn resulting in relatively higher dissolution at the boundaries." Thus, the no-flow boundary effect is not a numerical artifact and has previously been mentioned for the viscous fingering phenomenon where a less viscous fluid displaces a more viscous one (Morris and Ball 1990; Yang and Yortsos 1998). Although this explanation may be valid for the flow in a rectangular domain, it cannot be used to explain why the NAPL dissolution front propagates much faster along the top and bottom boundaries in the trapezoidal domain (i.e. $\theta = 20°$). Nevertheless, this phenomenon can be better explained using the proposed finger-amplitude growing theory associated with the corner effect at the entrance of a trapezoidal domain (in Sect. 10.3.1.4). In the case of the rectangular domain (where no corner effect exists) and $\bar{t} = 1600$ (in Fig. 10.15), the amplitude of the finger at either the top or the bottom boundary is of the same order of magnitude as that of the finger within the interior of the rectangular domain. However, in the case of the trapezoidal domain (where there exists the corner effect) and $\bar{t} = 1600$ (in Fig. 10.15), the amplitude of the finger at either the top or the bottom boundary is much greater than that of the finger within the interior of the trapezoidal domain. This demonstrated that the proposed finger-amplitude growing theory associated with the corner effect at the entrance of a trapezoidal domain is useful for correctly explaining why the finger at either the top or the bottom boundary grows much faster than that within the interior of the trapezoidal domain.

10.3.2.2 Mesh Discretization Effects of Computational Domains

When the finite element and finite difference methods are used to solve the NAPL dissolution problem in a fluid-saturated porous medium, the continuum domain of the problem needs to be discretized into many elements, so that the dimensionless governing equations (i.e. the partial differential equations) of the system can be approximately represented by a set of algebraic equations. This discretization process will cause the mesh discretization error, which may result in numerical diffusion during the numerical computation. Compared with the use of rectangular and square elements in the simulation of a rectangular domain, the use of general quadrilateral elements in the simulation of a trapezoidal domain can cause extra mesh discretization error (or extra numerical diffusion) due to the involvement of mesh inclination (i.e. grid orientation). For this reason, the total diffusion/dispersion tensor of the computational model for simulating the NAPL dissolution problem in a trapezoidal domain is equal to the sum of the physical diffusion/dispersion tensor and the numerical diffusion/dispersion tensor. Since the numerical diffusion/dispersion tensor is caused by the mesh discretization error, it is desirable to reduce the numerical diffusion/dispersion tensor so that the computational simulation result can converge to the physical solution.

To examine the effect of numerical diffusion/dispersion due to mesh discretization error on the computational simulation of a NAPL dissolution problem in a

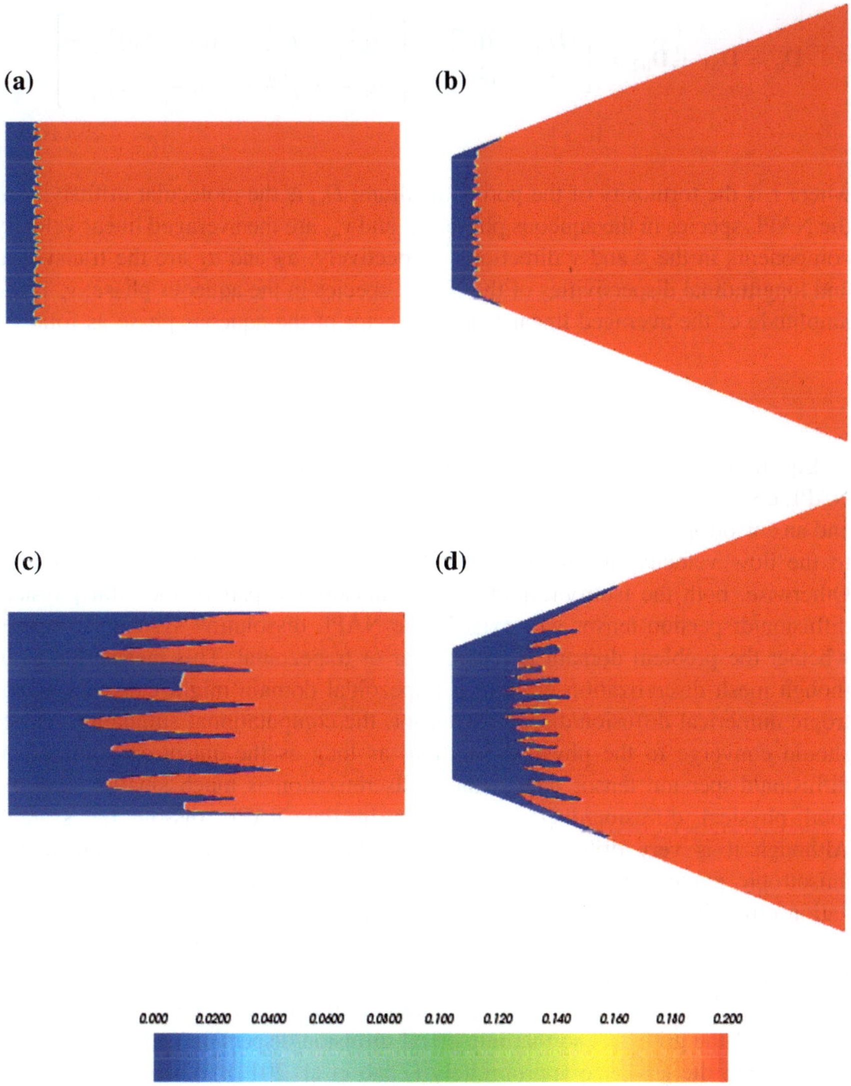

Fig. 10.15 Effects of corner and no-flow boundary on the evolution of NAPL dissolution fronts in the fluid-saturated porous medium (Supercritical system): **a** $\theta = 0, \bar{t} = 160$; **b** $\theta = 20°$, $\bar{t} = 160$; **c** $\theta = 0, \bar{t} = 1600$; **d** $\theta = 20°, \bar{t} = 1600$

trapezoidal domain, it is necessary to understand the constituents of the physical diffusion/dispersion tensor in a NAPL dissolution system. For this purpose, the physical diffusion/dispersion tensor in the NAPL dissolution system can be expressed in the following dimensional form:

$$\mathbf{D_h} = \mathbf{D_{h1}} + \mathbf{D_{h2}} = \begin{bmatrix} \tau D_m & 0 \\ 0 & \tau D_m \end{bmatrix} + \begin{bmatrix} \alpha_L \frac{v_{ax}^2}{v_a} + \alpha_T \frac{v_{ay}^2}{v_a} & (\alpha_L - \alpha_T) \frac{v_{ax} v_{ay}}{v_a} \\ (\alpha_L - \alpha_T) \frac{v_{ax} v_{ay}}{v_a} & \alpha_L \frac{v_{ay}^2}{v_a} + \alpha_T \frac{v_{ax}^2}{v_a} \end{bmatrix}, \tag{10.55}$$

where τ is the tortuosity of the porous medium; D_m is the molecular diffusivity of the NAPL species in the aqueous phase; v_{ax} and v_{ay} are the averaged linear velocity components in the x and y directions, respectively; α_T and α_L are the transversal and longitudinal dispersivities of the NAPL species in the aqueous phase; v_a is the amplitude of the averaged linear velocity vector of the aqueous phase as follows:

$$v_a = \sqrt{v_{ax}^2 + v_{ay}^2}. \tag{10.56}$$

Equation (10.55) indicates that the total physical diffusion/dispersion tensor in a NAPL dissolution system can be divided into two parts: an isotropic part (i.e. $\mathbf{D_{h1}}$) and an anisotropic part (i.e. $\mathbf{D_{h2}}$). Since the anisotropic part is directly proportional to the flow velocity, it can only vanish when there is no flow in the system. Otherwise, both the isotropic part and the anisotropic part of the total physical diffusion/dispersion tensor will exist in the NAPL dissolution system, no matter whether the problem domain is rectangular or trapezoidal. This means that even though mesh discretization error in a trapezoidal domain might cause an anisotropic numerical diffusion/dispersion tensor, the computational simulation results should converge to the physical solution, as long as the anisotropic numerical diffusion/dispersion tensor due to mesh discretization is much smaller than the total physical diffusion/dispersion tensor in the NAPL dissolution system. Although it is very difficult, if not impossible, to quantitatively evaluate the anisotropic numerical diffusion/dispersion tensor associated with mesh discretization error, it is possible to qualitatively judge its effect on the computational simulation through comparing the simulation results with the theoretical expectations, as discussed later.

In the computational simulation, the mesh discretization error decreases with the decrease of the grid size, so that numerical diffusion/dispersion can be minimized by choosing the grid size appropriately (Zhao et al. 2011). When the numerical results, which are obtained from the mesh of a given grid size, agree well with the corresponding analytical solutions and experimental results, the numerical diffusion/dispersion is considered to be minimum, compared with the physical diffusion/dispersion of the problem. To examine mesh discretization effects on the numerical results, we have simulated a subcritical NAPL dissolution system in the trapezoidal domain. In this case, the Zhao number used for the computational model (at the entrance of the flow) is reduced from 1.0 to 0.01, while the dimensionless time-step length is increased from 3.2 to 32. Figure 10.16 shows the corresponding computational simulation results. As expected theoretically (Tan and Homsy 1987), the NAPL dissolution front converges to the arc shape, which is controlled by the

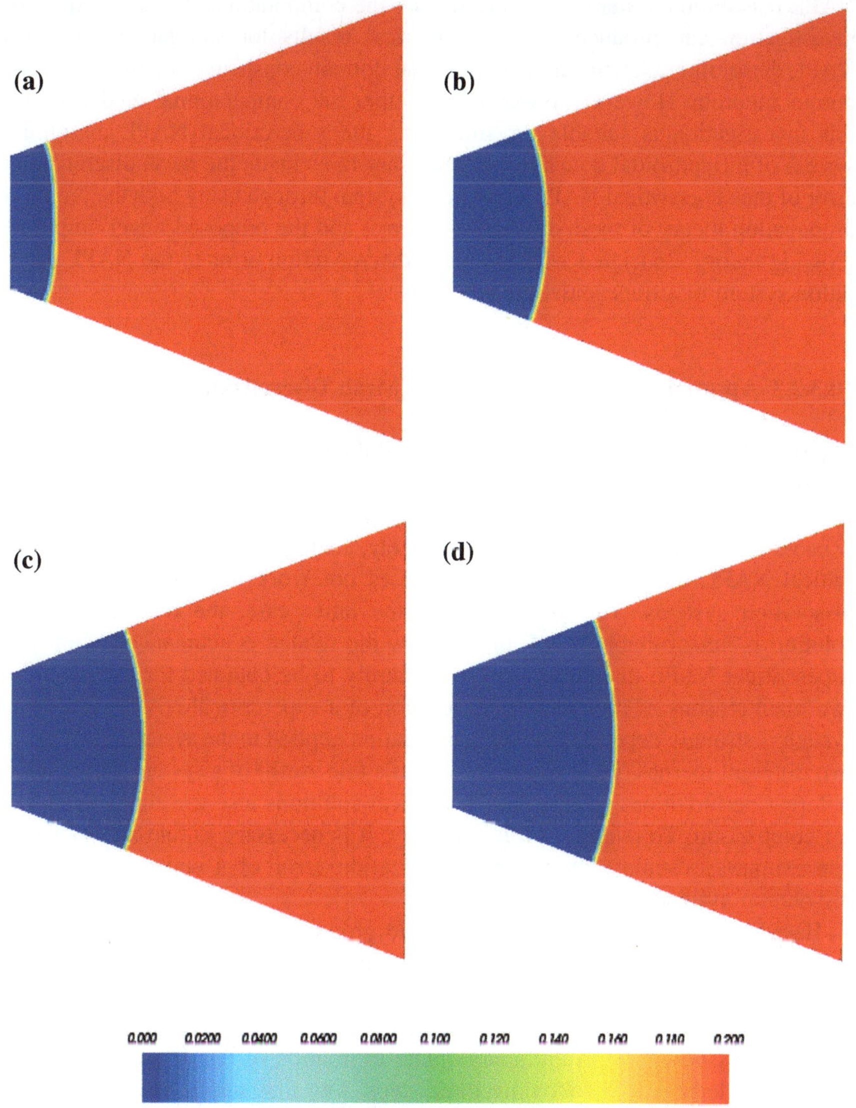

Fig. 10.16 The evolution of NAPL dissolution fronts in the trapezoidal domain consisting of the fluid-saturated porous medium (Subcritical system): **a** $\theta = 20°, \bar{t} = 320$; **b** $\theta = 20°, \bar{t} = 1600$; **c** $\theta = 20°, \bar{t} = 3200$; **d** $\theta = 20°, \bar{t} = 4800$

dynamic behaviour of the subcritical NAPL dissolution system. This is clearly evidenced in the computational simulation results. Since the numerical results have good agreement with the related theoretical expectation for the subcritical NAPL dissolution systems of trapezoidal domains, it can be concluded that the mesh discretization effect is negligible in the computational simulations of subcritical

NAPL dissolution systems. This means that the computational model used in this investigation can produce reliable numerical results for simulating subcritical NAPL dissolution systems in the trapezoidal domain consisting of a fluid-saturated porous medium. However, to examine whether the computational model used in this investigation is suitable for simulating the supercritical NAPL dissolution system of a trapezoidal domain, it is necessary to estimate the mesh discretization error of the supercritical NAPL dissolution system through using both the proposed propagation theory of mesh discretization error and the proposed finger-amplitude growing theory associated with the mesh discretization error in the NAPL dissolution system of a trapezoidal domain.

10.3.2.3 Approximate Estimation of the Mesh Discretization Error

In terms of estimating the discretization error of a numerical method, it is common practice to compare the numerical solution with the corresponding analytical solution (if any) or benchmark solution that is obtained using a very fine mesh (Zhao and Steven 1996a, b, c). Unfortunately, such practice works well for subcritical NAPL dissolution systems, but does not work for supercritical NAPL dissolution systems. This is because in the latter case, the related analytical solution is not available at all. In addition, the related benchmark solution of a supercritical NAPL dissolution system is hardly to be obtained for the following two main reasons: (1) the numerical solution of a supercritical NAPL dissolution system is strongly dependent on the perturbation applied to the system; and (2) any two different meshes (i.e. a coarse mesh and a fine mesh) of a system will result in different perturbations so that their numerical solutions lose the common ground for comparison. To overcome this difficulty, it is necessary to find a new way to approximately estimate the mesh discretization error of a supercritical NAPL dissolution system in this study.

Based on the proposed finger-amplitude growing theory associated with the mesh discretization error in the NAPL dissolution system of a trapezoidal domain, it is possible to approximately estimate the discretization error of the system through using Eq. (10.47) that is presented in the previous section. The specific procedure associated with the approximate estimation of the discretization error can be divided into the following three steps. The first step is to determine the dimensionless growth rate of the supercritical NAPL dissolution system. This can be done through considering the amplitudes of the fingers at two early computational simulation steps. For example, if the amplitudes of the finger are $A_{1Finger} = \delta_A e^{\bar{\omega}\bar{t}_1}$ and $A_{2Finger} = \delta_A e^{\bar{\omega}\bar{t}_2}$ at $\bar{t} = \bar{t}_1$ and $\bar{t} = \bar{t}_2$ respectively, then the ratio of these two amplitudes can be expressed as follows:

$$R_1 = \frac{A_{2Finger}}{A_{1Finger}} = e^{\bar{\omega}(\bar{t}_2 - \bar{t}_1)}. \tag{10.57}$$

From Eq. (10.57), the dimensionless growth rate of the supercritical NAPL dissolution system can be obtained as follows:

$$\bar{\omega} = \frac{\ln R_1}{\bar{t}_2 - \bar{t}_1}. \tag{10.58}$$

The second step is to determine the amplitude ratio of the equivalent initial perturbation (due to the discretization error) to the total initial perturbation in the supercritical NAPL dissolution system. In this situation, the discretization error is viewed as an equivalent initial perturbation, which can be determined from the computational simulation results of a supercritical NAPL dissolution system in which no any other initial perturbation is applied. On the other hand, the total initial perturbation, which is the sum of the equivalent initial perturbation (due to the discretization error) and the applied initial perturbation (due to some physical considerations), can be determined from the computational simulation results of a supercritical NAPL dissolution system in which an initial perturbation (due to some physical considerations) is applied. If the amplitudes of the fingers are $A_{1Finger} = \delta_{A1} e^{\bar{\omega}\bar{t}_1}$ (at $\bar{t} = \bar{t}_1$) for the supercritical NAPL dissolution system without any other initial perturbation applied and $A_{2Finger} = \delta_{A2} e^{\bar{\omega}\bar{t}_2}$ (at $\bar{t} = \bar{t}_2$) for the supercritical NAPL dissolution system with an initial perturbation applied, then the amplitude ratio of the equivalent initial perturbation (due to the discretization error) to the total initial perturbation can be evaluated under the condition of $A_{1Finger} = A_{2Finger}$. This can result in the following formula:

$$R_2 = \frac{\delta_{A1}}{\delta_{A2}} = e^{\bar{\omega}(\bar{t}_2 - \bar{t}_1)}, \tag{10.59}$$

where δ_{A1} is the amplitude of the equivalent initial perturbation due to the discretization error; δ_{A2} is the amplitude of the total initial perturbation.

The third step is to determine the amplitude of the equivalent initial perturbation due to the discretization error in the supercritical NAPL dissolution system. Since the initially applied perturbation is known, it can be represented by $\delta_{Ainitial}$, so that $\delta_{A2} = \delta_{Ainitial} + \delta_{A1}$. With consideration of Eq. (10.59), this leads to the following formula for evaluating δ_{A1}:

$$\delta_{A1} = \frac{R_2}{1 - R_2} \delta_{Ainitial}. \tag{10.60}$$

Based on the proposed procedure above, we have rerun the previously-used computational model of the supercritical NAPL dissolution system in the trapezoidal domain by setting the applied initial perturbation to be zero. Figure 10.17 shows the related computational simulation results, from which it can be observed that due to the dynamic behaviour of the supercritical NAPL dissolution system, the fingering phenomenon indeed takes place in the trapezoidal domain. This

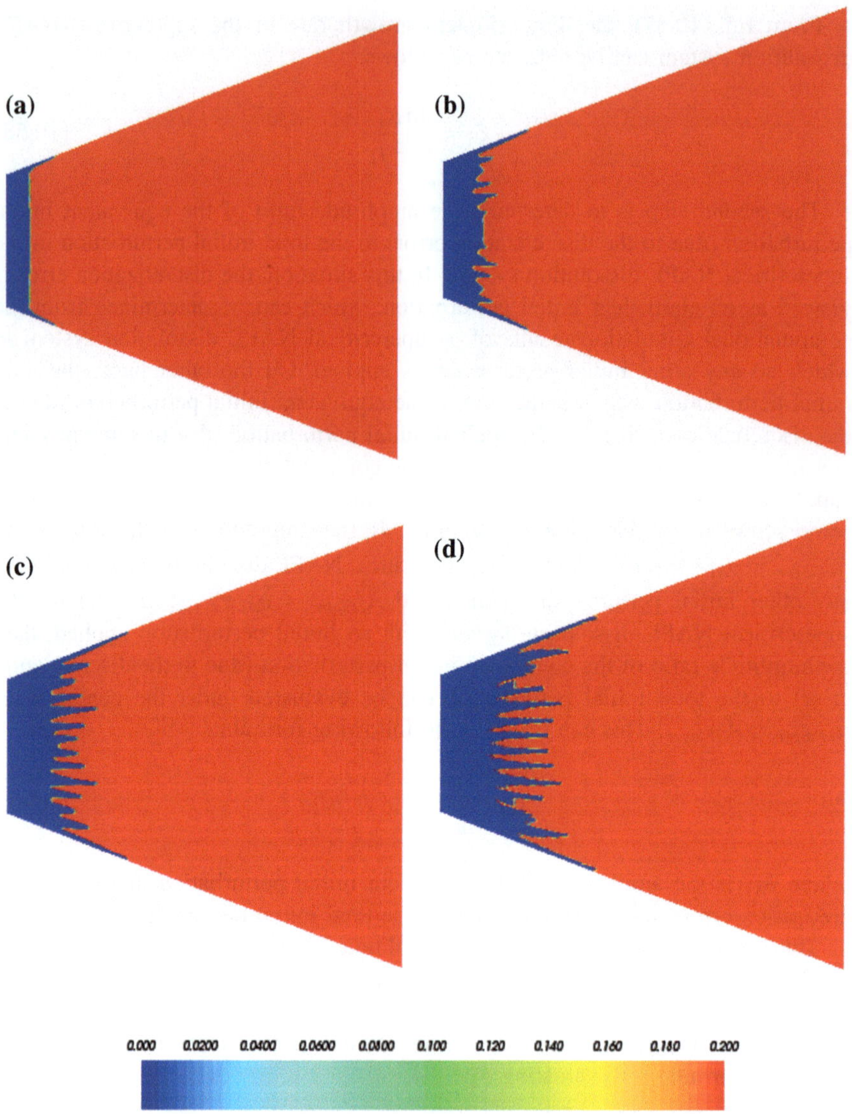

Fig. 10.17 The evolution of NAPL dissolution fronts in the trapezoidal domain consisting of fluid-saturated porous medium (Without applying initial perturbation): **a** $\theta = 20°, \bar{t} = 160$; **b** $\theta = 20°, \bar{t} = 320$; **c** $\theta = 20°, \bar{t} = 640$; **d** $\theta = 20°, \bar{t} = 1600$

indicates that the mesh discretization error, although it may be very small, can serve as a kind of small perturbation to trigger the fingering growth during the computational simulation of a supercritical NAPL dissolution system in the trapezoidal domain consisting of a fluid-saturated porous medium.

With two sets of computational simulation results of the supercritical NAPL dissolution system at hand, namely one obtained from the simulation with including an initial small perturbation and the other obtained from the simulation without including an initial small perturbation, it is possible to estimate the amplitude of the equivalent initial perturbation due to the discretization error in the supercritical NAPL dissolution system. As shown in Fig. 10.18, the ratio of the finger amplitude at $\bar{t} = 160$ to that at $\bar{t} = 128$ is equal to about 2.0 when the initial small perturbation, which has an amplitude of 0.002 (i.e. 0.01 S_{n0} as mentioned previously), is applied to the computational model of the NAPL dissolution system. This results in a dimensionless growth rate of 0.02166 (from Eq. (10.58)) for the NAPL dissolution system. On the other hand, the finger amplitude at $\bar{t} = 224$ when the initial small perturbation is not applied to the computational model is almost equal to that at $\bar{t} = 160$ when the initial small perturbation is applied to the computational model. From Eq. (10.59), this leads to a value of about 0.25 for R_2. Finally, substituting the value of R_2 into Eq. (10.60) yields a value of 0.00067 for the amplitude of the equivalent initial perturbation (i.e. δ_{A1}) due to the discretization error. Since δ_{A1} is smaller than 0.002 (i.e. $\delta_{A\text{initial}} = 0.002$), it is demonstrated that the effect of the total discretization error of the computational model is negligible in the computational simulations, compared with that of the initial small perturbation applied to the computational model. This also demonstrated that the proposed finger-amplitude growing theory associated with the mesh discretization error in the NAPL dissolution system of a trapezoidal domain in this study is useful for quantitatively assessing the correctness of computational simulations of NAPL dissolution-front instability problems in trapezoidal domains. If the amplitude of the equivalent initial perturbation due to the numerical discretization error is significantly smaller than the amplitude of the fingers generated by the numerical simulations of interest in the same geometry of the same finite element mesh, then one may safely neglect the numerical discretization effects on the numerical simulations of supercritical NAPL dissolution systems.

In summary, the propagation theory of the mesh discretization error associated with a NAPL dissolution system is first presented for a rectangular domain and then extended to a trapezoidal domain. This leads to the establishment of the finger-amplitude growing theory that is associated with both the corner effect and the mesh discretization effect in the NAPL dissolution systems of trapezoidal domains. This theory can be used to make the approximate error estimation of the corresponding computational simulation results. The related theoretical analysis and numerical results have demonstrated that: (1) both the corner effect and the mesh discretization effect can be quantitatively viewed as a kind of small perturbation so that they can have some considerable effects on the computational results of supercritical NAPL dissolution systems; (2) the proposed finger-amplitude growing theory associated with the corner effect at the entrance of a trapezoidal domain is useful for correctly explaining why the finger at either the top or the bottom boundary grows much faster than that within the interior of the trapezoidal domain; (3) the proposed finger-amplitude growing theory associated with the mesh discretization error in the NAPL dissolution system of a trapezoidal

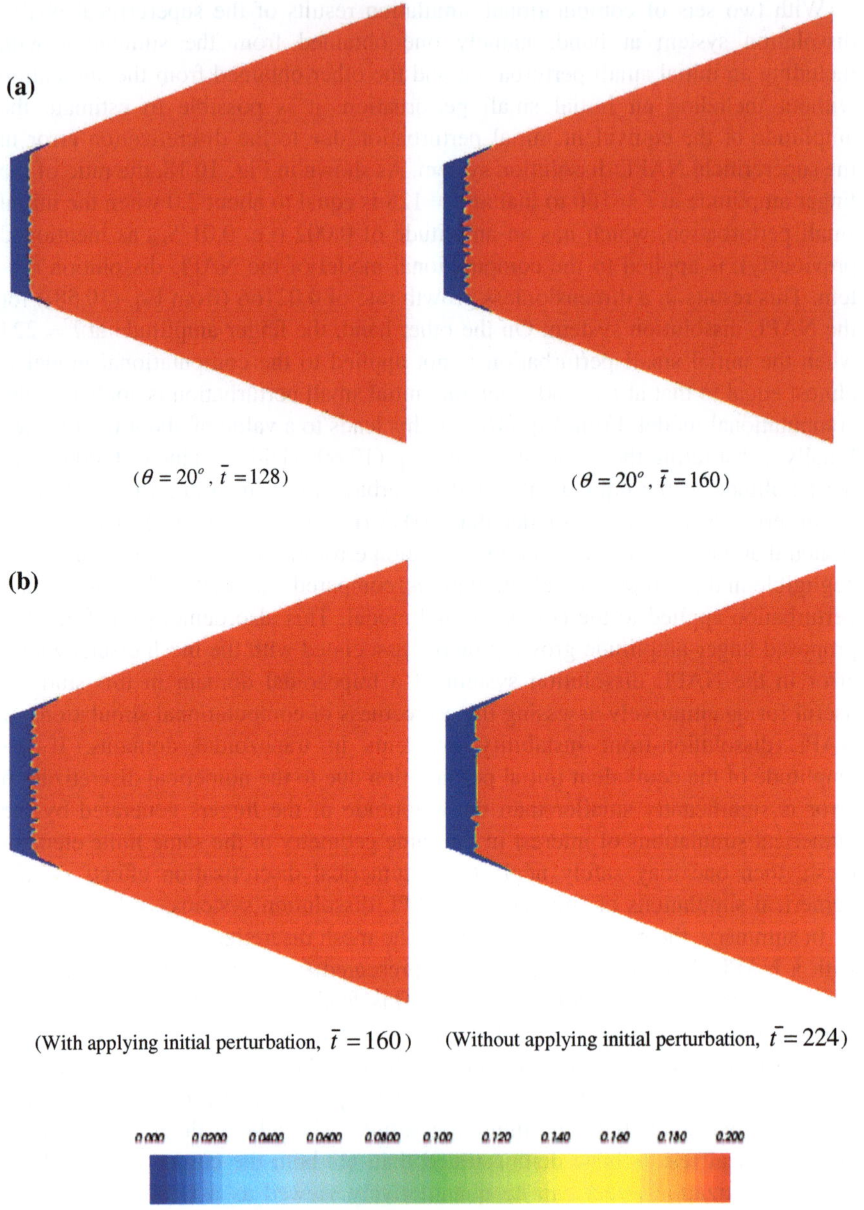

Fig. 10.18 Determination of the growth rate and equivalent initial perturbation in the supercritical NAPL dissolution system: **a** Determination of the growth rate; **b** Determination of the equivalent initial perturbation amplitude

domain can be used for quantitatively assessing the correctness of computational simulations of NAPL dissolution-front instability problems in trapezoidal domains, so that we can ensure that the computational simulation results are controlled by the physics of the NAPL dissolution system, rather than by the numerical artifacts.

References

Alt-Epping P, Smith L (2001) Computing geochemical mass transfer and water/rock ratios in submarine hydrothermal systems: implications for estimating the vigour of convection. Geofluids 1:163–181

Bear J (1972) Dynamics of fluids in porous media. American Elsevier Publishing Company, New York

Chadam J, Hoff D, Merino E, Ortoleva P, Sen A (1986) Reactive infiltration instabilities. IMA J Appl Math 36:207–221

Chadam J, Ortoleva P, Sen A (1988) A weekly nonlinear stability analysis of the reactive infiltration interface. IMA J Appl Math 48:1362–1378

Chen X, Jawitz JW (2008) Reactive tracer tests to predict nonaqueous phase liquid dissolution dynamics in laboratory flow chambers. Environ Sci Technol 42:5285–5291

Chen JS, Liu CW (2002) Numerical simulation of the evolution of aquifer porosity and species concentrations during reactive transport. Comput Geosci 28:485–499

Chen JS, Liu CW (2004) Interaction of reactive fronts during transport in a homogeneous porous medium with initial small non-uniformity. J Contam Hydrol 72:47–66

Chen JS, Liu CW, Lai GX, Ni CF (2009) Effects of mechanical dispersion on the morphological evolution of a chemical dissolution front in a fluid-saturated porous medium. J Hydrol 373:96–102

Christ JA, Ramsburg CA, Pennell KD, Abriola LM (2006) Estimating mass discharge from dense nonaqueous phase liquid source zones using upscaled mass transfer coefficients: an evaluation using multiphase numerical simulations. Water Resour Res 42:W11420. doi:10.1029/2006WR004886

Christ JA, Lemke LD, Abriola LM (2009) The influence of dimensionality on simulations of mass recovery from nonuniform dense non-aqueous phase liquid (DNAPL) source zones. Adv Water Resour 32:401–412

Daus AD, Frid EO, Sudicky EA (1985) Comparative error analysis in finite element formulations of the advection-dispersion equation. Adv Water Resour 8:86–95

DiFilippo EL, Carroll KC, Brusseau ML (2010) Impact of organic-liquid distribution and flow field heterogeneity on reduction in mass flux. J Contam Hydrol 115:14–25

Geller JT, Hunt JR (1993) Mass transfer from nonaqueous phase organic liquids in water-saturated porous media. Water Resour Res 29:833–845

Gerhard JI, Pang T, Kueper BH (2007) Time scales of DNAPL migration in sandy aquifers examined via numerical simulation. Ground Water 45:147–157

Grant GP, Gerhard JI (2007) Simulating the dissolution of a complex dense nonaqueous phase liquid zone: 1. Model to predict interfacial area. Water Resour Res 43:W12410. doi:10.1029/2007WR006038

Holzbecher EO (1998) Modeling density-driven flow in porous media. Springer, Berlin

Hong J, Hecker WC, Fletcher TH (1999) Predicting effectiveness factor for m-th order and langmuir rate equations in spherical coordinates. ACS Div Fuel Chem 44:1011–1015

Imhoff PT, Miller CT (1996) Dissolution fingering during the solubilization of nonaqueous phase liquids in saturated porous media: 1 Model predictions. Water Resour Res 32:1919–1928

Imhoff PT, Jaffe PR, Pinder GF (1994) An experimental study of complete dissolution of a nonaqueous phase liquid in saturated porous media. Water Resour Res 30:307–320

Imhoff PT, Thyrum GP, Miller CT (1996) Dissolution fingering during the solubilization of nonaqueous phase liquids in saturated porous media: 2 Experimental observations. Water Resour Res 32:1929–1942

Imhoff PT, Farthing MW, Gleyzer SN, Miller CT (2002) Evolving interface between clean and nonaqueous phase liquid (NAPL)-contaminated regions in two-dimensional porous media. Water Resour Res 38:1093–1106

Imhoff PT, Farthing MW, Miller CT (2003a) Modeling NAPL dissolution fingering with upscaled mass transfer rate coefficients. Adv Water Resour 26:1097–1111

Imhoff PT, Mann AS, Mercer M, Fitzpatrick M (2003b) Scaling DNAPL migration from the laboratory to the field. J Contam Hydrol 64:73–92

Kalia N, Balakotaiah V (2009) Effect of medium heterogeneities on reactive dissolution of carbonates. Chem Eng Sci 64:376–390

Maji R, Sudicky EA (2008) Influence of mass transfer characteristics for DNAPL source depletion and contaminant flux in a highly characterized glaciofluvial aquifer. J Contam Hydrol 102:105–119

Miller CT, Poirier-McNeil MM, Mayer AS (1990) Dissolution of trapped nonaqueous phase liquids: mass transfer characteristics. Water Resour Res 26:2783–2796

Miller CT, Christakos TG, Imhoff PT, McBride JF, Pedit JA, Trangenstein JA (1998) Multiphase flow and transport modeling in heterogeneous porous media: challenges and approaches. Adv Water Resour 21:77–120

Morris MI, Ball RC (1990) Renormalization of miscible flow functions. J Phys A: Math Gen 23:4199–4209

Ormond A, Ortoleva P (2000) Numerical modeling of reaction-induced cavities in a porous rock. J Geophys Res 105:16737–16747

Ortoleva P, Chadam J, Merino E, Sen A (1987) Geochemical self-organization II: The reactive-infiltration instability. Am J Sci 287:1008–1040

Parker JC, Park E (2004) Modeling field-scale dense nonaqueous phase liquid dissolution kinetics in heterogeneous aquifers. Water Resour Res 40:W05109. doi:10.1029/2003WR002807

Powers SE, Abriola LM, Weber WJ Jr (1994) An experimental investigation of nonaqueous phase liquid dissolution in saturated subsurface systems: transient mass transfer rates. Water Resour Res 30:321–332

Raffensperger JP, Garven G (1995) The formation of unconformity-type uranium ore deposits: coupled hydrochemical modelling. Am J Sci 295:639–696

Renard F, Gratier JP, Ortoleva P, Brosse E, Bazin B (1998) Self-organization during reactive fluid flow in a porous medium. Geophys Res Lett 25:385–388

Scheidegger AE (1961) General theory of dispersion in porous media. J Geophys Res 66:3273–3278

Seyedabbasi MA, Farthing MW, Imhoff PT, Miller CT (2008) The influence of wettability on NAPL dissolution fingering. Adv Water Resour 31:1687–1696

Soerens TS, Sabatini DA, Harwell JH (1998) Effects of flow bypassing and nonuniform NAPL distribution on the mass transfer characteristics of NAPL dissolution. Water Resour Res 34:1657–1673

Steefel CI, Lasaga AC (1990) Evolution of dissolution patterns: permeability change due to coupled flow and reaction. In: Melchior DC, Basset RL (eds) Chemical modeling in aqueous systems II, American Chemistry Society Symposium Series, vol 416, pp 213–225

Steefel CI, Lasaga AC (1994) A coupled model for transport of multiple chemical species and kinetic precipitation/dissolution reactions with application to reactive flow in single phase hydrothermal systems. Am J Sci 294:529–592

Tan CT, Homsy GM (1987) Stability of miscible displacements in porous media: radial source flow. Phys Fluids 30(5):1239–1245

Willson CS, Hall JL, Miller CT, Imhoff PT (1999) Factors affecting bank formation during surfactant-enhanced mobilization of residual NAPL. Environ Sci Technol 33:2440–2446

Yang Z, Yortsos YC (1998) Effect of no-flow boundaries on viscous fingering in porous media of large aspect ratio. Soc Petrol Eng J 3:285–292

Yeh GT, Tripathi VS (1991) A model for simulating transport of reactive multispecies components: model development and demonstration. Water Resour Res 27:3075–3094

Zhang C, Werth CJ, Webb AG (2007) Characterization of NAPL source zone architecture and dissolution kinetics in heterogeneous porous media using magnetic resonance imaging. Environ Sci Technol 41:3672–3678

Zhao C, Steven GP (1996a) A posteriori error estimator/corrector for natural frequencies of thin plate vibration problems. Comput Struct 59:949–963

Zhao C, Steven GP (1996b) An asymptotic formula for correcting finite element predicted natural frequencies of membrane vibration problems. Commun Numer Methods Eng 11:63–73

Zhao C, Steven GP (1996c) A practical error estimator for predicted natural frequencies of two-dimensional elastodynamic problems. Eng Comput 13:19–37

Zhao C, Hobbs BE, Hornby P, Ord A, Peng S, Liu L (2008a) Theoretical and numerical analyses of chemical-dissolution front instability in fluid-saturated porous rocks. Int J Numer Anal Meth Geomech 32:1107–1130

Zhao C, Hobbs BE, Ord A, Hornby P, Peng S (2008b) Effect of reactive surface areas associated with different particle shapes on chemical-dissolution front instability in fluid-saturated porous rocks. Transp Porous Media 73:75–94

Zhao C, Hobbs BE, Ord A, Hornby P, Peng S (2008c) Morphological evolution of three-dimensional chemical dissolution front in fluid-saturated porous media: a numerical simulation approach. Geofluids 8:113–127

Zhao C, Hobbs BE, Ord A (2009) Fundamentals of computational geoscience: numerical methods and algorithms. Springer, Berlin

Zhao C, Hobbs BE, Ord A, Peng S (2010a) Effects of mineral dissolution ratios on chemical-dissolution front instability in fluid-saturated porous media. Transp Porous Media 82:317–335

Zhao C, Hobbs BE, Ord A (2010b) Theoretical analyses of the effects of solute dispersion on chemical-dissolution front instability in fluid-saturated porous rocks. Transp Porous Media 84:629–653

Zhao C, Hobbs BE, Ord A (2010c) Theoretical analyses of nonaqueous-phase-liquid dissolution induced instability in two-dimensional fluid-saturated porous media. Int J Numer Anal Meth Geomech 34:1767–1796

Zhao C, Hobbs BE, Regenauer-Lieb K, Ord A (2011) Computational simulation for the morphological evolution of nonaqueous-phase-liquid dissolution fronts in two-dimensional fluid-saturated porous media. Comput Geosci 15:167–183

Zhao C, Hobbs BE, Ord A (2012) Effects of domain shapes on the morphological evolution of nonaqueous-phase-liquid dissolution fronts in fluid-saturated porous media. J Contam Hydrol 138–139:123–140

Zienkiewicz OC (1977) The finite element method. McGraw-Hill, London

Chapter 11
Fundamental Theory for Acidization Dissolution-Front Instability in Fluid-Saturated Carbonate Rocks

Instability of acidization dissolution fronts in carbonate rocks is an important mechanism of the karst formation that is commonly observed in nature. This mechanism has been successfully used to increase the oil production in petroleum industry through injecting hydrochloric acid into carbonate rocks in the surroundings of drilling wells (Fredd and Fogler 1998; Golfier et al. 2002; Panga et al. 2005; Kalia and Balakotaiah 2007, 2009; Cohen et al. 2008). Due to the rapid rates of acidization dissolution, the dissolution front may become unstable under certain conditions, so that preferential flow channels such as wormholes can be generated. Generally, the acidization dissolution process of carbonate rocks involves the following three steps: (1) the acid is transported by diffusion and advection to the solid particle surfaces of the porous rock; (2) the chemical dissolution reaction takes place at the solid particle surfaces of the porous rock; and (3) the products of the chemical dissolution reaction are transported away from the solid particle surfaces of the porous rock. This means that the pore-fluid flow, which transports both the acid to and the chemical product away from the solid particle surfaces of the porous rock through advection, plays a critical role in the acidization dissolution process of carbonate rocks.

Under the stimulus of developing a new technology to increase oil production from carbonate reservoirs, a large amount of experimental and theoretical research has been carried out to investigate the effects of various factors, such as the acid injecting rate, acid volume, rock permeability, reaction kinetics and so forth, on the formation of the wormholes (Fredd and Fogler 1998; Golfier et al. 2002; Panga et al. 2005; Kalia and Balakotaiah 2007, 2009; Cohen et al. 2008). The existing experimental results have displayed the following five possible dissolution structures: the face or compact structure, the conical wormhole structure, the dominant wormhole structure, the ramified wormhole structure and the uniform dissolution structure. Although the instability phenomena of acidization dissolution fronts in carbonate rocks have been observed in many laboratory experiments and computational simulations (Golfier et al. 2002; Panga et al. 2005; Kalia and Balakotaiah 2007, 2009; Cohen et al. 2008), a theoretical criterion, which can be used to assess

C. Zhao, *Physical and Chemical Dissolution Front Instability in Porous Media*, Lecture Notes in Earth System Sciences,
DOI: 10.1007/978-3-319-08461-9_11

the onset of acidization dissolution-front instability, is still lacking. Since the laboratory experiments were conducted using carbonate rocks of remarkable heterogeneity, it is difficult to distinguish whether the wormhole structure is caused either by the rock heterogeneity or by the instability of the acidization dissolution front. It needs to be pointed out that the rock heterogeneity reflects some geometrical characteristic of a system, while the instability of the acidization dissolution front reflects the overall dynamic characteristic of the system. From the linear stability theory, the system under consideration should be ideally comprised of a homogeneous porous medium with a small perturbation, namely a carbonate rock of slight heterogeneity, for dealing with the instability of acidization dissolution fronts in fluid-saturated carbonate rocks.

In terms of the stability analysis of acidization dissolution fronts in fluid-saturated porous rocks, the current status of theoretical study is that in order to avoid mathematical difficulties, the porosity of the rock is assumed to be constant during the acidization dissolution reaction, but the permeability of the rock is allowed to have a significant change (Sherwood 1987; Hinch and Bhatt 1990). Due to this assumption, the related theoretical results may be valid for fluid-saturated sandstones, but are definitely not valid for fluid-saturated carbonate rocks. In the case of considering the acidization dissolution of fluid-saturated sandstones, Sherwood (1987), Hinch and Bhatt (1990) assumed that the rock itself does not dissolve, and the injected acid can only etch clays and other particles that may block the pore throats. As a result, the porosity of the sandstone varies little, but the permeability of the sandstone may be increased by perhaps an order of magnitude. However, when the acidization dissolution of a fluid-saturated carbonate rock is considered, the rock itself can be completely dissolved, resulting in no resistance to pore-fluid flow. This means that both the porosity and permeability of the carbonate rock can have a dramatic change during the acidization dissolution. For this reason, the existing theoretical results obtained by Sherwood (1987), Hinch and Bhatt (1990) are not suitable for describing the dynamic behavior associated with the propagation of acidization dissolution fronts in a fluid-saturated carbonate rock, to say nothing of providing a theoretical criterion that can be used to assess the instability of acidization dissolution fronts in the fluid-saturated carbonate rock.

When the chemical dissolution rate is very slow and the mineral dissolution ratio is small enough, a considerable amount of theoretical research and computational simulation have been conducted to investigate the instability of chemical dissolution fronts in fluid-saturated porous media (Chadam et al. 1986, 1988; Ortoleva et al. 1987; Ormond and Ortoleva 2000; Chen and Liu 2002, 2004; Zhao et al. 2008; Chen et al. 2009; Szymczak and Ladd 2009). In particular, Zhao et al. (2009) have proposed a comprehensive dimensionless number, known as the Zhao number, to represent the dynamic characteristic of the chemical dissolution system that is comprised of a fluid-saturated porous medium. Using the proposed Zhao number and its critical value, a theoretical criterion is derived for assessing the instability of chemical dissolution fronts in fluid-saturated porous media (Zhao et al. 2008). Although the acidization dissolution of a carbonate rock can be viewed as a special kind of chemical dissolution problem in the fluid-saturated porous medium, there exist some significant differences between the acidization dissolution problem and the slow-rate chemical dissolution problem. First, for the acidization dissolution problem, the reactant (i.e. acid) concentration is assumed to be constant in the upstream direction of the dissolution front, while for the slow-rate chemical dissolution problem, the dissolved mineral concentration is assumed to be constant in the downstream direction of the dissolution front. Second, the reactant (i.e. acid) concentration in the downstream direction of the dissolution front is assumed to be zero for the acidization dissolution problem, while the dissolved mineral concentration in the upstream direction of the dissolution front is assumed to be zero for the slow-rate chemical dissolution problem. Despite these differences, similar mathematical methods can be used to conduct the theoretical analysis for both the acidization dissolution problem and the slow-rate chemical dissolution problem because they have similar mathematical models.

11.1 Mathematical Analysis of the Acidization Dissolution Problem

To facilitate the mathematical analysis of the acidization dissolution-front instability in a fluid-saturated carbonate rock, the Darcy scale mathematical model (Golfier et al. 2002) is used with the following assumptions: (1) Darcy's law can be used to describe pore-fluid flow in the carbonate rock, even the carbonate rock is completely dissolved; (2) the acidization of the carbonate rock is a fast process; (3) the pore-fluid is incompressible; (4) the Carman-Kozeny law (Detournay and Cheng 1993; Scheidegger 1974; Nield and Bejan 1992) can be used to express the relationship between the porosity and permeability of the carbonate rock; (5) the propagation speed of the acidization dissolution front is much slower than the Darcy velocity of the pore-fluid flow; (6) the mass exchange between the acid and carbonate rock during acidization dissolution is proportional to both the acid concentration and the difference between the porosity and its final value of the carbonate rock, so that the acidization dissolution stops when either the acid is completely depleted (i.e. the acid concentration is zero) or the carbonate rock is completely dissolved (i.e. the porosity is equal to the final porosity). Under these assumptions, the governing equations and related formulas of the acidization dissolution-front instability problem in a fluid-saturated carbonate rock can be written as follows:

$$\vec{u} = -\frac{k(\phi)}{\mu}\nabla p, \tag{11.1}$$

$$\frac{\partial \phi}{\partial t} + \nabla \cdot \vec{u} = 0, \tag{11.2}$$

$$\frac{\partial}{\partial t}(\phi C) + \nabla \cdot [C\vec{u}] = \nabla \cdot [\phi D \nabla C] - \alpha_0(\phi_f - \phi)C, \tag{11.3}$$

$$\frac{\partial \phi}{\partial t} = \frac{\chi \alpha_0}{\rho_s}(\phi_f - \phi)C, \tag{11.4}$$

$$\alpha = \alpha_0(\phi_f - \phi), \tag{11.5}$$

where $\vec{u}$ is the Darcy velocity vector within the carbonate rock; p and C are the pore-fluid pressure and the acid concentration (with a unit of moles per pore-fluid volume); μ is the dynamic viscosity of the pore-fluid; ϕ is the porosity of the carbonate rock; D is the effective diffusivity/dispersivity of the acid; $k(\phi)$ is the permeability of the porous medium; χ is the stoichiometric coefficient of the carbonate rock; α is the mass exchange rate between the acid and the carbonate rock; ρ_s is the molar density (i.e. moles per volume) of the carbonate rock; ϕ_f is the final (i.e. maximum) porosity when the carbonate rock is completely dissolved.

Equation (11.5) indicates that the mass exchange rate between the acid and the carbonate rock is directly proportional to both the porosity of the carbonate rock and the mass exchange rate coefficient (i.e. α_0) between the acid and the carbonate rock. Once the carbonate rock is completely dissolved, the porosity of the carbonate rock is equal to its final porosity (i.e. ϕ_f), so that the mass exchange rate between the acid and the carbonate rock is equal to zero. This implies that the dissolution reaction can be automatically stopped in the region where the carbonate rock is completely dissolved.

If the acid reacts with the carbonate (e.g. calcite or dolomite), then the corresponding acidization dissolution can be expressed as follows:

$$CaCO_3 + 2H^+ \rightarrow Ca^{2+} + H_2O + CO_2,$$

$$CaMg(CO_3)_2 + 4H^+ \rightarrow Ca^{2+} + Mg^{2+} + 2H_2O + 2CO_2. \tag{11.6}$$

Equation (11.6) indicates that one mole acid can dissolve 0.5 mol calcite, so that the stoichiometric coefficient of the dissolved calcite is 0.5. Similarly, since one mole acid can dissolve 0.25 mol dolomite, the stoichiometric coefficient of the dissolved dolomite is 0.25. Suppose the weight density of the carbonate rock is 2700 kg/m^3, the molar density of calcite is about 27 mol/l, while the molar density of dolomite is about 14.7 mol/l.

To consider the permeability change caused by a change in porosity, the Carman-Kozeny law (Detournay and Cheng 1993; Scheidegger 1974; Nield and Bejan 1992) is used to calculate permeability k, for a given porosity ϕ.

$$k(\phi) = \frac{k_0(1 - \phi_0)^2 \phi^3}{\phi_0^3 (1 - \phi)^2}, \tag{11.7}$$

where ϕ_0 and k_0 are the initial reference porosity and permeability of the carbonate rock respectively.

Mathematically, the governing equations and related formulas of the acidization dissolution-front instability problem in a fluid-saturated carbonate rock can be rewritten into the following form:

$$\frac{\partial \phi}{\partial t} - \nabla \cdot [\psi(\phi)\nabla p] = 0, \tag{11.8}$$

$$\frac{\partial}{\partial t}(\phi C) - \nabla \cdot [C\psi(\phi)\nabla p] - \nabla \cdot [\phi D \nabla C] + \alpha_0(\phi_f - \phi)C = 0, \tag{11.9}$$

$$\frac{\partial \phi}{\partial t} - \frac{\chi \alpha_0}{\rho_s}(\phi_f - \phi)C = 0, \tag{11.10}$$

$$\psi(\phi) = \frac{k(\phi)}{\mu}. \tag{11.11}$$

11.1.1 Determination of the Propagation Speed of the Acidization Dissolution Front in a Fluid-Saturated Carbonate Rock

When a planar acidization dissolution-front is considered to propagate in the full space, analytical solutions can be obtained for both the propagation speed of the acidization dissolution front and the downstream pressure gradient of the pore-fluid in the fluid-saturated carbonate rock. In this special case, a planar dissolution front is assumed to propagate in the positive x direction, so that all quantities are independent of the transverse coordinates y and z. For this reason, Eqs. (11.8)–(11.10) can be rewritten as follows:

$$\frac{\partial \phi}{\partial t} - \frac{\partial}{\partial x}[\psi(\phi)\frac{\partial p}{\partial x}] = 0, \tag{11.12}$$

$$\frac{\partial}{\partial t}(\phi C) - \frac{\partial}{\partial x}[\phi D\frac{\partial C}{\partial x} + C\psi(\phi)\frac{\partial p}{\partial x}] + \alpha_0(\phi_f - \phi)C = 0, \tag{11.13}$$

$$\frac{\partial \phi}{\partial t} - \frac{\chi \alpha_0}{\rho_s}(\phi_f - \phi)C = 0. \tag{11.14}$$

If the acid is injected at the location of x approaching negative infinity, then the boundary conditions of the problem can be expressed as

$$\lim_{x \to -\infty} C = C_0, \quad \lim_{x \to -\infty} \frac{\partial p}{\partial x} = p'_{fx} \quad \text{(upstream boundary)}, \tag{11.15}$$

$$\lim_{x \to \infty} C = 0, \quad \lim_{x \to \infty} \frac{\partial p}{\partial x} = p'_{0x} \quad \text{(downstream boundary)}, \tag{11.16}$$

where C_0 is the injected acid concentration in the carbonate rock; p'_{fx} is the pore-fluid pressure gradient as x approaching negative infinity in the upstream of the pore-fluid flow; p'_{0x} is the unknown pore-fluid pressure gradient as x approaching positive infinity in the downstream of the pore-fluid flow. Because p'_{fx} drives the pore-fluid flow continuously along the positive x direction, it has a negative algebraic value (i.e. $p'_{fx} < 0$) in the mathematical analysis.

The initial condition for this problem is: $\phi(x, 0) = \phi_0$ expect at the negative infinity, where $\lim_{\bar{x} \to -\infty} \phi(x, 0) = \phi_f$. Note that ϕ_0 is the initial porosity of the carbonate rock.

To transform a moving boundary problem of the acidization dissolution front (in an $x - t$ coordinate system) into a steady-state boundary problem of the acidization dissolution front (in a $\xi - t$ coordinate system), the following coordinate mapping can be used.

$$\xi = x - v_{front} t, \tag{11.17}$$

where v_{front} is the propagation speed of the planar acidization dissolution-front in the fluid-saturated carbonate rock.

From the mathematical point of view, the following relationships exist between the partial derivatives with respect to ξ and t and those with respect to x and t (Turcotte and Schubert 1982):

$$\left(\frac{\partial}{\partial t}\right)_\xi = \left(\frac{\partial}{\partial t}\right)_x + \frac{\partial}{\partial x}\frac{\partial x}{\partial t} = \left(\frac{\partial}{\partial t}\right)_x + v_{front}\frac{\partial}{\partial x}, \quad \left(\frac{\partial}{\partial \xi}\right)_t = \left(\frac{\partial}{\partial x}\right)_t, \tag{11.18}$$

where derivatives are taken with x or t held constant as appropriate.

When Eq. (11.17) is considered, the following relationship exists:

$$u_\xi = u_x - v_{front}, \quad u_\xi = \frac{\partial \xi}{\partial t}, \quad u_x = \frac{\partial x}{\partial t} \tag{11.19}$$

where u_ξ is the relative velocity to the $\xi - t$ coordinate system; u_x is the absolute velocity in the $x - t$ coordinate system; v_{front} is the moving velocity of the $\xi - t$ coordinate system relative to the $x - t$ coordinate system (i.e. the convected velocity of the $\xi - t$ coordinate system).

Since Darcy's law is valid in both the $x - t$ coordinate system and the $\xi - t$ coordinate system, Eq. (11.19) can be further expressed as follows:

$$-\frac{k(\phi)}{\mu}\frac{\partial p}{\partial \xi} = -\frac{k(\phi)}{\mu}\frac{\partial p}{\partial x} - v_{front}, \quad \frac{\partial p}{\partial \xi} = \frac{\partial p}{\partial x} + \frac{\mu}{k(\phi)}v_{front}, \tag{11.20}$$

where $\partial p/\partial \xi$ is the pore-fluid pressure gradient in the $\xi - t$ coordinate system; $\partial p/\partial x$ is the pore-fluid pressure gradient in the $x - t$ coordinate system.

Equation (11.20) indicates that the pore-fluid pressure gradient measured in the $\xi - t$ coordinate system is different from that in the $x - t$ coordinate system as a result of the motion of the $\xi - t$ coordinate system. This means that the pore-fluid pressure measured in the $\xi - t$ coordinate system is different from that in the $x - t$ coordinate system unless the convected motion of the $\xi - t$ coordinate system is slow enough. However, both the pore-fluid pressure gradient and the Darcy velocity are transformed from the $\xi - t$ coordinate system into the $x - t$ coordinate system using the following formulas:

$$\frac{\partial p}{\partial x} = \frac{\partial p}{\partial \xi} - \frac{\mu}{k(\phi)}v_{front}, \quad u_x = u_\xi + v_{front}. \tag{11.21}$$

This means that if the Darcy velocity, pore-fluid pressure and its gradient in the $x - t$ coordinate system are of interest, u_x and $\partial p/\partial x$ can be directly used to replace u_ξ and $\partial p/\partial \xi$ when the theoretical analysis is conducted in the $\xi - t$ coordinate system, so that the results in the $x - t$ coordinate system can be obtained without a need to carry out the transformation expressed in Eq. (11.21). For this reason, $(\partial/\partial \xi)_t = (\partial/\partial x)_t$ is used in the following theoretical analysis.

Since the transformed acidization-dissolution system in the $\xi - t$ coordinate system is in a steady state, the following equation can be derived from Eq. (11.18):

$$\left(\frac{\partial}{\partial t}\right)_x = -v_{front}\frac{\partial}{\partial \xi}, \quad \left(\frac{\partial}{\partial \xi}\right)_t = \left(\frac{\partial}{\partial x}\right)_t. \tag{11.22}$$

Substituting Eq. (11.22) into Eqs. (11.12)–(11.14) yields the following equations:

$$\frac{\partial}{\partial \xi}\left[\psi(\phi)\frac{\partial p}{\partial \xi} + v_{front}\phi\right] = 0, \tag{11.23}$$

$$\frac{\partial}{\partial \xi}\left[\phi D\frac{\partial C}{\partial \xi} + C\psi(\phi)\frac{\partial p}{\partial \xi} + v_{front}\left(C + \frac{\rho_s}{\chi}\right)\phi\right] = 0, \tag{11.24}$$

$$v_{front}\frac{\partial \phi}{\partial \xi} + \frac{\chi\alpha_0}{\rho_s}(\phi_f - \phi)C = 0. \tag{11.25}$$

Integrating Eqs. (11.23) and (11.24) from negative infinity to positive infinity and using the boundary conditions (i.e. Eqs. (11.15) and (11.16)) yield the following equations:

$$C_0\psi(\phi_f)p'_{fx} + v_{front}\phi_f(C_0 + \frac{\rho_s}{\chi}) - v_{front}\phi_0\frac{\rho_s}{\chi} = 0, \tag{11.26}$$

$$\psi(\phi_0)p'_{0x} + v_{front}\phi_0 - \psi(\phi_f)p'_{fx} - v_{front}\phi_f = 0. \tag{11.27}$$

Solving Eqs. (11.26) and (11.27) simultaneously results in the following analytical solutions:

$$\begin{aligned} v_{front} &= \frac{-\psi(\phi_f)p'_{fx}C_0}{\phi_f C_0 + (\phi_f - \phi_0)\frac{\rho_s}{\chi}} = \frac{-\psi(\phi_0)p'_{0x}C_0}{\phi_0 C_0 + (\phi_f - \phi_0)\frac{\rho_s}{\chi}} \\ &= \frac{u_{0x}C_0}{\phi_0 C_0 + (\phi_f - \phi_0)\frac{\rho_s}{\chi}}, \end{aligned} \tag{11.28}$$

$$p'_{0x} = \frac{\psi(\phi_f)[\phi_0 C_0 + (\phi_f - \phi_0)\frac{\rho_s}{\chi}]}{\psi(\phi_0)[\phi_f C_0 + (\phi_f - \phi_0)\frac{\rho_s}{\chi}]}p'_{fx}, \tag{11.29}$$

where u_{0x} is the Darcy velocity in the far downstream of the pore-fluid flow as x approaching positive infinity. Using Darcy's law, u_{0x} can be expressed as

$$u_{0x} = \frac{\phi_0 C_0 + (\phi_f - \phi_0)\frac{\rho_s}{\chi}}{\phi_f C_0 + (\phi_f - \phi_0)\frac{\rho_s}{\chi}}u_{fx}, \tag{11.30}$$

where u_{fx} is the Darcy velocity in the far upstream of the pore-fluid flow as x approaching negative infinity.

11.1.2 Instability Analysis of the Acidization Dissolution Front in a Fluid-Saturated Carbonate Rock

To facilitate the theoretical analysis, Eqs. (11.8)–(11.10) are changed into the following dimensionless equations:

$$\frac{\partial\phi}{\partial\bar{t}} - \nabla\cdot[\psi^*(\phi)\nabla\bar{p}] = 0, \tag{11.31}$$

$$\frac{\partial}{\partial\bar{t}}(\phi\bar{C}) - \nabla\cdot[D^*(\phi)\nabla\bar{C} + \bar{C}\psi^*(\phi)\nabla\bar{p}] + \frac{\partial\phi}{\varepsilon\partial\bar{t}} = 0, \tag{11.32}$$

$$\frac{\partial\phi}{\partial\bar{t}} - (\phi_f - \phi)\bar{C} = 0, \tag{11.33}$$

where

$$\bar{x} = \frac{x}{L^*}, \quad \bar{y} = \frac{y}{L^*}, \quad \bar{C} = \frac{C}{C_0}, \quad \bar{p} = \frac{p}{p^*}, \quad \bar{\vec{u}} = \frac{\vec{u}}{u^*}, \quad \bar{t} = \frac{t}{t^*} \tag{11.34}$$

$$t^* = \frac{\rho_s}{\alpha_0 \chi C_0}, \quad L^* = \sqrt{\phi_f D t^*}, \quad p^* = \frac{\phi_f D}{\psi(\phi_f)}, \quad u^* = \frac{\phi_f D}{L^*},$$

$$\varepsilon = \frac{\chi C_0}{\rho_s}, \quad D^*(\phi) = \frac{\phi D}{\phi_f D}, \quad \psi^*(\phi) = \frac{\psi(\phi)}{\psi(\phi_f)}. \tag{11.35}$$

Note that ε is the mineral dissolution ratio of a carbonate rock. It is directly proportional to the ratio of the molar concentration of the consumed acid to the molar density of the dissolved carbonate rock during the acidization dissolution. If the acid concentration is 1 mol/l, then the mineral dissolution ratio of calcite (i.e. $CaCO_3$) is about 0.019, while the mineral dissolution ratio of dolomite (i.e. $CaMg(CO_3)_2$) is about 0.017. However, if the pH value of the solution is equal to 4 (in the case of a weak acid), then the mineral dissolution ratios of calcite and dolomite are 0.19×10^{-5} and 0.17×10^{-5} respectively. Since the geometrical domain shape of the theoretical acidization dissolution problem is a full space, L^*, t^* and p^* can be defined as the intrinsic length, time and pore-fluid pressure of the acidization dissolution problem.

The corresponding boundary conditions can be expressed in a dimensionless form as follows:

$$\lim_{\bar{x} \to -\infty} \bar{C} = 1, \quad \lim_{\bar{x} \to -\infty} \frac{\partial \bar{p}}{\partial \bar{x}} = \bar{p}'_{fx} \quad \text{(upstream boundary)}, \tag{11.36}$$

$$\lim_{\bar{x} \to \infty} \bar{C} = 0, \quad \lim_{\bar{x} \to \infty} \frac{\partial \bar{p}}{\partial \bar{x}} = \bar{p}'_{0x}, \quad \text{(downstream boundary)}. \tag{11.37}$$

In this case, the initial condition for this theoretical problem is: $\phi(\bar{x}, 0) = \phi_0$ expect at the negative infinity, where $\lim_{\bar{x} \to -\infty} \phi(\bar{x}, 0) = \phi_f$.

The acidization dissolution front divides the problem domain into two regions, an upstream region and a downstream region, relative to the acidization dissolution front. Across this front, the porosity undergoes a jump from its initial value into its final value. Thus, the acidization dissolution-front propagation problem can be considered as a Stefan moving boundary problem. This means that the dimensionless governing equations of the acidization dissolution-front propagation problem in both the upstream region and the downstream region can be expressed in the following forms:

$$\phi_f \bar{v}_{front} \frac{\partial \bar{C}}{\partial \bar{x}} + \nabla \cdot (\nabla \bar{C} + \bar{C} \nabla \bar{p}) = 0, \quad \nabla^2 \bar{p} = 0, \quad \phi = \phi_f \quad \text{(in the upstream region)}, \tag{11.38}$$

$$\bar{C} = 0, \quad \nabla^2 \bar{p} = 0, \quad \phi = \phi_0 \quad \text{(in the downstream region)}. \tag{11.39}$$

If the acidization dissolution front is denoted by $S(\bar{x}, \bar{t}) = 0$, then the dimensionless pore-fluid pressure, acid concentration and mass fluxes of both the acid and the pore-fluid should be continuous on $S(\bar{x}, \bar{t}) = 0$. This leads to the following interface conditions for this moving dissolution front problem.

$$\lim_{S \to 0^-} \bar{C} = \lim_{S \to 0^+} \bar{C}, \quad \lim_{S \to 0^-} \bar{p} = \lim_{S \to 0^+} \bar{p} \tag{11.40}$$

$$\lim_{S \to 0^-} \frac{\partial \bar{C}}{\partial \bar{n}} = -\frac{\bar{v}_{front}}{\varepsilon}(\phi_f - \phi_0), \quad \lim_{S \to 0^-} \frac{\partial \bar{p}}{\partial \bar{n}} = \frac{\psi(\phi_0)}{\psi(\phi_f)} \lim_{S \to 0^+} \frac{\partial \bar{p}}{\partial \bar{n}} - \bar{v}_{front}(\phi_f - \phi_0), \tag{11.41}$$

where $\bar{n}$ is the dimensionless unit normal vector of the moving dissolution front.

Note that since the propagation speed of the acidization dissolution front is usually much slower than the Darcy velocity of the pore-fluid flow, the term, $\bar{v}_{front}(\phi_f - \phi_0)$, can be neglected from Eq. (11.41) hereafter. When the planar acidization dissolution-front is under stable conditions, the base solutions for the acidization dissolution problem can be derived from Eqs. (11.38) and (11.39) with the related boundary conditions (i.e. Eqs. (11.36) and (11.37)) and interface conditions (i.e. Eqs. (11.40) and (11.41)). As a result, the following base solutions are obtained.

$$\bar{C}(\xi) = 1 - \exp[-(\bar{p}'_{fx} + \phi_f \bar{v}_{front})\xi], \quad \bar{p}(\xi) = \bar{p}'_{fx}\xi + \bar{p}_{C2}$$

$$\phi = \phi_f \quad \text{(in the upstream region)}, \tag{11.42}$$

$$\bar{C}(\xi) = 0, \quad \bar{p}(\xi) = \bar{p}'_{0x}\xi + \bar{p}_{C1}, \quad \phi = \phi_0, \quad \text{(in the downstream region)}, \tag{11.43}$$

where $\bar{p}_{C1}$ and $\bar{p}_{C2}$ are two constants to be determined. For example, $\bar{p}_{C1}$ can be determined by setting the dimensionless pore-fluid pressure $\bar{p}(\xi)$ to be a constant at a prescribed location of the downstream region, while $\bar{p}_{C2}$ can be determined using the pore-fluid pressure continuity condition at the interface between the upstream and downstream regions.

Substituting Eq. (11.42) into Eq. (11.41) yields the following equation:

$$\bar{v}_{front} = -\frac{\varepsilon \bar{p}'_{fx}}{\varepsilon \phi_f + (\phi_f - \phi_0)}. \tag{11.44}$$

If the related dimensionless quantities are used, Eq. (11.28) can be changed into the dimensionless form, in which the dimensionless propagation speed of the acidization dissolution front is exactly the same as that expressed in Eq. (11.44). This can demonstrates that the above mathematical deduction is correct.

Next, the linear stability analysis (Chadam et al. 1986, 1988; Zhao et al. 2008) is conducted to derive the critical condition, under which the acidization dissolution front becomes unstable during its propagation. For this purpose, a small time-dependent perturbation is added on the planar acidization dissolution-front, so that the total solution of the acidization dissolution system is equal to the sum of the base solution and the perturbed solution of the system.

$$S(\xi, \bar{y}, t) = \xi - \delta \exp(\bar{\omega}\, \bar{t}) \cos(\overline{my}), \tag{11.45}$$

$$\bar{p}_{total}(\xi, \bar{y}, \bar{t}) = \bar{p}(\xi, \bar{t}) + \delta \hat{p}(\xi) \exp(\bar{\omega}\, \bar{t}) \cos(m\bar{y}), \tag{11.46}$$

$$\bar{C}_{total}(\xi, \bar{y}, \bar{t}) = \bar{C}(\xi, \bar{t}) + \delta \hat{C}(\xi) \exp(\bar{\omega}\, \bar{t}) \cos(m\bar{y}), \tag{11.47}$$

where $\bar{\omega}$ is the dimensionless growth rate of the small perturbation; $\bar{m}$ is the dimensionless wavenumber of the small perturbation; δ is the amplitude of the small perturbation and $\delta << 1$ by the definition of a linear stability analysis.

Because $S(\xi, \bar{y}, \bar{t})$ is a function of coordinates ξ and $\bar{y}$, the following derivatives exist mathematically:

$$\left(\frac{\partial}{\partial \xi}\right)_{\xi} = \frac{\partial S}{\partial \xi}\frac{\partial}{\partial S} = \left(\frac{\partial}{\partial \xi}\right)_{S}, \tag{11.48}$$

$$\left(\frac{\partial}{\partial \bar{y}}\right)_{\xi} = \frac{\partial S}{\partial \bar{y}}\frac{\partial}{\partial S} + \left(\frac{\partial}{\partial \bar{y}}\right)_{S} = \frac{\partial S}{\partial \bar{y}}\left(\frac{\partial}{\partial \xi}\right)_{S} + \left(\frac{\partial}{\partial \bar{y}}\right)_{S}, \tag{11.49}$$

$$\left(\frac{\partial^2}{\partial \xi^2}\right)_{\xi} = \left(\frac{\partial^2}{\partial \xi^2}\right)_{S}, \tag{11.50}$$

$$\left(\frac{\partial^2}{\partial \bar{y}^2}\right)_{\xi} = \frac{\partial^2 S}{\partial \bar{y}^2}\frac{\partial}{\partial \xi} + \left(\frac{\partial S}{\partial \bar{y}}\right)^2 \frac{\partial^2}{\partial \xi^2} + 2\frac{\partial S}{\partial \bar{y}}\frac{\partial^2}{\partial \xi \partial \bar{y}} + \left(\frac{\partial^2}{\partial \bar{y}^2}\right)_{S}. \tag{11.51}$$

Substituting Eqs. (11.46)–(11.51) into Eqs. (11.38) and (11.39) yields the first-order perturbation equations of the acidization dissolution system as follows:

$$\hat{C} = 0, \quad \frac{\partial^2 \hat{p}}{\partial \xi^2} - \bar{m}^2 \hat{p} + \bar{m}^2 \bar{p}'_{0x} = 0 \quad \text{(in the downstream region)}, \tag{11.52}$$

$$\frac{\partial^2 \hat{C}}{\partial \xi^2} + (\phi_f \bar{v}_{front} + \bar{p}'_{fx}) \frac{\partial \hat{C}}{\partial \xi} - \bar{m}^2 \hat{C} + \bar{m}^2 (\phi_f \bar{v}_{front} + \bar{p}'_{fx}) \exp[-(\phi_f \bar{v}_{front} + \bar{p}'_{fx})\xi]$$
$$+ (\phi_f \bar{v}_{front} + \bar{p}'_{fx}) \exp[-(\phi_f \bar{v}_{front} + \bar{p}'_{fx})\xi] \frac{\partial \hat{p}}{\partial \xi} = 0,$$
$$\frac{\partial^2 \hat{p}}{\partial \xi^2} - \bar{m}^2 \hat{p} + \bar{m}^2 \bar{p}'_{fx} = 0 \qquad \text{(in the upstream region).} \tag{11.53}$$

The corresponding boundary conditions of the first-order perturbation problem are:

$$\hat{C} = 0, \quad \lim_{x \to \infty} \frac{\partial \hat{p}}{\partial \xi} = 0 \qquad \text{(downstream boundary)}, \tag{11.54}$$

$$\lim_{x \to -\infty} \hat{C} = 0, \quad \lim_{x \to -\infty} \frac{\partial \hat{p}}{\partial \xi} = 0, \qquad \text{(upstream boundary)}. \tag{11.55}$$

The interface conditions for this first-order perturbation problem can be expressed as follows:

$$\hat{C} = 0, \quad \lim_{S \to 0^-} \hat{p} = \lim_{S \to 0^+} \hat{p} \tag{11.56}$$

$$\lim_{S \to 0^-} \frac{\partial \hat{C}}{\partial \bar{n}} = -\frac{\omega}{\varepsilon}(\phi_f - \phi_0), \quad \lim_{S \to 0^-} \frac{\partial \hat{p}}{\partial \bar{n}} = \frac{\psi(\phi_0)}{\psi(\phi_f)} \lim_{S \to 0^+} \frac{\partial \hat{p}}{\partial \bar{n}}. \tag{11.57}$$

Solving Eqs. (11.52) and (11.53) with the boundary and interface conditions (i.e. Eqs. (11.54)–(11.57)) yields the following analytical results:

$$\hat{C} = 0, \quad \hat{p}(\xi) = \bar{p}'_{0x}[1 - \frac{1-\beta}{1+\beta}\exp(-\bar{m}\xi)], \quad \text{(in the downstream region)}, \tag{11.58}$$

$$\hat{C}(\xi) = (\phi_f \bar{v}_{front} + \bar{p}'_{fx})\{\exp[-(\phi_f \bar{v}_{front} + \bar{p}'_{fx})\xi] - \frac{(1+\beta)\phi_f \bar{v}_{front} + 2\bar{p}'_{fx}}{(\phi_f \bar{v}_{front} + \bar{p}'_{fx})(1+\beta)} \exp(\sigma\xi)\}$$
$$+ \frac{1-\beta}{1+\beta} \bar{p}'_{fx} \exp\{[\bar{m} - (\phi_f \bar{v}_{front} + \bar{p}'_{fx})]\xi\},$$

$$\hat{p}(\xi) = \bar{p}'_{fx}[1 + \frac{1-\beta}{1+\beta}\exp(\bar{m}\xi)] \quad \text{(in the upstream region)}, \tag{11.59}$$

where

$$\beta = \frac{\psi(\phi_0)}{\psi(\phi_f)} = \frac{k(\phi_0)}{k(\phi_f)} = \left(\frac{\phi_0}{\phi_f}\right)^3 \left(\frac{1-\phi_f}{1-\phi_0}\right)^2, \tag{11.60}$$

$$\sigma = \frac{\sqrt{(\phi_f \bar{v}_{front} + \bar{p}'_{fx})^2 + 4\bar{m}^2} - (\phi_f \bar{v}_{front} + \bar{p}'_{fx})}{2}. \tag{11.61}$$

The characteristic equation for the dimensionless growth rate of the perturbed acidization dissolution system can be derived from the consideration of both Eqs. (11.57) and (11.59). This yields the following equation for the first-order derivative of the dimensionless concentration of the acid:

$$\begin{aligned} \left.\frac{\partial \hat{C}}{\partial \xi}\right|_{\xi=0} &= -\frac{1}{(1+\beta)}\{(\phi_f \bar{v}_{front} + \bar{p}'_{fx})^2(1+\beta) + [(1+\beta)\phi_f \bar{v}_{front} + 2\bar{p}'_{fx}]\sigma \\ &\quad - (1-\beta)\bar{p}'_{fx}[\bar{m} - (\phi_f \bar{v}_{front} + \bar{p}'_{fx})]\} \\ &= -\frac{1}{(1+\beta)}\{[2\bar{p}'_{fx} + (1+\beta)\phi_f \bar{v}_{front}]\sigma + 2(\bar{p}'_{fx})^2 + (1+\beta)(\phi_f \bar{v}_{front})^2 \\ &\quad - [(1-\beta)\bar{m} - (3+\beta)\phi_f \bar{v}_{front}]\bar{p}'_{fx}\}. \end{aligned} \tag{11.62}$$

To reflect the overall dynamic characteristic of an acidization dissolution system, a dimensionless number, known as the Zhao number, can be defined as follows:

$$Zh = -p'_{fx} = -\frac{p'_{fx}L^*}{p^*} = -\frac{k(\phi_f)L^* p'_{fx}}{\phi_f \mu D} = \frac{v_{flow}}{\sqrt{\phi_f D}}\sqrt{\frac{\rho_s}{\alpha_0 \chi C_0}}. \tag{11.63}$$

The physical meaning of each term in the Zhao number can be explained using the following equation:

$$Zh = F_{Advection} F_{Diffusion} F_{Chemical}, \tag{11.64}$$

where $F_{Advection}$ is a term to represent the solute advection; $F_{Diffusion}$ is a term to represent the solute diffusion/dispersion; $F_{Chemical}$ is a term to represent the chemical kinetics of the dissolution reaction. They can be expressed as follows:

$$F_{Advection} = v_{flow}, \tag{11.65}$$

$$F_{Diffusion} = \frac{1}{\sqrt{\phi_f D(\phi_f)}} \tag{11.66}$$

$$F_{Chemical} = \sqrt{\frac{\rho_s}{\alpha_0 \chi C_0}}. \tag{11.67}$$

Equation (11.64) clearly indicates that the Zhao number is a comprehensive dimensionless number, which can be used to simultaneously represent the interaction between the advection, diffusion/dispersion and chemical kinetics in the acidization dissolution system. This means that the Zhao number and its critical value can be used to assess the instability of the acidization dissolution front in a fluid-saturated carbonate rock. For this reason, the main purpose of this section is to derive the critical Zhao number of the acidization dissolution system. Toward this end, the dimensionless growth rate of the perturbed acidization dissolution system can be expressed as follows:

$$\begin{aligned}\bar{\omega}(\bar{m}) = {} & \frac{\varepsilon}{(1+\beta)(\phi_f - \phi_0)}\{[(1+\beta)\phi_f \bar{v}_{front} - 2Zh]\sigma + 2Zh^2 + (1+\beta)(\phi_f \bar{v}_{front})^2 \\ & + [(1-\beta)\bar{m} - (3+\beta)\phi_f \bar{v}_{front}]Zh\}.\end{aligned} \tag{11.68}$$

Letting $\bar{\omega}(\bar{m}) = 0$ yields the following characteristic equation for the critical Zhao number of the acidization dissolution system:

$$\begin{aligned}[(1+\beta)\phi_f \bar{v}_{front} - 2Zh_{critical}]\sigma_1 + 2(Zh_{critical})^2 + (1+\beta)(\phi_f \bar{v}_{front})^2 \\ + [(1-\beta)\bar{m} - (3+\beta)\phi_f \bar{v}_{front}]Zh_{critical} = 0,\end{aligned} \tag{11.69}$$

where

$$\sigma_1 = \frac{\sqrt{(\phi_f \bar{v}_{front} - Zh_{critical})^2 + 4\bar{m}^2} - (\phi_f \bar{v}_{front} - Zh_{critical})}{2}, \tag{11.70}$$

$$\phi_f \bar{v}_{front} = \frac{\varepsilon \phi_f Zh_{critical}}{\varepsilon \phi_f + (\phi_f - \phi_0)} = R Zh_{critical}, \quad R = \frac{\varepsilon \phi_f}{(1+\varepsilon)\phi_f - \phi_0}. \tag{11.71}$$

To facilitate the solution of Eq. (11.69), it can be rewritten as follows:

$$\begin{aligned}& [(1+\beta)\phi_f \bar{v}_{front} - 2Zh_{critical}]\frac{\sqrt{(\phi_f \bar{v}_{front} - Zh_{critical})^2 + 4\bar{m}^2}}{2} \\ & = [(1+\beta)\phi_f \bar{v}_{front} - 2Zh_{critical}]\frac{\phi_f \bar{v}_{front} - Zh_{critical}}{2} \\ & \quad - 2(Zh_{critical})^2 - (1+\beta)(\phi_f \bar{v}_{front})^2 - [(1-\beta)\bar{m} - (3+\beta)\phi_f \bar{v}_{front}]Zh_{critical}.\end{aligned} \tag{11.72}$$

After some mathematical manipulations are made, Eq. (11.72) can be written in the following form:

$$[(1+\beta)\phi_f\bar{v}_{front} - 2Zh_{critical}]\sqrt{(\phi_f\bar{v}_{front} - Zh_{critical})^2 + 4\bar{m}^2}$$
$$= -2(Zh_{critical})^2 - [2(1-\beta)\bar{m} - (3+\beta)\phi_f\bar{v}_{front}]Zh_{critical} - (1+\beta)(\phi_f\bar{v}_{front})^2. \tag{11.73}$$

The square of the left hand side of Eq. (11.73) leads to the following equation:

$$\begin{aligned}
&[(1+\beta)\phi_f\bar{v}_{front} - 2Zh_{critical}]^2[(\phi_f\bar{v}_{front} - Zh_{critical})^2 + 4\bar{m}^2] \\
&= 4(Zh_{critical})^4 - 4(3+\beta)\phi_f\bar{v}_{front}(Zh_{critical})^3 \\
&\quad + \{[12 + 8\beta + (1+\beta)^2](\phi_f\bar{v}_{front})^2 + 16m^2\}(Zh_{critical})^2 \\
&\quad - \{[4(1+\beta) + 2(1+\beta)^2](\phi_f\bar{v}_{front})^3 + 16(1+\beta)\bar{m}^2\phi_f\bar{v}_{front}\}Zh_{critical} \\
&\quad + (1+\beta)^2(\phi_f\bar{v}_{front})^4 + 4(1+\beta)^2\bar{m}^2(\phi_f\bar{v}_{front})^2.
\end{aligned} \tag{11.74}$$

Similarly, the square of the right hand side of Eq. (11.73) yields the following equation:

$$\begin{aligned}
&\{-2(Zh_{critical})^2 - [2(1-\beta)\bar{m} - (3+\beta)\phi_f\bar{v}_{front}]Zh_{critical} - (1+\beta)(\phi_f\bar{v}_{front})^2\}^2 \\
&= 4(Zh_{critical})^4 + 4[2(1-\beta)\bar{m} - (3+\beta)\phi_f\bar{v}_{front}](Zh_{critical})^3 \\
&\quad + \{4(1-\beta)^2\bar{m}^2 - 4(1-\beta)(3+\beta)\bar{m}\,\phi_f\bar{v}_{front}\}(Zh_{critical})^2 \\
&\quad + \{[(3+\beta)^2 + 4(1+\beta)](\phi_f\bar{v}_{front})^2\}(Zh_{critical})^2 \\
&\quad + [4(1-\beta^2)\bar{m}(\phi_f\bar{v}_{front})^2 - 2(3+\beta)(1+\beta)(\phi_f\bar{v}_{front})^3]Zh_{critical} \\
&\quad + (1+\beta)^2(\phi_f\bar{v}_{front})^4
\end{aligned} \tag{11.75}$$

Using the equating condition between Eqs. (11.74) and (11.75) leads to the standard third-power algebraic equation as follows:

$$a_1 Zh^3_{critical} + b_1 Zh^2_{critical} + c_1 Zh_{critical} + d_1 = 0, \tag{11.76}$$

where

$$a_1 = 8(1-\beta)\bar{m}, \tag{11.77}$$

$$b_1 = -4(1-\beta)(3+\beta)\bar{m}\phi_f\bar{v}_{front} - (12 + 8\beta - 4\beta^2)\bar{m}^2, \tag{11.78}$$

$$c_1 = 4(1-\beta)(1+\beta)\bar{m}(\phi_f \bar{v}_{front})^2 + 16(1+\beta)\bar{m}^2 \phi_f \bar{v}_{front} \quad (11.79)$$

$$d_1 = -4(1+\beta)^2 \bar{m}^2 (\phi_f \bar{v}_{front})^2. \quad (11.80)$$

To further reduce the power of the algebraic equation, Eq. (11.71) is used to replace the related terms in Eq. (11.76). As a result, the following standard linear equation is obtained:

$$a_2 Zh_{critical} + b_2 = 0, \quad (11.81)$$

where

$$a_2 = 8(1-\beta)\bar{m} - 4(1-\beta)(3+\beta)\bar{m}R + 4(1-\beta^2)\bar{m}R^2, \quad (11.82)$$

$$b_2 = -4(3-\beta)(1+\beta)\bar{m}^2 + 16(1+\beta)\bar{m}^2 R - 4(1+\beta)^2 \bar{m}^2 R^2. \quad (11.83)$$

Since Eq. (11.81) is the standard linear algebraic equation, its analytical solution for the critical Zhao number of the acidization dissolution system can be obtained in an explicit manner.

$$Zh_{critical} = \frac{[(3-\beta)(1+\beta) - 4(1+\beta)R + (1+\beta)^2 R^2]\bar{m}}{2(1-\beta) - (1-\beta)(3+\beta)R + (1-\beta^2)R^2}. \quad (11.84)$$

If the mineral dissolution ratio of a carbonate rock (i.e. ε) is very small, then the value of R is very small. In the limit case of R approaching zero, Eq. (11.84) approaches the following equation (for $\bar{m} = 1$):

$$Zh_{critical} = \frac{(3-\beta)(1+\beta)}{2(1-\beta)}. \quad (11.85)$$

Since Eq. (11.85) is exactly the same as the previous theoretical solution for the chemical dissolution of a much smaller mineral dissolution ratio (Zhao et al. 2008), it can be further demonstrated that the mathematical deduction carried out in this study is correct.

Using Eqs. (11.63) and (11.84), a theoretical criterion can be established to judge the instability of an acidization dissolution front in the fluid-saturated carbonate rock. If $Zh > Zh_{critical}$, then the acidization dissolution front in the fluid-saturated carbonate rock becomes unstable, while if $Zh < Zh_{critical}$, then the acidization dissolution front in the fluid-saturated carbonate rock is stable. If $Zh = Zh_{critical}$, then the acidization dissolution front in the fluid-saturated carbonate rock is neutrally unstable.

11.2 Theoretical Understanding of Some Fundamental Characteristics of Acidization Dissolution-Front Instability Problems

The proposed theory in this study can be used to understand the following fundamental characteristics of acidization dissolution-front instability problems in fluid-saturated carbonate rocks: (1) both the intrinsic time and length scales of an acidization dissolution system; (2) the asymptotic behavior of an acidization dissolution system under two limit conditions; (3) effects of several factors on the critical Zhao number of an acidization dissolution system; (4) effects of several factors on the propagation speed of an acidization dissolution front in fluid-saturated carbonate rocks.

11.2.1 The Intrinsic Time and Length Scales of an Acidization Dissolution System

The derived intrinsic time and length of an acidization dissolution system can be used to determine the time scale at which a dissolution front can be formed and the length scale at which the instability of the acidization dissolution front can be initiated. Since the intrinsic time (i.e. $t^* = 1/\alpha_0\varepsilon)$) is inversely proportional to both the mass exchange rate coefficient (i.e. α_0) and the mineral dissolution ratio (i.e. ε), the time scale at which a dissolution front can be formed should be equal to the intrinsic time of the acidization dissolution system. For an acidization dissolution system used to increase the oil production in petroleum engineering (Golfier et al. 2002), α_0 and ε can be assumed to be 10 (1/s) and 0.1 approximately, resulting in the intrinsic time of one second for the acidization dissolution system. However, for an acidization dissolution system involved in the generation of the karst formation, α_0 and ε may be assumed to be 1×10^{-3} and 1×10^{-5} (1/s) approximately, resulting in the intrinsic time of 1×10^8 s for the acidization dissolution system. The intrinsic length of an acidization dissolution system can be determined using the following equation:

$$L^* = \sqrt{\phi_f D t^*} = \sqrt{\frac{\phi_f D}{\alpha_0 \varepsilon}}, \tag{11.86}$$

where D is the effective diffusivity/dispersivity of the acid; t^* is the intrinsic time of the acidization dissolution system; α_0 is the mass exchange rate coefficient; ε is the mineral dissolution ratio; φ_f is the final (i.e. maximum) porosity when the carbonate rock is completely dissolved.

If $\phi_f = 1$ and $D = 1 \times 10^{-9}$ m^2/s, which are the same as those used in the previous numerical simulation of an experiment (Golfier et al. 2002), the intrinsic length of the acidization dissolution system, which is commonly used to increase the oil production in petroleum engineering, is about 3.16×10^{-5} m, while the intrinsic length of the acidization dissolution system involved in the generation of the karst formation is about 3.16×10^{-1} m. This means that the instability of the acidization dissolution front can occur at the pore scale for the acidization dissolution system that is used to increase the oil production in petroleum engineering, while it can occur at the meter scale for the acidization dissolution system involved in the generation of the karst formation. To ensure the occurrence of the instability, the length in the perpendicular direction of the dissolution front propagation should be large enough to accommodate a single wavelength of the fundamental mode. This implies that the length scale at which the instability of the acidization dissolution front can be initiated should be equal to $2\pi L^*$. It is interesting to note that for a very fast dissolution reaction, if α_0 is assumed to be $1 \times 10^9 (1/\mathrm{s})$ in the acidization dissolution system (with $\varepsilon = 0.1$), then the intrinsic length of the acidization dissolution system is decreased to about 3.16×10^{-9} m, indicating that the instability of the acidization dissolution front can take place at the nanometer scale. The understanding of the intrinsic length scale of an acidization dissolution problem is important for the computational simulation of the problem. To capture the fundamental characteristics of an acidization dissolution system, the mesh sizes should not be larger than the intrinsic length of the system if the finite element method is used to simulate the instability of the acidization dissolution front in fluid-saturated carbonate rocks.

11.2.2 The Asymptotic Behavior of an Acidization Dissolution System under Two Limit Conditions

Equation (11.84) indicates that the critical Zhao number of an acidization dissolution system is a function of β, R and $\bar{m}$, where β is dependent on both the initial and final porosities of the acidization dissolution system (see Eq. (11.60)) and R is dependent on the mineral dissolution ratio (i.e. ε), the initial porosity and the final porosity of the acidization dissolution system (see Eq. (11.71)). This means that for a given acidization dissolution system, the critical Zhao number is a positive finite number. On the other hand, from Eq. (11.63), the Zhao number of the acidization dissolution system can be rewritten in the following form.

$$Zh = \frac{v_{flow}}{\sqrt{\phi_f D}} \sqrt{\frac{1}{\alpha_0 \varepsilon}}. \tag{11.87}$$

Since v_{flow}, D and ϕ_f are all positive finite numbers, it is possible to theoretically investigate the asymptotic behaviors of the acidization dissolution system in the following two limit cases. In the first limit case, the mass exchange rate coefficient (i.e. α_0) is a constant but the mineral dissolution ratio (i.e. ε)

approaches zero, so that the Zhao number of the acidization dissolution system tends to infinity. This indicates that the acidization dissolution front is always unstable in the first limit case. On contrary, in the second limit case, the mineral dissolution ratio (i.e. ε) is a constant but the mass exchange rate coefficient (i.e. α_0) approaches infinity, so that the Zhao number of the acidization dissolution system tends to zero. This means that the acidization dissolution front is always stable unless the critical Zhao number of the system is equal to zero, which requires that the dimensionless wavenumber (i.e. $\bar{m}$) be zero in the acidization dissolution system. The implication of the second limit case is that if the acidization dissolution reaction takes place instantaneously, the acidization dissolution front may become stabilized in the acidization dissolution system.

11.2.3 Effects of Several Factors on the Critical Zhao Number of an Acidization Dissolution System

From the theoretical analysis in this study, the critical Zhao number (see Eq. (11.84)) of an acidization dissolution system can be expressed as a function of the dimensionless wavenumber (i.e. $\bar{m}$), the mineral dissolution ratio (i.e. ε), the initial porosity (i.e. ϕ_0) and the final porosity (i.e. ϕ_f) of the acidization dissolution system. Since the critical Zhao number is directly proportional to the dimensionless wavenumber, it is only necessary to investigate how other three factors, namely the mineral dissolution ratio (i.e. ε), the initial porosity (i.e. ϕ_0) and the final porosity (i.e. ϕ_f), affect the critical Zhao number of the acidization dissolution system. For this purpose, the dimensionless wavenumber is assumed to be unity in the following analysis.

Figure 11.1 shows the variation of the critical Zhao number, which is marked as Zh_Critical, with the mineral dissolution ratio due to different initial and final porosities. In Fig. 11.1a, the initial porosity is fixed to be 0.1 (i.e. $\phi_0 = 0.1$), so that the effect of both the mineral dissolution ratio and the final porosity on the critical Zhao number can be investigated, while in Fig. 11.1b, the final porosity is fixed to be 0.5 (i.e. $\phi_f = 0.5$), so as to investigate the effect of both the mineral dissolution ratio and the initial porosity on the critical Zhao number of the acidization dissolution system. It can be observed from Fig. 11.1a, b that with the increase of the mineral dissolution ratio, the critical Zhao number of the acidization dissolution system increases gradually. This implies that the increase of the mineral dissolution ratio can stabilize the acidization dissolution front in fluid-saturated carbonate rocks. For a given mineral dissolution ratio in Fig. 11.1a, it is noted that the critical Zhao number of the acidization dissolution system decreases with the increase of the final porosity, which is marked as ϕ_f in this figure. However, for a given mineral dissolution ratio in Fig. 11.1b, the critical Zhao number of the acidization dissolution system increases with the increase of the initial porosity, which is marked as ϕ_0 in this figure. This means that the increase

Fig. 11.1 Variation of the critical Zhao number with the mineral dissolution ratio due to different initial and final porosities

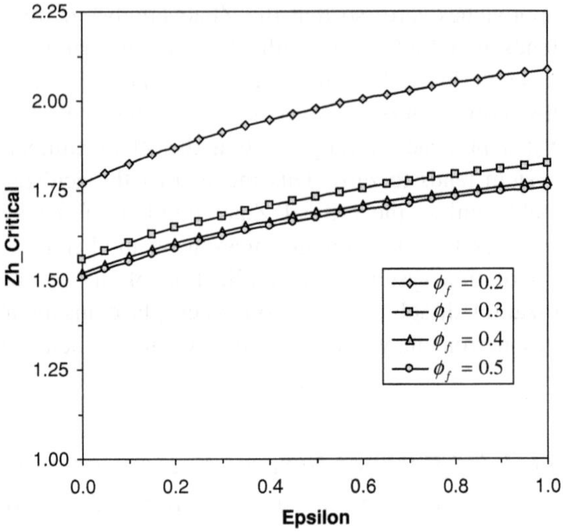

A (Initial porosity=0.1)

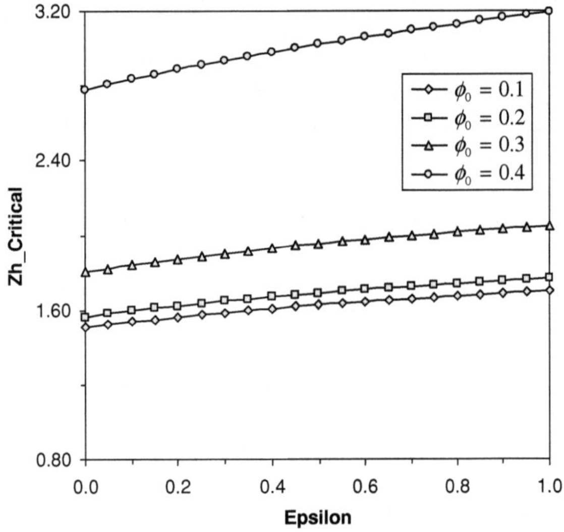

B (Final porosity=0.5)

of the final porosity of the carbonate rock can destabilize the acidization dissolution front, while the increase of the initial porosity can stabilize the acidization dissolution front in fluid-saturated carbonate rocks.

11.2.4 Effects of Several Factors on the Propagation Speed of an Acidization Dissolution Front

The proposed theory indicates that the following four factors can affect the dimensionless propagation speed of an acidization dissolution front in fluid-saturated carbonate rocks: (1) the mineral dissolution ratio (i.e. ε); (2) the initial porosity (i.e. ϕ_0); (3) the final porosity (i.e. ϕ_f); and (4) the Zhao number (i.e. Zh) of the acidization dissolution system. Since the dimensionless propagation speed of an acidization dissolution front is directly proportional to the Zhao number of the acidization dissolution system, only two special values of the Zhao number, namely $Zh = Zh_{critical}$ and $Zh = \phi_f$, are considered to investigate the effects of the other three factors on the dimensionless propagation speed of an acidization dissolution front in fluid-saturated carbonate rocks.

Figure 11.2 shows the variation of the dimensionless propagation speed of the acidization dissolution front with the mineral dissolution ratio due to different initial and final porosities in the case of $Zh = Zh_{critical}$. The general variation trend is that with the increase of the mineral dissolution ratio, the dimensionless propagation speed of the acidization dissolution front increases remarkably. This indicates that with the increase of the mineral dissolution ratio, the acidization dissolution front propagates faster in the acidization dissolution system. For a given mineral dissolution ratio in the case of the initial porosity being equal to 0.1 (see Fig. 11.2a), an increase in the final porosity can result in a decrease in the dimensionless propagation speed of the acidization dissolution front. However, for a given mineral dissolution ratio in the case of the final porosity being equal to 0.5 (see Fig. 11.2b), an increase in the initial porosity can result in an increase in the dimensionless propagation speed of the acidization dissolution front. This means that the increase of the initial porosity can enable the acidization dissolution front to propagate faster, while the increase of the final porosity can enable the acidization dissolution front to propagate slower in the acidization dissolution system.

During the theoretical analysis in this study, a coefficient (namely R) is introduced to facilitate the mathematical deduction. The physical meaning of this coefficient is that it represents the dimensionless propagation speed of the acidization dissolution front in the special case of $Zh = \phi_f$. Figure 11.3 shows the variation of the dimensionless propagation speed of the acidization dissolution front with the mineral dissolution ratio due to different initial and final porosities in this special case. Compared with the results shown in Fig. 11.2, it is noted that the general trends of the effects of the mineral dissolution ratio, initial porosity and final porosity on the dimensionless propagation speed of the acidization dissolution front in the case of $Zh = \phi_f$ (see Fig. 11.3) are very similar to those in the case of $Zh = Zh_{critical}$ (see Fig. 11.2), although the specific values of the dimensionless propagation speed are different in these two cases.

Fig. 11.2 Variation of the dimensionless speed of the acidization dissolution front with the mineral dissolution ratio due to different initial and final porosities ($Zh = Zh_{critical}$)

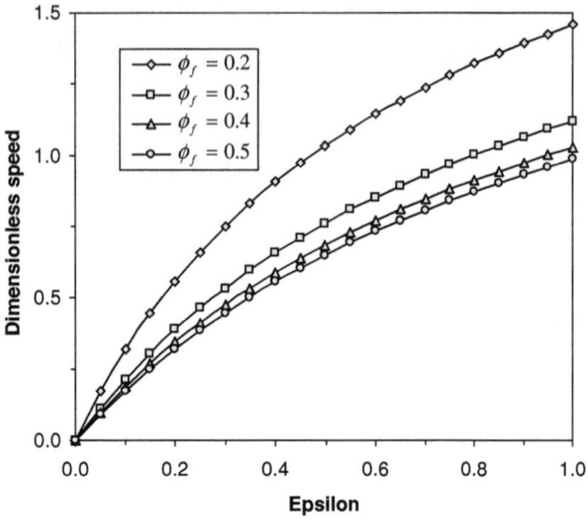

A (Initial porosity=0.1)

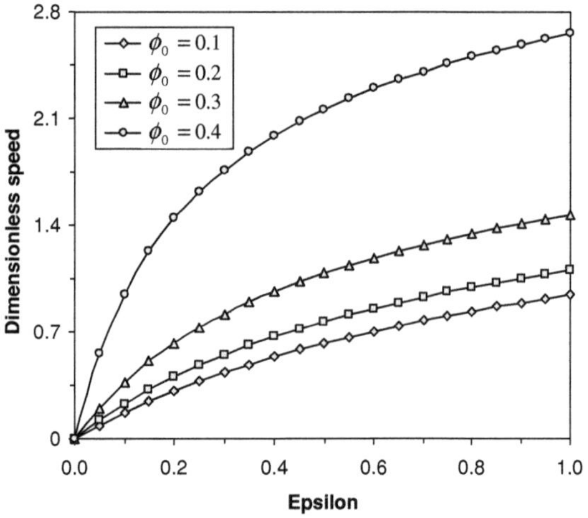

B (Final porosity=0.5)

11.3 Application of the Proposed Theory

To demonstrate how to use the proposed theoretical criterion to assess the instability of an acidization dissolution front propagating in fluid-saturated carbonate rocks, it is necessary to obtain the laboratory experimental data or field observation information. In terms of the experimental data available, Golfier et al. (2002) conducted a series of laboratory experiments to simulate the instability phenomena of an acidization dissolution front at the centimeter length scale. In particular, they used the numerical method to reproduce the instability phenomena observed

Fig. 11.3 Variation of the dimensionless speed of the acidization dissolution front with the mineral dissolution ratio due to different initial and final porosities ($Zh = \phi_f$)

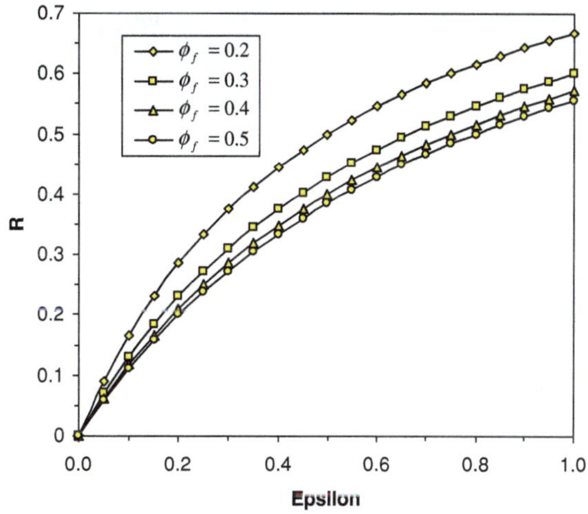

A (Initial porosity=0.1)

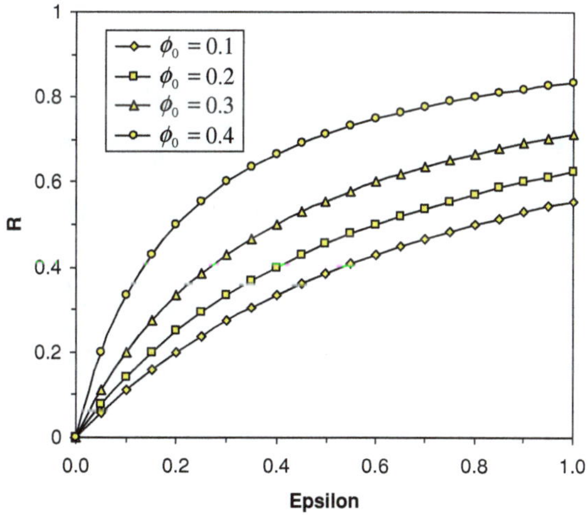

B (Final porosity=0.5)

during the laboratory experiments, so that both the simulation results and the detailed data were presented in their paper (Golfier et al. 2002). This means that their results and data can be employed to illustrate how to use the proposed theoretical criterion to assess the instability of an acidization dissolution front when it propagates in fluid-saturated carbonate rocks.

Figures 11.4 and 11.5 show the simulated experimental sample and the related simulation results in the work of Golfier et al. (2002). To identify the different modes of the acidization dissolution front, it is useful to introduce the previous conclusions that were drawn from the study on the evolution of NAPL dissolution

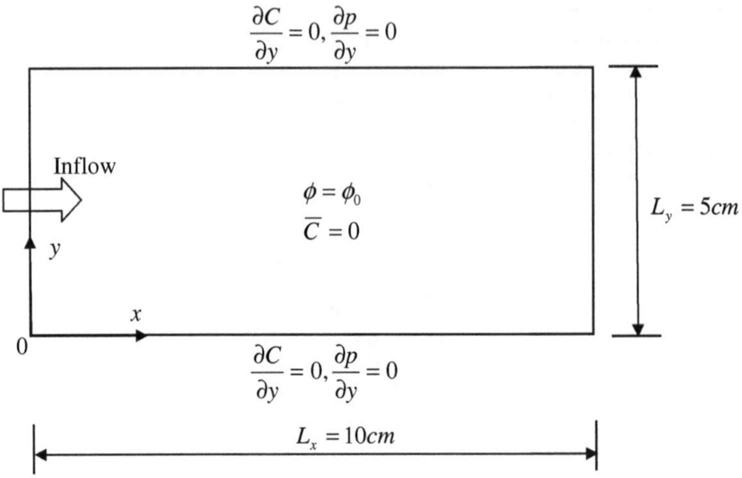

Fig. 11.4 The simulated sample in the paper of Golfier et al. (2002)

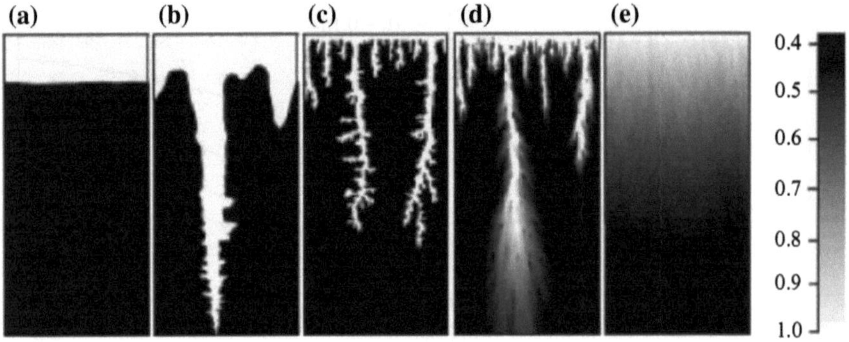

Fig. 11.5 The simulation results in the paper of Golfier et al. (2002): **a** $Pe = 8.32 \times 10^{-4}$; **b** $Pe = 4.14 \times 10^{-3}$; **c** $Pe = 1.66$; **d** $Pe = 83.2$; **e** $Pe = 832$

fronts in fluid-saturated porous media (Zhao et al. 2011). If the Zhao number of a dissolution system is of the same order of magnitude as the critical Zhao number of the system, then the fundamental mode is predominant. If the Zhao number is one order of magnitude higher than the critical Zhao number, then the (normal) fingering mode is the predominant pattern of the dissolution front. If the Zhao number is two orders of magnitude higher than the critical Zhao number, then the fractal mode is predominant for the dissolution front. Compared with the previous results of Zhao et al. (2011), the acidization dissolution fronts shown in Fig. 11.4a–c are very similar to the fundamental mode, the (normal) fingering mode and the fractal mode observed in the NAPL dissolution front propagation system respectively.

The conventional numerical procedures based on the finite element and/or finite difference methods performed well for dealing with dispersion-dominated transport problems (in the small Peclet number cases), but suffered from excessive artificial oscillation when they are applied to the simulation of advection-dominated transport problems (in the large Peclet number cases). To overcome this difficulty, a time-splitting method, which splits the mass transport equation into a hyperbolic part (consisting of advective terms) and an elliptic part (consisting of diffusive/dispersive terms), was used to obtain numerical results in the paper of Golfier et al. (2002). A scheme proposed by Takacs (1985) is employed for solving the hyperbolic part so as to reduce the numerical (i.e. artificial) diffusion/dispersion. This scheme is a TVD (total variation diminishing), oscillation-free and unconditional stable scheme. The elliptic part is solved using the conventional implicit discretization scheme. The basic property of the TVD scheme is that the total variation of the numerical solution does not increase during the solution process (Gottlieb and Shu 1998). By using the above-mentioned time-splitting method, it is possible to dealing with advection-dominated transport problems (in the very large Peclet number cases), such as those considered in the paper of Golfier et al. (2002).

The following parameters were used in the paper of Golfier et al. (2002): the concentration of injected acid is 150 kg/m^3; the density of the carbonate rock is 2700 kg/m^3; the initial and final porosities are 0.38 and 1.0 respectively; the effective diffusivity/dispersivity of the acid is 1×10^{-9} m^2/s; the dynamic viscosity of the pore-fluid is 1×10^{-3} (Pa · s); the mass exchange rate coefficient is 10 (1/s); the permeability of the carbonate rock is 1×10^{-11} m^2; the stoichiometric coefficient of the carbonate rock in terms of the weight density (i.e. weight per volume) is 1.37; the width and height of the simulated sample are 5 and 10 cm respectively.

To assess the instability of an acidization dissolution front, it is necessary to evaluate both the Zhao number and the critical Zhao number of the acidization dissolution system. Firstly, we explain how to evaluate the critical Zhao number of the acidization dissolution system. Based on the above parameters, the mineral dissolution ratio of the acidization dissolution system can be evaluated as follows:

$$\varepsilon = \frac{\chi C_0}{\rho_s} = \frac{1.37 \times 150}{2700} \approx 0.076. \tag{11.88}$$

This can result in the following intrinsic time of the acidization dissolution system:

$$t^* = \frac{1}{\alpha_0 \varepsilon} = \frac{1}{10 \times 0.076} \approx 1.316(\mathrm{s}). \tag{11.89}$$

Similarly, the intrinsic length of the acidization system can be determined below:

$$L^* = \sqrt{\phi_f D t^*} = \sqrt{1.0 \times 10^{-9} \times 1.316} \approx 3.628 \times 10^{-5} (\mathrm{m}). \tag{11.90}$$

To determine the dimensionless wavenumber of the acidization dissolution system, it is necessary to evaluate the dimensionless width of the acidization dissolution system as follows:

$$\bar{L}_y = \frac{L_y}{L^*} = \frac{0.05}{3.628 \times 10^{-5}} \approx 1378. \tag{11.91}$$

This leads to the following dimensionless wavenumber of the acidization dissolution system:

$$\bar{m} = \frac{2\pi}{\bar{L}_y} = \frac{2 \times 3.142}{1378} \approx 4.56 \times 10^{-3}. \tag{11.92}$$

In the case of $\phi_f = 1.0$,

$$\beta = 0 \text{ (from Equation (11.60))}, \tag{11.93}$$

$$R = \frac{\varepsilon}{(1+\varepsilon) - \phi_0} = \frac{0.076}{1.076 - 0.38} \approx 0.11 \text{ (from Equation (11.70))}. \tag{11.94}$$

Consequently, the critical Zhao number of the acidization dissolution system can be determined as follows:

$$\begin{aligned} Zh_{critical} &= \frac{(3 - 4R + R^2)\bar{m}}{2 - R + R^2} = \frac{(3 - 4 \times 0.11 + 0.11^2) \times 4.56 \times 10^{-3}}{2 - 0.11 + 0.11^2} \\ &\approx 6.17 \times 10^{-3}. \end{aligned} \tag{11.95}$$

Next, we explain how to evaluate the Zhao number of the acidization dissolution system. For this purpose, it is necessary to obtain the Darcy velocity used in the paper of Golfier et al. (2002). However, Golfier et al. (2002) did not directly give the Darcy velocity, but gave the Peclet number instead of the Darcy velocity. In the case of evaluating the Peclet number, they used $l = \sqrt{K} = \sqrt{10^{-11}} \approx 3.16 \times 10^{-6} (\mathrm{m})$ and $D = 1 \times 10^{-9}$ (m^2/s) in their paper. This means that the Darcy velocity used in the paper of Golfier et al. (2002) can be evaluated using the following formula:

$$v_{flow} = \frac{PeD}{l} = \frac{10^{-9} Pe}{3.16 \times 10^{-6}} \approx 3.16 \times 10^{-4} Pe (\mathrm{m/s}), \tag{11.96}$$

where Pe is the Peclet number used in the paper of Golfier et al. (2002).

When $Pe = 8.32 \times 10^{-4}$, 4.14×10^{-3}, 1.66, 83.2 and 832, the corresponding values of the Darcy velocity are 2.63×10^{-7}, 1.31×10^{-6}, 5.25×10^{-4}, 2.63×10^{-2} and 26.3 m/s respectively.

Finally, the Zhao number of the acidization dissolution system can be determined as follows:

$$Zh = \frac{v_{flow}}{\sqrt{\phi_f D}} \sqrt{\frac{1}{\alpha_0 \varepsilon}} = v_{flow} \sqrt{\frac{1}{1.0 \times 10^{-9} \times 10 \times 0.076}} \approx 3.63 \times 10^4 v_{flow}. \tag{11.97}$$

From Eq. (11.97), the Zhao number of the acidization dissolution system is 9.55×10^{-3}, 4.76×10^{-2}, 1.91×10, 9.55×10^2 and 9.55×10^3 for $Pe = 8.32 \times 10^{-4}$, 4.14×10^{-3}, 1.66, 83.2 and 832 respectively. Since all the Zhao numbers are greater than the corresponding critical Zhao number (i.e. $Zh_{critical} \approx 6.17 \times 10^{-3}$), the acidization dissolution fronts considered in all the five cases in the paper of Golfier et al. (2002) are unstable. This demonstrates that the proposed instability theory can be used to identify the fundamental mode and other modes of the acidization dissolution front in the fluid-saturated carbonate rock.

In summary, the proposed theory includes two fundamental concepts, namely the intrinsic time and length of an acidization dissolution system, and a theoretical criterion that involves the comparison of the Zhao number and its critical value of the acidization dissolution system. The intrinsic time is used to determine the time scale at which the acidization dissolution front is formed, while the intrinsic length is used to determine the length scale at which the instability of the acidization dissolution front can be initiated. Under the assumption that the acidization dissolution reaction is a fast process, the critical Zhao number, which is used to assess the instability likelihood of an acidization dissolution front propagating in fluid-saturated carbonate rocks, has been derived in a strictly mathematical manner. Based on the proposed instability theory of a propagating acidization dissolution front, it has been theoretically recognized that: (1) the increase of the mineral dissolution ratio can stabilize the acidization dissolution front in fluid-saturated carbonate rocks; (2) the increase of the final porosity of the carbonate rock can destabilize the acidization dissolution front, while the increase of the initial porosity can stabilize the acidization dissolution front in fluid-saturated carbonate rocks; (3) the increase of the mineral dissolution ratio can cause an increase in the dimensionless propagation speed of the acidization dissolution front; and (4) the increase of the initial porosity can enable the acidization dissolution front to propagate faster, while the increase of the final porosity can enable the acidization dissolution front to propagate slower in the acidization dissolution system. In addition, an application example has been employed to illustrate how to use the proposed theory to assess the instability likelihood of an acidization dissolution front in the fluid-saturated carbonate rock.

References

Chadam J, Hoff D, Merino E, Ortoleva P, Sen A (1986) Reactive infiltration instabilities. IMA J Appl Math 36:207–221

Chadam J, Ortoleva P, Sen A (1988) A weekly nonlinear stability analysis of the reactive infiltration interface. IMA J Appl Math 48:1362–1378

Chen JS, Liu CW (2002) Numerical simulation of the evolution of aquifer porosity and species concentrations during reactive transport. Comput Geosci 28:485–499

Chen JS, Liu CW (2004) Interaction of reactive fronts during transport in a homogeneous porous medium with initial small non-uniformity. J Contam Hydrol 72:47–66

Chen JS, Liu CW, Lai GX, Ni CF (2009) Effects of mechanical dispersion on the morphological evolution of a chemical dissolution front in a fluid-saturated porous medium. J Hydrol 373:96–102

Cohen CE, Ding D, Quintard M, Bazin B (2008) From pore scale to wellbore scale: impact of geometry on wormhole growth in carbonate acidization. Chem Eng Sci 63:3088–3099

Detournay E, Cheng AHD (1993) Fundamentals of poroelasticity: comprehensive rock engineering. : In: Hudson JA, Fairhurst C (ed) Analysis and design methods, vol. 2. Pergamon Press, New York

Fredd CN, Fogler HS (1998) Influence of transport and reaction on wormhole formation in porous media. AIChE J 44:1933–1949

Golfier F, Zarcone C, Bazin B, Lenormand R, Lasseux D, Quintard M (2002) On the ability of a Darcy-scale model to capture wormhole formation during the dissolution of a porous medium. J Fluid Mech 457:213–254

Gottlieb S, Shu CW (1998) Total variation diminishing Runge-Kutta schemes. Math Comput 67:73–85

Hinch EJ, Bhatt BS (1990) Stability of an acid front moving through porous rock. J Fluid Mech 212:279–288

Kalia N, Balakotaiah V (2007) Modeling and analysis of wormhole formation in reactive dissolution in carbonate rocks. Chem Eng Sci 62:919–928

Kalia N, Balakotaiah V (2009) Effect of medium heterogeneities on reactive dissolution of carbonates. Chem Eng Sci 64:376–390

Nield DA, Bejan A (1992) Convection in porous media. Springer, New York

Ormond A, Ortoleva P (2000) Numerical modeling of reaction-induced cavities in a porous rock. J Geophys Res 105:16737–16747

Ortoleva P, Chadam J, Merino E, Sen A (1987) Geochemical self-organization II: the reactive-infiltration instability. Am J Sci 287:1008–1040

Panga MKR, Ziauddin M, Balakotaiah V (2005) Two-scale continuum model for simulation of wormholes in carbonate acidization. AIChE J 51:3231–3248

Scheidegger AE (1974) The physics of flow through porous media. University of Toronto Press, Toronto

Sherwood JD (1987) Stability of a plane reaction front in a porous medium. Chem Eng Sci 42:1823–1829

Szymczak P, Ladd AJC (2009) Wormhole formation in dissolving fractures. J Geophys Res 114:B06203. doi:10.1029/2008JB006122

Takacs L (1985) A two-step scheme for the advection equation with minimized dissipaption and dispersion errors. Mon Weather Rev 113:1050–1065

Turcotte DL, Schubert G (1982) Geodynamics: applications of continuum physics to geological problems. Wiley, New York

Zhao C, Hobbs BE, Hornby P, Ord A, Peng S, Liu L (2008) Theoretical and numerical analyses of chemical-dissolution front instability in fluid-saturated porous rocks. Int J Numer Anal Meth Geomech 32:1107–1130

Zhao C, Hobbs BE, Ord A (2009) Fundamentals of computational geoscience: numerical methods and algorithms. Springer, Berlin

Zhao C, Hobbs BE, Regenauer-Lieb K, Ord A (2011) Computational simulation for the morphological evolution of nonaqueous-phase-liquid dissolution fronts in two-dimensional fluid-saturated porous media. Comput Geosci 15:167–183

Summary Statements

Physical and chemical dissolution front instability problems exist ubiquitously in many scientific and engineering fields. In geoenvironmental engineering, the remediation of contaminated sites using fresh water to flush the contaminated soils involves propagation of the dissolved contaminant front in the water-saturated porous medium. In mineral mining engineering, the extraction of minerals in the deep Earth using the in-situ leaching technique may result in propagation of the dissolved mineral front in the fluid-saturated porous medium. In petroleum industry, the secondary recovery of oil by acidifying the oil field to uniformly increase porosity and hence the yield of oil is associated with propagation of the acid-dissolved material front in porous rocks. Thus, a systematical study on the dynamic mechanisms of physical and chemical dissolution-front instability phenomena is beneficial not only for understanding ore forming mechanisms, which are imperative to develop advanced techniques for exploring new ore deposits in the deep Earth, but also for understanding dissolved contaminant transport, which is important to develop innovative techniques for rehabilitating contaminated soils. Toward this end, both theoretical analyses and computational simulations have been extensively carried out in such a systematical study. As a result, the following conclusions have been drawn from the research work reported in the monograph.

(1) To solve chemical dissolution-front propagation problems, it is necessary to deal with a coupled system between porosity, pore-fluid pressure and reactive chemical-species transport in fluid-saturated porous media. Due to the morphological instability of a chemical dissolution front, this problem needs to be solved mathematically and numerically. Mathematical methods are used to establish a theoretical criterion for assessing whether or not a chemical dissolution system under consideration is in a critical or supercritical state. A segregated algorithm based on a combination of the finite element and finite difference methods has been proposed for

C. Zhao, *Physical and Chemical Dissolution Front Instability in Porous Media*, Lecture Notes in Earth System Sciences,
DOI: 10.1007/978-3-319-08461-9

simulating the morphological evolution of chemical dissolution fronts in chemical dissolution systems of critical and supercritical Zhao numbers. A set of analytical solutions have been derived for a benchmark problem to verify the proposed numerical procedure. Not only can the derived analytical solutions be used to verify any numerical method before it is used to solve this kind of chemical dissolution-front propagation problem, but also they can be used to understand the fundamental mechanisms behind the morphological instability of a chemical dissolution front during its propagation within fluid-saturated porous media of critical and supercritical Zhao numbers. The related numerical results have demonstrated that the proposed segregated algorithm and related numerical procedure are useful for and capable of simulating the morphological instability of chemical dissolution fronts within fluid-saturated porous media.

(2) The related theoretical and numerical results from investigating the effects of particle reactive surface areas have demonstrated that: first, since the shape coefficient of spherical grains is greater than that of cubic grains, the chemical dissolution system consisting of spherical grains is more unstable than that consisting of cubic grains; second, the instability likelihood of a natural porous medium, which is comprised of irregular grains, is smaller than that of an idealized porous medium, which is comprised of regular spherical grains; third, reactive surface areas associated with different particle shapes can have a significant influence on the morphological evolution of an unstable chemical-dissolution front within the fluid-saturated porous medium.

(3) The related theoretical results from investigating the effects of mineral dissolution ratios have revealed that the mineral dissolution ratio plays an important role in controlling the propagation speed of a planar chemical dissolution-front in the fluid-saturated porous medium. An increase in the value of the mineral dissolution ratio can result in a remarkable decrease in the value of the dimensionless propagation speed of a planar chemical dissolution-front. On the other hand, the related computational simulation results have demonstrated that the mineral dissolution ratio has a considerable influence on the evolution pattern of a planar chemical dissolution-front during its propagation in the fluid-saturated porous medium. An increase in the mineral dissolution ratio can reduce the likelihood for a planar chemical dissolution-front to evolve from the initial planar shape into different morphologies within the fluid-saturated porous medium of finite size.

(4) The theoretical results from examining the effects of solute dispersion have led to the following findings. First, the propagation speed of a planar chemical dissolution-front in the case of considering solute dispersion effects is exactly the same as that when solute dispersion effects are neglected. This indicates that solute dispersion does not affect the propagation speed of the planar chemical dissolution-front in a fluid-saturated porous medium. Second, consideration of solute dispersion can cause a significant increase in the critical Zhao number, which is used to

judge whether or not a planar chemical dissolution-front may become unstable in the fluid-saturated porous medium. This means that the consideration of solute dispersion can stabilize a planar chemical dissolution-front because an increase in the critical Zhao number reduces the likelihood of the planar chemical dissolution-front instability in a fluid-saturated porous medium. Third, for both a given solute dispersion ratio and a given longitudinal dispersivity, an increase in the final porosity value destabilizes the chemical dissolution-front so that it becomes easier for a planar chemical dissolution-front to evolve into different morphologies. In addition, the present results can be used as benchmark solutions for verifying numerical methods employed to simulate detailed morphological evolution processes of chemical dissolution fronts in two-dimensional fluid-saturated porous media.

(5) The related theoretical and numerical results from investigating the effects of medium permeability anisotropy have clearly demonstrated that a decrease in the medium anisotropic permeability factor (or ratio), which is defined as the ratio of the principal permeability in the transversal direction to that in the longitudinal direction parallel to the pore-fluid inflow direction, can stabilize the chemical dissolution front so that it becomes more difficult for a planar chemical dissolution-front to evolve into different morphologies in the chemical dissolution system. On the other hand, the medium anisotropic permeability ratio can have significant effects on the morphological evolution of the chemical dissolution front. When the Zhao number of the chemical dissolution system is greater than its critical value, the greater the medium anisotropic permeability ratio, the faster the irregular chemical dissolution-front grows.

(6) The related theoretical and numerical results from examining the effects of pore-fluid and medium compressibility have led to the following findings. First, not only can pore-fluid compressibility affect the propagation speeds of chemical dissolution fronts in both subcritical and supercritical chemical dissolution systems, but also it can affect the growth and amplitudes of irregular chemical dissolution fronts in supercritical chemical dissolution systems. Second, medium compressibility may have a little influence on the propagation speeds of chemical dissolution fronts, but it can have significant effects on the growth and amplitudes of irregular chemical dissolution-fronts in supercritical chemical dissolution systems. Third, both medium and pore-fluid compressibility may stabilize irregular chemical dissolution-fronts in supercritical chemical dissolution systems.

(7) To simulate the chemical dissolution-front evolution in a three-dimensional fluid-saturated porous medium, a combined numerical procedure consisting of the finite difference and finite element methods has been proposed. Since the problem belongs to a complex system science problem, a small randomly-generated perturbation of porosity is added into the initial porosity of a three-dimensional homogeneous domain to trigger the instability of a planar chemical dissolution-front during its propagation

within the fluid-saturated porous medium. To test the correctness and accuracy of the proposed numerical procedure, a three-dimensional benchmark problem has been constructed and the related analytical solution has been derived. This enables the proposed numerical procedure to be used for simulating the morphological evolution of a three-dimensional chemical dissolution-front from a stable, planar state into an unstable, fingering state. The related numerical results have demonstrated that the proposed numerical procedure is useful for and capable of simulating the morphological evolution of a three-dimensional chemical dissolution-front within the fluid-saturated porous medium.

(8) A theoretical criterion for assessing the instability of planar NAPL dissolution-fronts in two-dimensional fluid-saturated porous media of finite domains has been established. Not only can the present theoretical results be used for theoretical understanding of the effect of solute dispersion on the instability of a NAPL dissolution front in the fluid-saturated porous medium of either a finite domain or an infinite domain, but also they can be used as benchmark solutions for verifying numerical methods employed to simulate the detailed morphological evolution processes of NAPL dissolution fronts in two-dimensional fluid-saturated porous media. The related simulation results have revealed that: (i) the proposed numerical procedure is useful and applicable for simulating the morphological evolution of NAPL dissolution fronts in two-dimensional fluid-saturated porous media of finite domains; (ii) if the Zhao number of a NAPL dissolution system is in the lower range of the supercritical Zhao numbers, the fundamental mode is predominant; (iii) if the Zhao number is in the middle range of the supercritical Zhao numbers, the (normal) fingering mode is the predominant pattern of the NAPL dissolution front; and (iv) if the Zhao number is in the higher range of the supercritical Zhao numbers, the fractal mode is predominant for the NAPL dissolution front.

(9) The related numerical simulation results from investigating the effects of domain shapes have demonstrated that: (i) domain shapes have a significant effect on both the propagation speed and the morphological evolution pattern of a NAPL dissolution front in the fluid-saturated porous medium; (ii) an increase in the divergent angle of a trapezoidal domain can lead to a decrease in the propagation speed of the NAPL dissolution front; (iii) the morphological evolution pattern of the NAPL dissolution front in a rectangular domain is remarkably different from that in a trapezoidal domain of a large divergent angle; (iv) for a rectangular domain, the simplified dispersion model, which is commonly used in the theoretical analysis and numerical simulation, is valid for solving NAPL dissolution instability problems in fluid-saturated porous media; and (v) compared with diverging flow (when the trapezoidal domain is inclined outward), converging flow (when the trapezoidal domain is inclined inward) can enhance the growth of NAPL fingers, indicating that pump-and-treat systems by extracting contaminated groundwater might enhance NAPL

dissolution fingering and lead to less uniform dissolution fronts.

(10) The propagation theory of mesh discretization errors associated with a NAPL dissolution system is first presented for a rectangular domain and then extended to a trapezoidal domain. This leads to the establishment of the finger-amplitude growing theory associated with both the corner effect and the mesh discretization effect in the NAPL dissolution systems of trapezoidal domains. This theory can be used to make the approximate error estimation of the corresponding computational simulation results. The related theoretical analysis and numerical results have demonstrated that: (i) both the corner effect and the mesh discretization effect can be quantitatively viewed as a kind of small perturbation so that they can have some considerable effects on the computational results of supercritical NAPL dissolution systems; (ii) the proposed finger-amplitude growing theory associated with the corner effect at the entrance of a trapezoidal domain is useful for correctly explaining why the finger at either the top or the bottom boundary grows much faster than that within the interior of the trapezoidal domain; and (iii) the proposed finger-amplitude growing theory associated with the mesh discretization error in the NAPL dissolution system of a trapezoidal domain can be used for quantitatively assessing the correctness of computational simulations of NAPL dissolution-front instability problems in trapezoidal domains, so that it can be ensured that the computational simulation results are controlled by the physics of the NAPL dissolution system, rather than by the numerical artifacts.

(11) The intrinsic time is used to determine the time scale at which the acidization dissolution front is formed, while the intrinsic length is used to determine the length scale at which the instability of the acidization dissolution front can be initiated. Under the assumption that the acidization dissolution reaction is a fast process, the critical Zhao number, which is used to assess the instability likelihood of an acidization dissolution-front propagation in fluid-saturated carbonate rocks, has been derived in a strictly mathematical manner. Based on the proposed instability theory of a propagating acidization dissolution front, it has been theoretically recognized that: (i) the increase of the mineral dissolution ratio can stabilize the acidization dissolution front in fluid-saturated carbonate rocks; (ii) the increase of the final porosity of the carbonate rock can destabilize the acidization dissolution front, while the increase of the initial porosity can stabilize the acidization dissolution front in fluid-saturated carbonate rocks; (iii) the increase of the mineral dissolution ratio can cause an increase in the dimensionless propagation speed of the acidization dissolution front; and (iv) the increase of the initial porosity can enable the acidization dissolution front to propagate faster, while the increase of the final porosity can enable the acidization dissolution front to propagate slower in the acidization dissolution system.

(12) At the end of this monograph, it needs to be pointed out that there remains much to be done in this particular research field. For instance, only physical and chemical dissolution fronts of planar shapes are treated theoretically in this field. Physical and chemical dissolution fronts of other different shapes, such as circular, cylindrical and spherical shapes, have not been theoretically considered so far. In addition, temperature effects in non-isothermal dissolution systems have also neglected in the current research. To solve energy shortage problems facing the mankind, physical and chemical dissolution-front instability phenomena should be employed to develop advanced and innovative techniques for extracting unconventional hydrocarbons, such as gas hydrates, heavy oil, and shale oil/gas, from the reservoirs that are comprised of low porosity rocks.

Index

C. Zhao, *Physical and Chemical Dissolution Front Instability in Porous Media*, Lecture Notes in Earth System Sciences,
DOI: 10.1007/978-3-319-08461-9

MIX
Papier aus verantwortungsvollen Quellen
Paper from responsible sources
FSC® C105338

If you have any concerns about our products,
you can contact us on
ProductSafety@springernature.com

In case Publisher is established outside the EU,
the EU authorized representative is:
Springer Nature Customer Service Center GmbH
Europaplatz 3, 69115 Heidelberg, Germany

Printed by Libri Plureos GmbH
in Hamburg, Germany